AF304964

A HISTORY OF THE SKY

IN 300 INFOGRAPHICS

A HISTORY OF THE SKY

IN 300 INFOGRAPHICS

PIERRE LÉNA &
CHRISTIAN GRATALOUP

with
LÉNA HESPEL
PHILIPPE PAJOT
LÉGENDES CARTOGRAPHIE

Contents

To Mohammad Heydari-Malayeri (1947–2023)

A talented astrophysicist, linguist and man of peace, he
worked tirelessly to create a new Farsi vocabulary of words
for modern astronomical terms. Just as this book aims to
do, he dived into thousands of years of history to bring
knowledge to the general public of today.

The human sky

Christian Grataloup and Pierre Léna

It is hard to imagine what our earliest ancestors felt when they looked up at the star-studded sky. How did this mix of emotions and questions, which we also share, first emerge in the human mind? The stars, the phases of the Moon, the movement of the Sun: for thousands of years, every human society has wondered about the sky that shines above them. Many have used the stars to find their place on Earth and navigate across it. Most have sought to understand what connections may exist between the Earth as humans see it and the mysteries of the sky. Myths, legends and stories have explained its origin, peopled it with a pantheon of gods, or imbued it with unattainable perfection.

Anyone who closely observes the sky will start to recognize certain patterns and regular occurrences, and a few irregular ones. Discoveries like these are what led to the birth of science. The study of the skies is probably the most important source of ideas, counting systems, explanatory models and geometric shapes in history. We can still see many traces of these early discoveries in our daily lives. If there is one scientific field whose origins can be traced back many millennia, it is astronomy. Without it, there would be no accurate cartography, and therefore no atlases and no precise calendars. It developed across many different cultures – Babylonian and Chinese, Mayan and Polynesian – before a cohesive system of astronomy emerged in Greece, spreading across Europe and Asia and developing over time. The pace of progress increased five centuries ago in Europe, when a formalized field of science began to emerge, gradually breaking free from religious doctrines and the practice of astrology. From the 18th century onwards, it started to spread across the world.

It is only recently that we humans, who have been gazing at the same sky for thousands of years, have ventured beyond our earthly home. As yet, we have only travelled a stone's throw away, to our planet's nearest neighbour, the Moon. Now we dream of colonizing other planets. Already, our probes travel across the Solar System and our radio signals are beamed across the galaxy, sending messages to the universe.

Exploring this long and beautiful history teaches us to appreciate the greatness of humanity's potential and to embrace our fragility with humility and patience.

A tale of surprises

Here we must contend with a paradox: humans today, and particularly children, are becoming much less aware of the heavens above us. Especially in Western countries, it is harder than ever to see the night sky. Artificial lighting in urban areas makes it difficult for us to appreciate the sight of this source of contemplation, meditation and delight. And the entire world is fast becoming urbanized. Just to look at a star, we often have to travel far away from human habitation – and sometimes this is no longer possible. For many people, even the Moon is no longer a familiar friend. We read about the conquest of space in the media, but we only look at space through a screen. The aim of this book, therefore, is to combine human history with the history of the universe, in all its vastness but also its historicity, to inspire our readers to go out on a clear night and look at what our ancestors were able to see shining in the darkness.

The history of the sky is fascinating, but often little known. It was written by men and women whose stories we celebrate. Alexander, Genghis Khan and Sundiata Keita changed the world, as did Euclid, Ptolemy and Kepler, and so many others who have now been forgotten also played their part in driving this story forward, transforming how we thought about the universe. Our senses, which allow us to perceive the world and which feed our imaginations, are limited and often mislead us. Our minds are able to look beyond the false evidence of our senses, but only thanks to a huge amount of labour, often anonymous. This labour and its fruits are the main lessons conveyed by this volume.

The goal we have set out to achieve is a difficult one. This book seeks to interest two readerships that are usually seen as distinct: lovers of astronomy and students of history. But beyond that, it is for all those who are too nervous to ask basic questions and query the vocabulary of modern science. It shows that what we see as simple today was sometimes very complicated in the past. It answers the question: how do we know what we know about the sky?

There are already plenty of beautiful astronomy books in the world. The purpose of this one is to bring together an interest in the history of knowledge about the stars and an interest in the history of human societies. Lovers of science need to understand the diversity of its roots, the slow pace of its development and its moments of wonder.

A history in three eras, and beyond

The 16th and 17th centuries in Europe were a time of great change: over the course of a few decades, knowledge built up over thousands of years suddenly bore new fruit. There was a time before and after Copernicus, Kepler, Galileo and Newton, to take a few turning points. But equally, everything changed in the early 20th century with relativity and quantum physics. At first, the universe was small, widely believed to have the Earth at its centre, and was estimated to have existed for a few thousand years. In the time from Galileo to Einstein, humans learned that they were located in the Milky Way galaxy and understood that this information came from very far away and a long time ago: the sky had become history. For a century now, we have known that there is a whole universe of galaxies with a history going back many billions of years: space and time, filled with matter and energy, which together form a single continuum: spacetime.

This book includes three chronological sections: up to the 16th century; from the 17th to the 20th century; and after 1900. These are preceded by an overview of what the human eye can see without the help of any scientific instruments, the uses that societies made of this knowledge and the variety of ways that people sought to understand and impose order on the night sky. The tale of the slow, and then increasingly fast, development of our current understanding of the sky culminates with a final chapter on the conquest of space.

For humans, seeking to understand the heavens is both a deeply humbling and wildly ambitious task.

Visualizing history

In the past, globes came in pairs: the terrestrial globe had a matching celestial globe. The spectacular globes made by Coronelli, exhibited at the Bibliothèque Nationale de France in Paris, are a perfect example of this. This book follows in the footsteps of this tradition. It is therefore not surprising that it contains a number of maps of the celestial vault. These views of the sky from Earth are rooted in specific societies, which are also explored. This book explores the range of responses from communities that were looking up at the same sky, but from their own distinct perspective. Today, it is only possible to briefly summarize the vast swathe of knowledge about the universe produced by an international community of scientists.

Between these two conceptions of the universe lies an entire history, the story of the development of a scientific discipline, which required a great deal of ingenuity, technical skill and advances in mathematics. In order to explore these fully, it was necessary to record the intellectual speculations of the ancients as well as long-forgotten instruments, right up to the technical and scientific tools of the last few decades, with telescopes installed on satellites, interplanetary probes, exploratory devices on the Moon, Mars and beyond. A geographical history in pictures: celestial and terrestrial maps, of course, alongside a vast range of images.

This would not be the book it is today if it had not benefited from the invaluable teamwork of everyone at our publishers, as well as a number of other supporters, who are thanked at the end of the book. We must give a special mention to Léna Hespel, a young science journalist who, over the course of more than a year, worked closely with us on research, interviews and the writing of this book. Without her, it would not exist, and we would like to offer her our deepest thanks for her hard work.

1

The sky with the naked eye

We are able to observe the sky because it is lit by the Sun and the Moon, which give a pattern to the passage of time. The regular cycles of these two celestial bodies impact us just as they impacted most human societies throughout history. At night, the sight of the Milky Way and the stars, the repeated rhythm of their rising and setting, still evokes an emotional response. Faced with these regular but mysterious events, every civilization came up with their own myths and tales in an attempt to explain the skies. They saw in them the presence of gods and sought to connect celestial sights with the fates of individual humans. When a phenomenon such as an eclipse seemed to disrupt the established order, fear and awe were the response. Using no instruments other than the naked eye, human beings looked beyond the Earth on which they stood and sought new knowledge.

The night sky

Gazing at the starry sky on a cloudless night has been a shared experience since the earliest human civilizations. It has given rise to myths, poetry, art and science.

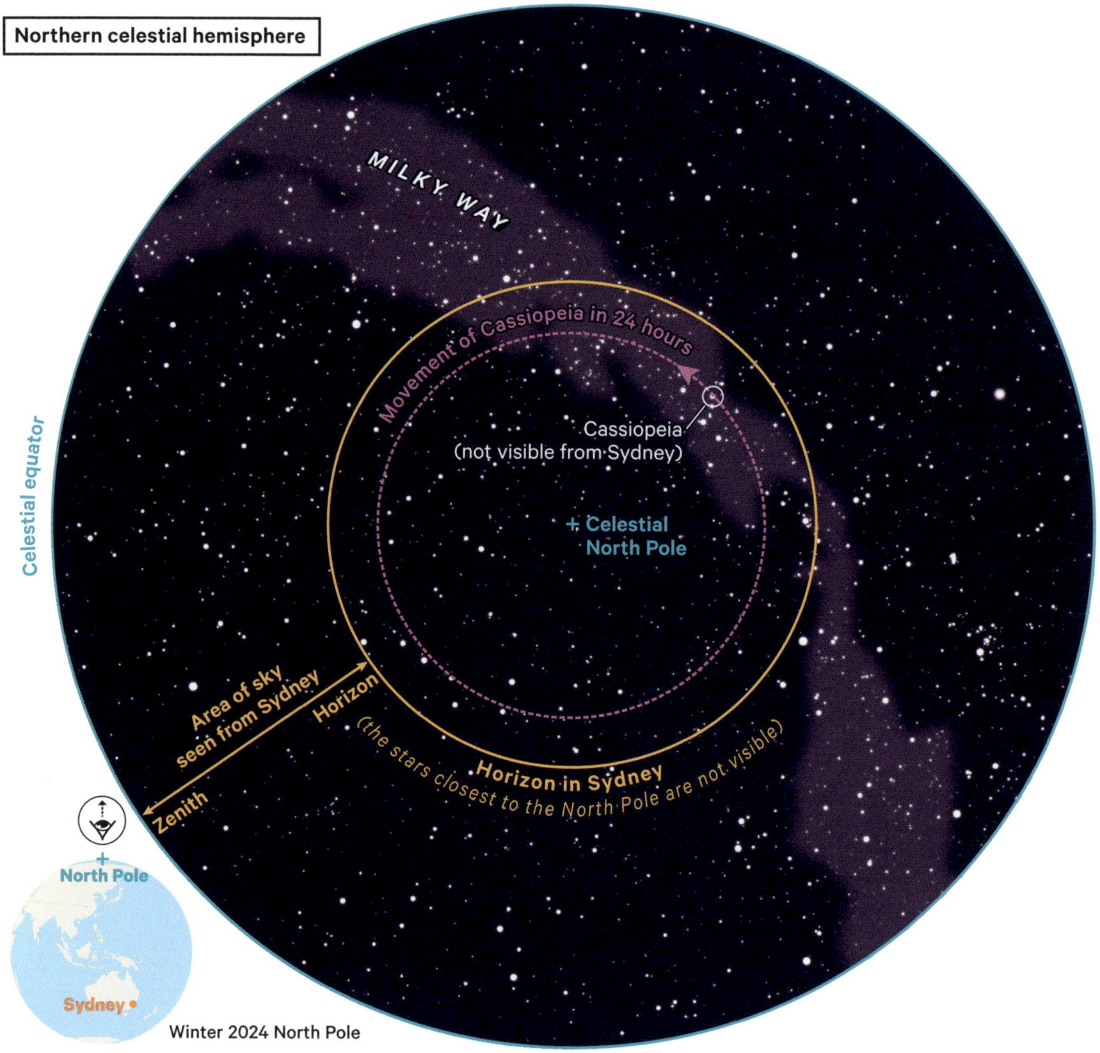

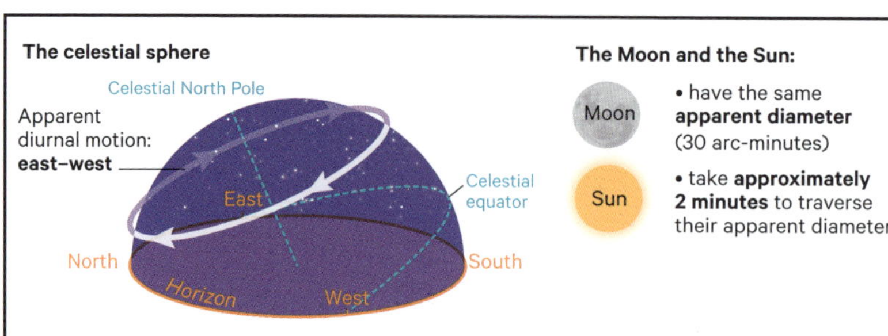

The celestial sphere

The Moon and the Sun:

- have the same **apparent diameter** (30 arc-minutes)

- take **approximately 2 minutes** to traverse their apparent diameter

A celestial sphere that holds the stars

The sky seems to turn slowly as time goes by, and the stars appear to move in circles around the celestial North or South Pole. This apparent motion gave rise to the theory of a 'celestial sphere', on which the stars were fixed in place.

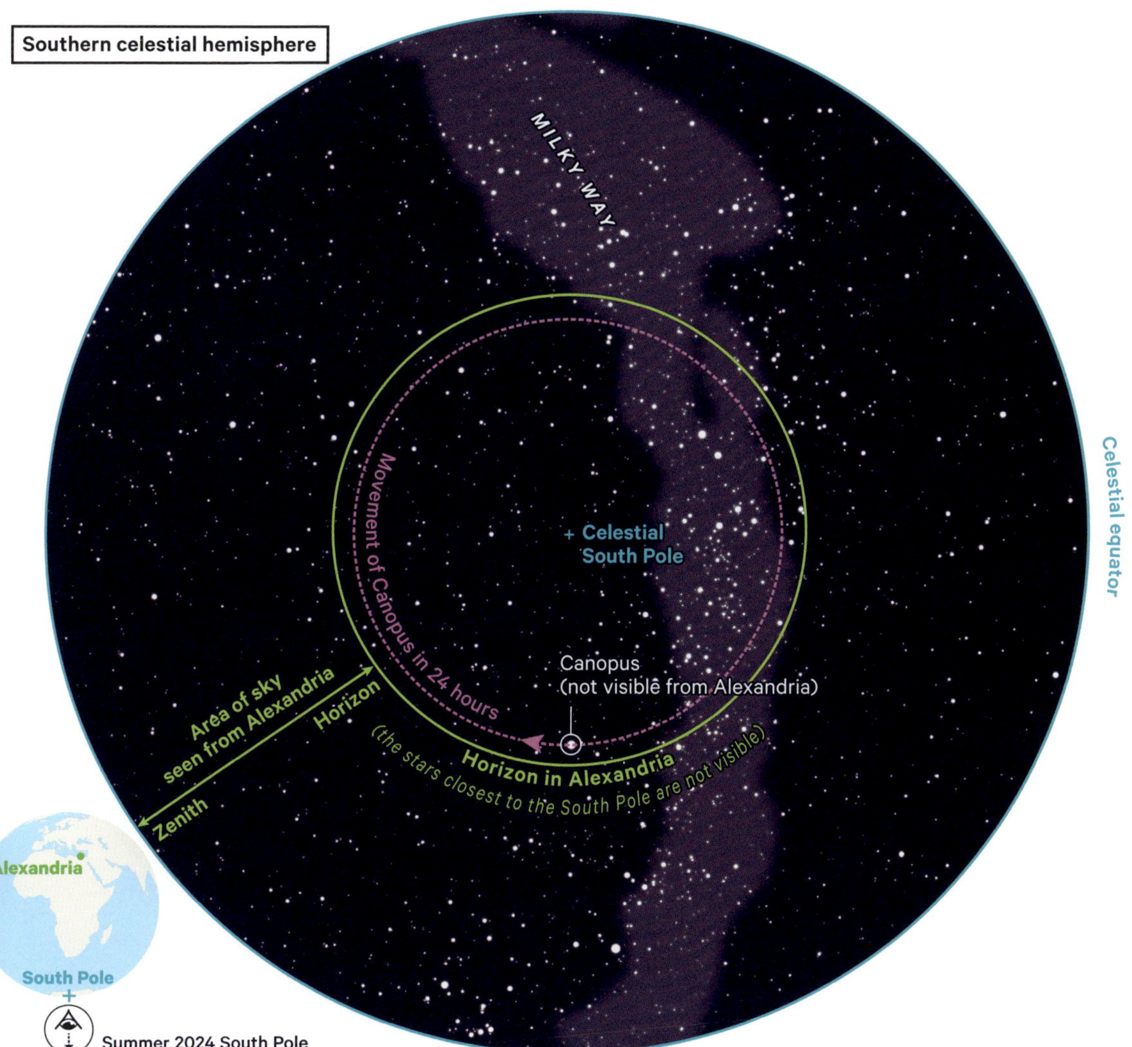

The two celestial hemispheres

Aside from the equator, there is no place on Earth where it is possible to see the entire celestial sphere. From the South Pole, only the southern celestial hemisphere is visible, and from the North Pole, only the northern celestial hemisphere. Which portion is visible depends on the latitude of the observer's location. Some stars that are close to the celestial poles are visible throughout the year at night. Others are only visible during certain seasons. Every night, the stars rise in the east around four minutes earlier than the previous night, which means that the night sky changes over time. After 365 days, the same night sky is visible once again.

Light and the human eye

The result of five hundred million years of evolution, the human eye is an amazing organ. As well as seeing in sunlight, it is also able to see in the dimmer light of the Moon and stars, and can discern colours and tiny details, depending on its distance from the object being observed.

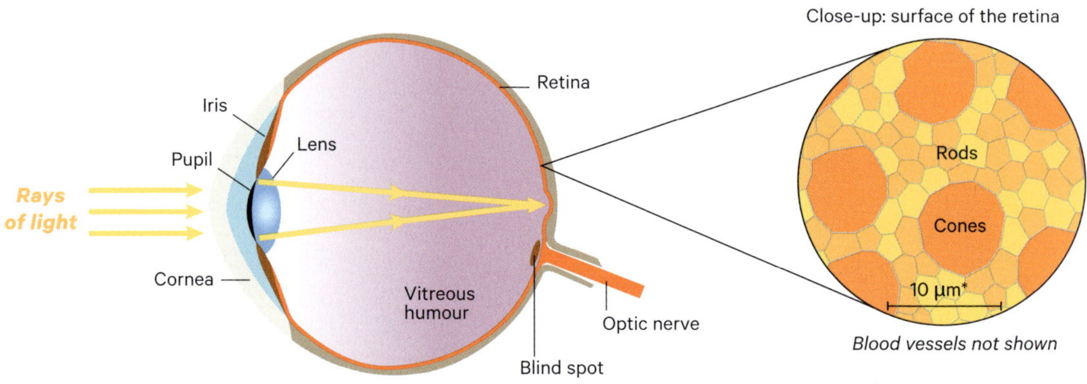

▲ The retina of the human eye

The eye is a spherical organ controlled by muscles. Light enter through the transparent cornea, passes through a lens, which focuses it, and forms an inverted image on a layer of photosensitive cells called the retina. These cells are divided into cones and rods: the cones detect colour and sharp detail, while the rods are more sensitive in dim light and at night, but are not as capable of perceiving fine details. The signals received by the retina are transmitted to the brain, which interprets them, flips them the right way up and creates the image that we perceive. The eye can detect even the faintest traces of light.

▼ Discerning details

The eye's ability to make out the details of an object depends on the distribution of the light-sensitive cells (cones and rods) in the retina. The acuity of the human eye can be expressed as its angular resolution: the smallest angle between two points that it is able to discern. Binoculars and telescopes are capable of discerning much smaller angles, so the images they produce are more detailed. The Moon's angular diameter is half a degree, and the eye can easily make out craters on the Moon's surface. The angular diameter of the Sun is the same, although it is important to avoid looking directly at the Sun.

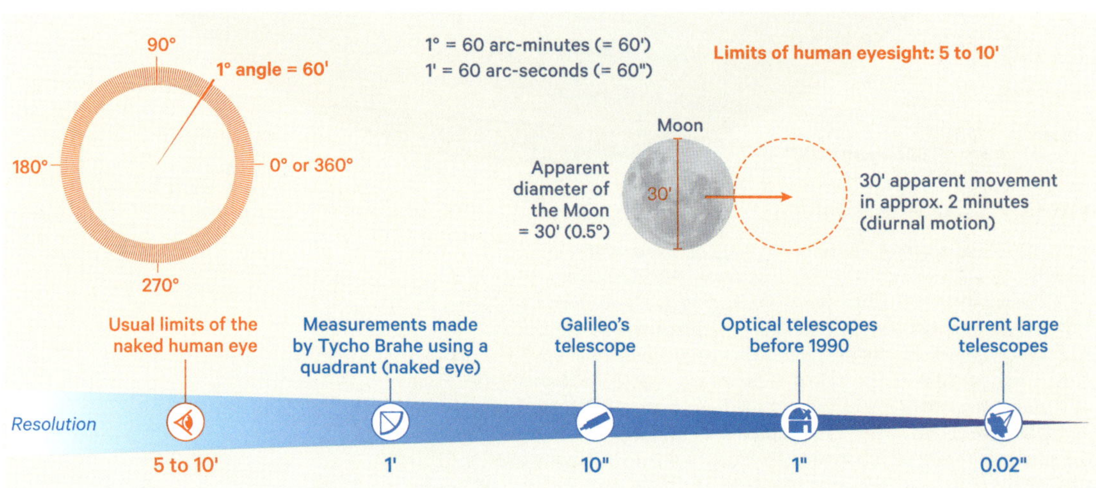

Seven sisters in the sky

The Pleiades (from the Greek for 'many') are a small group (asterism) of seven bright stars, forming an open cluster that occupies a space around four times the size of the Moon, in the constellation of Taurus. In Greek myth, the Pleiades were the daughters of the giant Atlas. They are 450 light-years away from Earth and also include a large number of less bright stars. As they are fairly close to the celestial equator, this striking asterism can be observed by the naked eye by many cultures at different latitudes. For example, since the Neolithic era, the Pleiades' rising (in November) has marked the beginning of the harvest and a time of plenty in Polynesia.

When looking at this cluster of stars on a moonless night, at first the eye sees a small blurry smudge. Then, as the eye becomes used to the dark, it can make out the stars and count them. Children, whose eyes are sharper, can often make them out more easily than adults. This test of sight was used by Darius (6th century BCE) and Alexander (4th century BCE) to select archers for their armies.

The two photographs on the left demonstrate the resolution of the human eye. A star has such a small apparent diameter that it should appear as nothing more than a speck. However, even in the photograph below, the stars are slightly blurred, because the camera is not perfect. This has been exaggerated in the top image in order to show what a person with average vision would see, at a resolution of 8 arc-minutes.

The Milky Way

A vast swathe of light that looks as if it has been painted across the sky, the Milky Way is visible throughout the year. Many cultures, from the ancient Greeks to the Chinese and Native Americans, have myths or folktales associated with it. The longstanding theory that it was made up of individual stars that could not be distinguished by the naked eye was confirmed by Galileo's observations.

EAST

A sleeping goddess

In Greek myth, Heracles was the son of a mortal woman, Alcmene, and Zeus, the king of the gods. Although his father was a god, the baby would not be immortal unless he fed from the breast of Hera, Zeus's wife. She refused to feed the baby, but while she was sleeping, Zeus brought the young Heracles to her and the baby bit her breast, causing a few drops of milk to spurt out, which became this ribbon of light across the sky.

Star-crossed lovers

In a Chinese folktale, the cowherd Niu Lang and the fairy Zhi Nu fell in love. The girl's jealous mother snatched her away and Niu Lang set off in pursuit. To stop him, the mother took a gold pin from her hair and tore a hole in the sky. A river of light spilled out of the gap and separated the lovers, who became two bright stars. Once a year, the guilty mother allows a flock of magpies to build a bridge where the lovers can meet.

The Way of St James

Since the 11th century, Christian pilgrims have crossed Europe, walking to the shrine of St James the Great in Santiago de Compostela in northern Spain. On summer nights, the pilgrims walked southwest, using the Milky Way to navigate. This swathe of light above their heads therefore came to be called the 'Way of St James'.

Navajo souls

While Haashch'ééshzhiní, the Black God, was placing the stars in the sky, Ma'ii, the coyote trickster, took the remaining stars from his buckskin pouch and emptied them out. According to the Navajo people, this is how the Milky Way was born. The souls of the dead follow this path of light to the afterlife and their campfires are the stars that run alongside it.

The living waters of Tane

To the people of Polynesia, the Milky Way is the 'living waters' of the god who created humans, known as Tane to the Tahitians, and Kane to the Hawaiians. In the Maori language, it is called Ika-roa ('big fish') or Ika-o-te-rangi ('heavenly fish'), after the gigantic fish that is believed to have given birth to the stars.

The night's backbone

Among the San people of southern Africa, the Milky Way is known as the 'night's backbone', the 'spine of the sky' or even the 'back of God'. Some of the San believe it was created by a girl from an ancient race, who threw a handful of ashes from a fire into the sky. Thus the path of light was born, and it guides people as they make their way home at night.

View of the sky from the mountain of Cerro Armazones in the Atacama Desert, Chile: Latitude 24°35' S, longitude 70°11' W, altitude 3,064 m, May 2024, around 10 pm. The lights on the horizon are from traffic on the Pan-American Highway.

The Sun, a star that moves

The movement of the Sun across the sky marks the passing of time, giving us days and seasons. Unlike the stars, which seem to be fixed in place on the celestial sphere, its position changes over the course of the year. In any given place on Earth, the Sun rises more or less in the east and sets in the west, and reaches its highest point at around midday – towards the south in the northern hemisphere and vice versa.

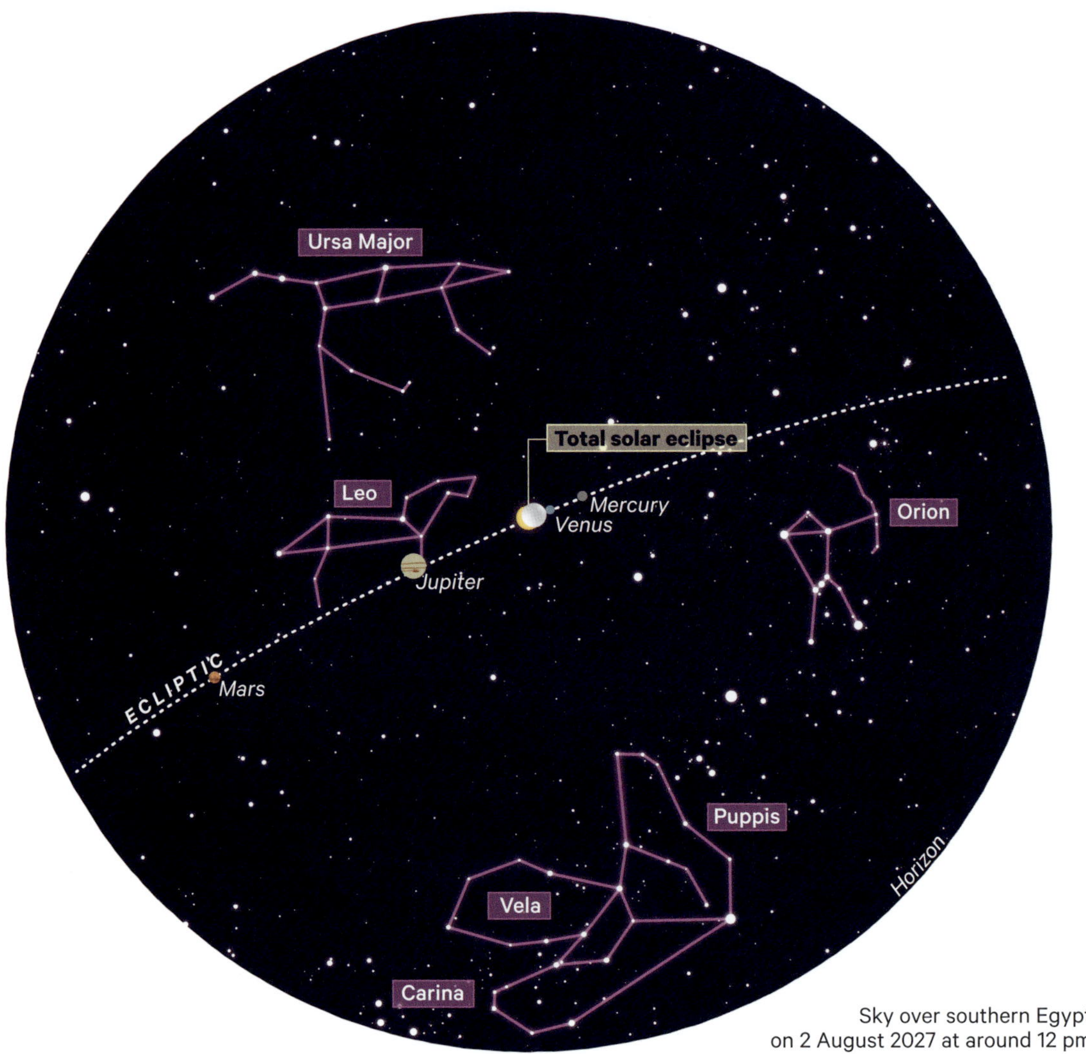

Sky over southern Egypt
on 2 August 2027 at around 12 pm

A journey across the celestial sphere

The idea that the Sun is moving through the stars on the celestial sphere can only be understood after it has set and the stars have appeared. During a total eclipse of the Sun, when it is hidden by the Moon, the sky grows dark, planets and stars appear, and we are able to see the Sun's position among them. On 2 August 2027, a solar eclipse will occur in southern Egypt, and the position of the Sun among the constellations will be visible. Because different stars appear during different seasons, this suggests that the Sun moves across the celestial sphere. The diagram shows the ecliptic, the path across this sphere that the Sun follows over the course of a year.

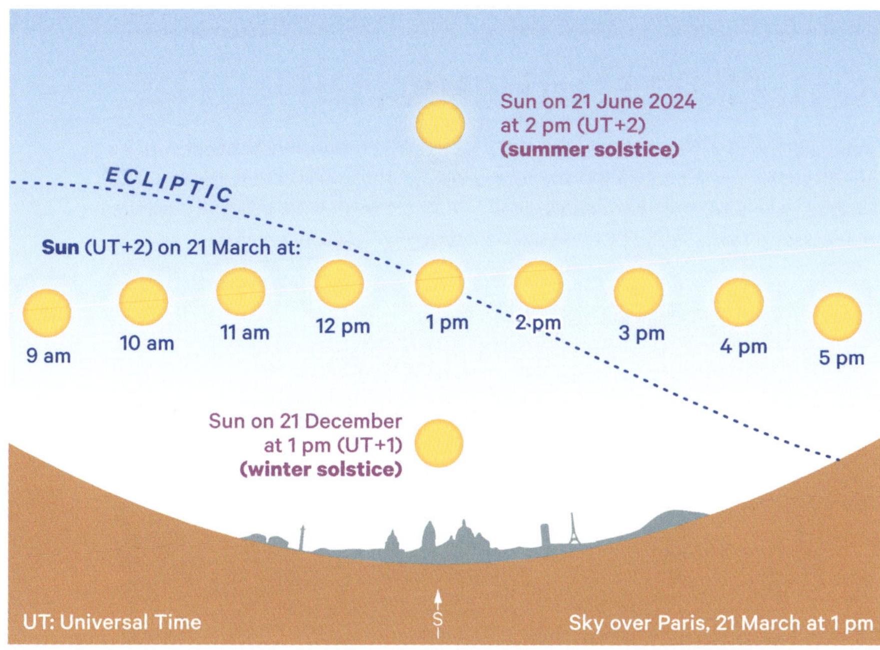

Solar noon

During the day, the Sun seems to move across the sky (its diurnal motion). When the Sun reaches its highest point above the southern horizon (in the northern hemisphere) or the northern horizon (in the southern hemisphere), this marks solar noon in that place. Its height changes from one day to the next; the Sun is at its highest at the summer solstice and at its lowest at the winter solstice, and is between these two positions at the spring and autumn equinoxes.

From sunrise to sunset

On the celestial sphere shown below, only the celestial North Pole is visible. Like the stars, the Sun seems to move around the axis of the poles. When viewed from Earth, it seems to trace a circle (the orange arrows below) that is only partly above the horizon. This part corresponds to the time between sunrise and sunset. As the position of the Sun on the sphere changes over the course of the year, sometimes above, sometimes below the celestial equator, the length of days changes. The days are at their longest at the summer solstice, and shortest at the winter solstice. There are 12 hours of daylight at the equinoxes, when the Sun is on the celestial equator, tracing a great circle on the sphere, rising in the east and setting in the west.

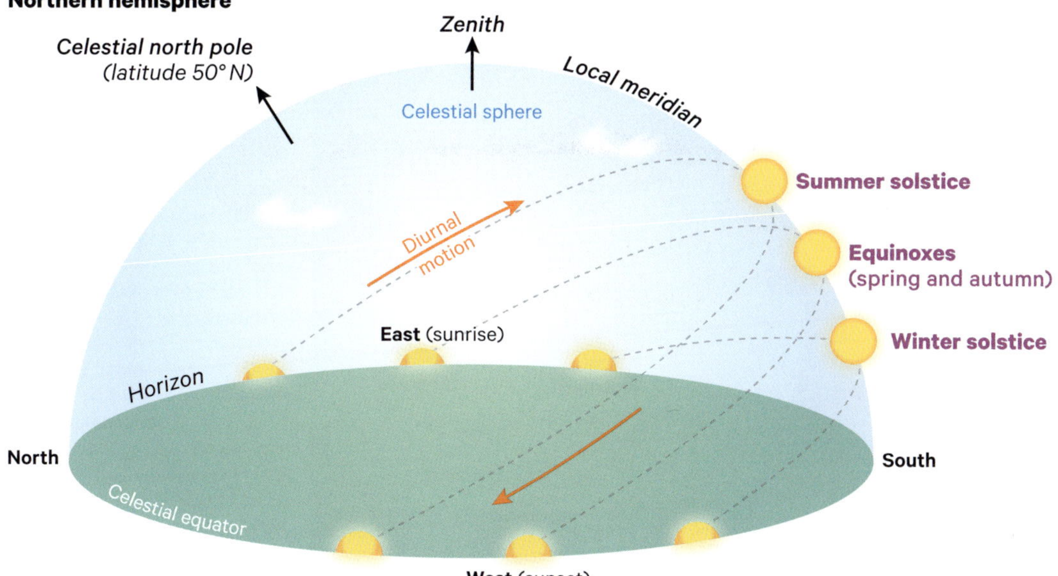

The Moon, constantly changing

The Moon is probably the celestial object that aroused the most curiosity among ancient peoples, due to the clear changes in its appearance over a short period of time, its shifting position in the sky, the fact that it can be seen by day as well as by night, and its fairly frequent eclipses. The influence of the Moon can be seen across multiple aspects of human civilization, particularly calendars, poetry and science.

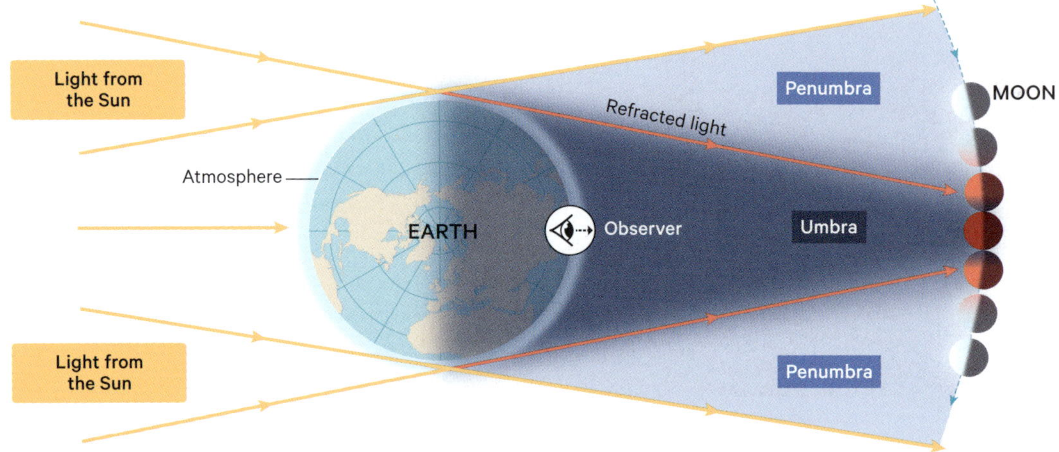

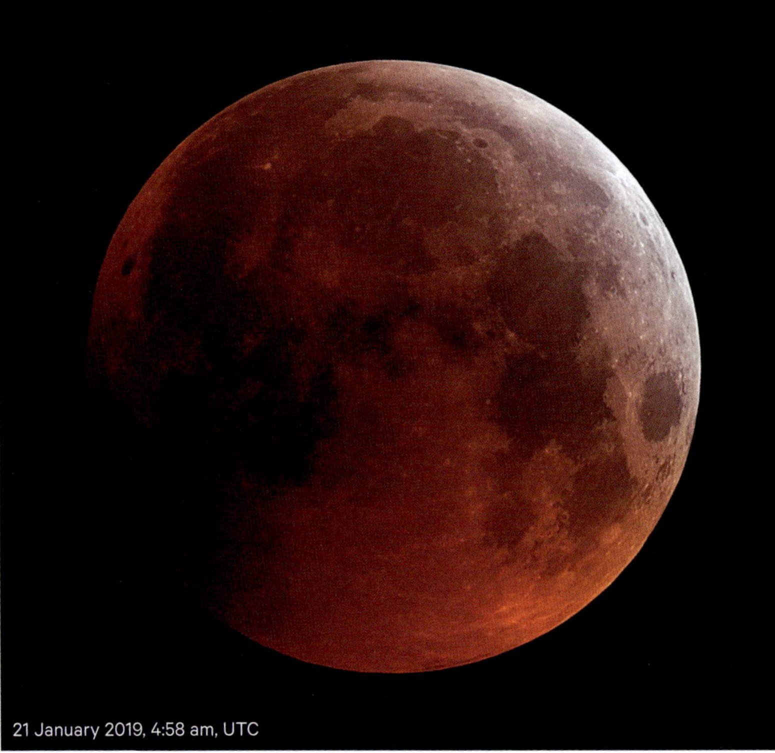
21 January 2019, 4:58 am, UTC

Lunar eclipses

A lunar eclipse occurs when the Sun, the Earth (orbiting the Sun) and the Moon (orbiting the Earth) are aligned. If the Moon's orbit were on the same plane as the Earth's, there would be a lunar eclipse every lunar month. As the two planes are slightly different, eclipses occur less often, but still fairly frequently. A total lunar eclipse happens when the Moon passes entirely into the Earth's shadow. Other eclipses are partial and can be hard to see if the Moon only passes into the Earth's penumbra (outer shadow). The Moon can remain in the Earth's shadow for almost two hours. The part of the Moon in shadow takes on an orange hue, because the Sun's rays are curved by the refraction of the Earth's atmosphere, which absorbs blue light. The shadow on the Moon has a curved edge, due to the shape of the Earth.

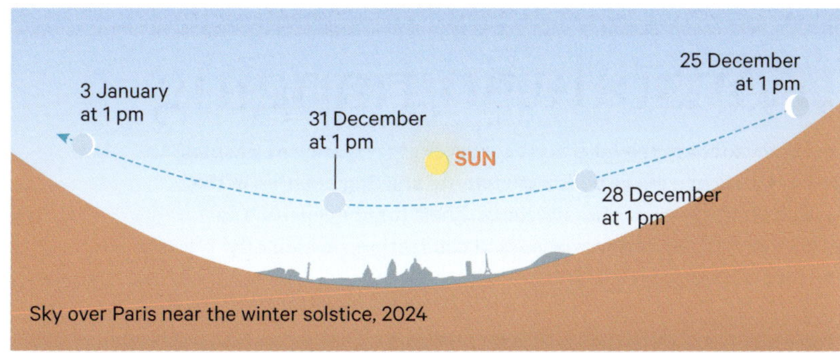

Sky over Paris near the winter solstice, 2024

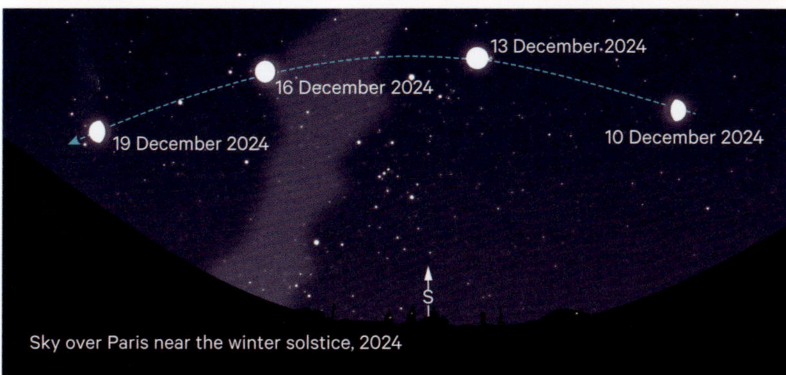

Sky over Paris near the winter solstice, 2024

By day and by night

Like the Sun, the Moon moves across the celestial sphere, relative to the stars. It takes around 28 days to orbit the Earth, while the Earth rotates on its axis once a day. Viewed from the Earth, the Moon is sometimes close to the Sun (a new moon) and sometimes in the opposite direction to the Sun (full moon). The above left image shows the positions and phases of the Moon every three days as seen from Paris at the time when the Sun is at its highest in the south. The image below shows the Moon at night, two weeks earlier, its position also marked every three days, when it is almost directly opposite the Sun. Day by day, it seems to slowly move from west to east across the celestial sphere, while travelling from east to west in its diurnal motion across the sky.

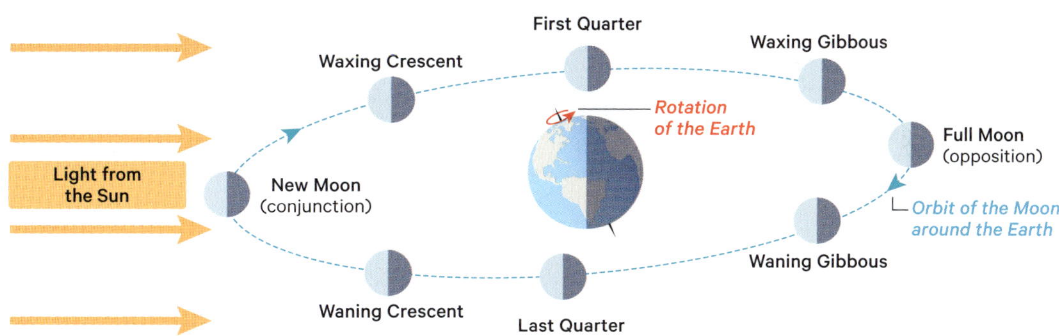

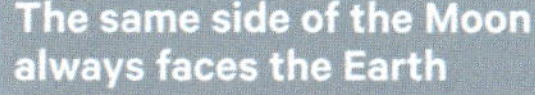

| New Moon | Waxing Crescent | First Quarter | Waxing Gibbous | Full Moon | Waning Gibbous | Last Quarter | Waning Crescent |

The phases of the Moon

The hemisphere of the Moon that can be seen from Earth is always the same. The far hemisphere had never been seen until it was photographed by a satellite launched from Earth in 1959. As the Moon orbits the Earth in around 28 days, and the Earth orbits the Sun in around 365 days, the combination of these two movements means that an observer on Earth sees regular phases of the Moon: sometimes its face is completely illuminated (full moon), sometimes it is entirely in shadow (new moon), and then there are all the phases in between. This monthly cycle, combined with the fact that the Moon is visible from everywhere on Earth, led to the invention of lunar calendars in many cultures.

The planets, wandering stars

Saturn, Mars, Jupiter, Mercury and Venus: five of the Solar System's planets are visible to the naked eye. Then there are Uranus and Neptune, which can be seen with binoculars. Planets are striking because they are constantly changing position relative to the stars. Their trajectories, like that of the Moon, remain within the same band of sky as the Sun's trajectory. Many cultures identify the planets with heavenly beings. The word 'planet' comes from the Greek word for 'wanderer'.

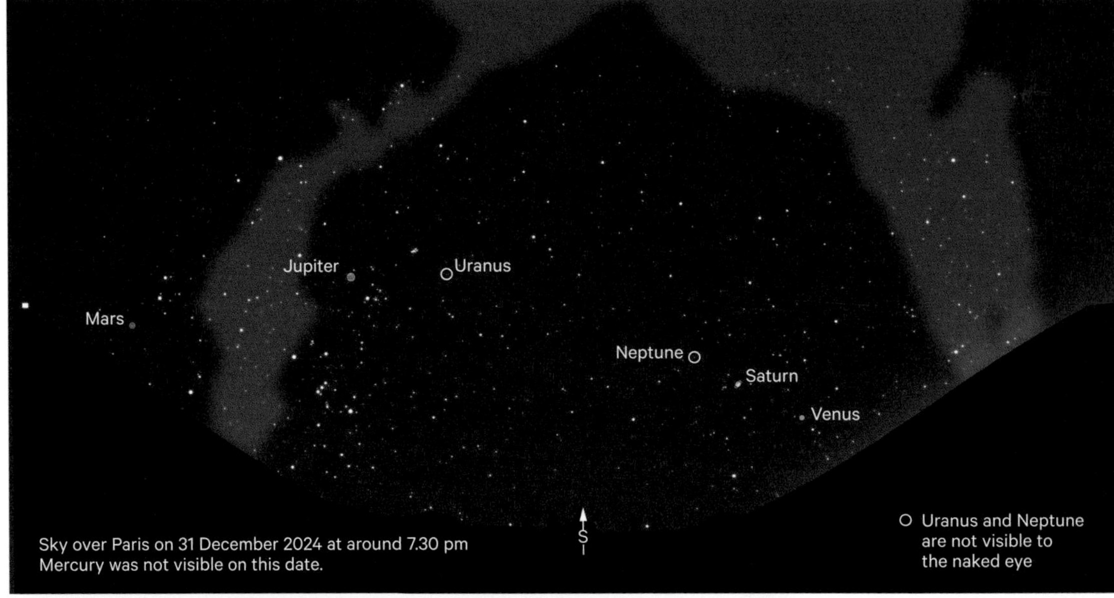

Sky over Paris on 31 December 2024 at around 7.30 pm
Mercury was not visible on this date.

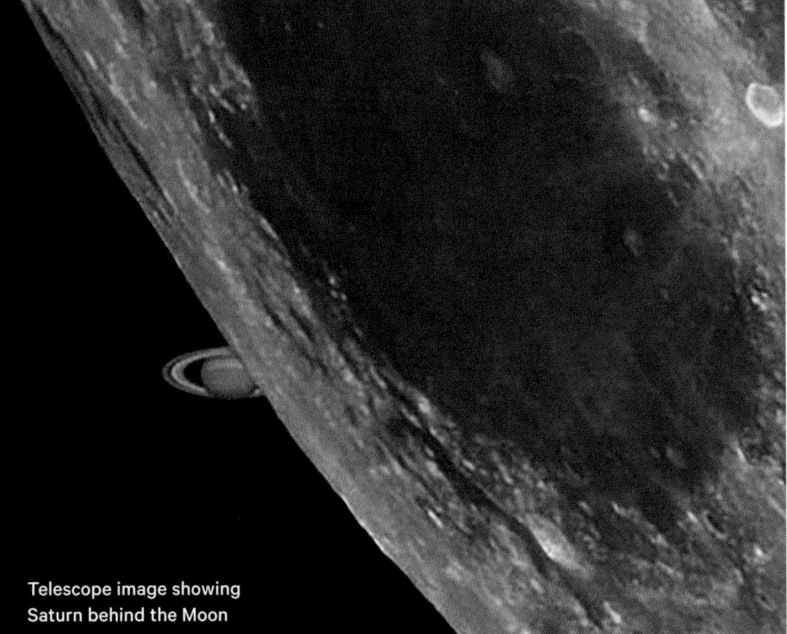

Telescope image showing
Saturn behind the Moon

A meeting of the planets

On the night of December 2024, shortly after sunset, six of the planets in the Solar System were aligned along the ecliptic. Only Uranus and Neptune were not visible to the naked eye, and Mercury had already set. It is a rare occurrence for these travellers across the sky to be seen together. Venus is brighter than the brightest star. Mars is most notable for its red colour, as well as the changes to its brightness. The periods of Jupiter and Saturn are extremely long: many decades pass before they are once again seen in the same constellation. Sometimes they are hidden by the Moon, proving that they are situated beyond its orbit.

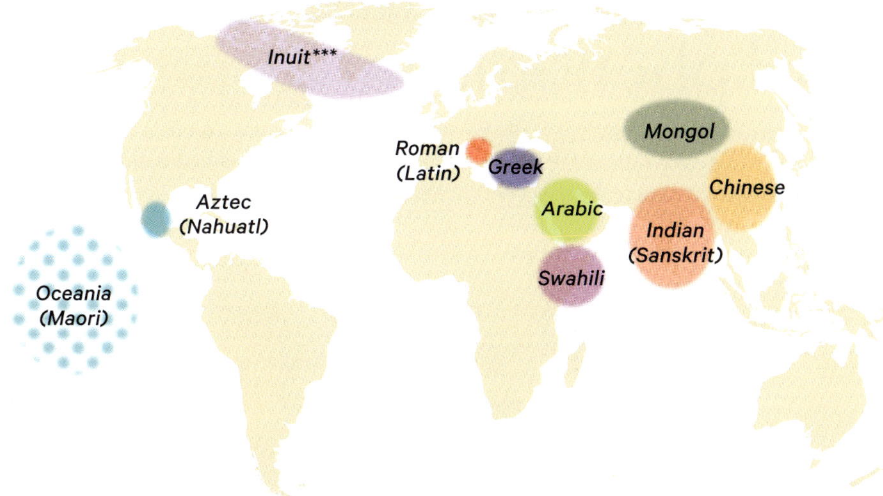

Sun	
Latin:	*Sol*
Greek:	*Helios*
Sanskrit:	*Sûrya*
Mongol:	*Nar*
Chinese:	*Tàiyáng*
Arabic:	*Chams*
Maori:	*Rā (Tama-nui-te-rā)*
Nahuatl:	*Tonatiuh*
Swahili:	*Jua*
Inuit:	*Siqniq*

Moon	
Latin:	*Luna*
Greek:	*Selene*
Sanskrit:	*Chandra*
Mongol:	*Sar*
Chinese:	*Yuèliang*
Arabic:	*Qamar*
Maori:	*Marama*
Nahuatl:	*Meztli*
Swahili:	*Mwezi*
Inuit:	*Taqqiq*

Saturn  378 days

Saturn	
Latin:	*Saturnus*
Greek:	*Phaenon*
Sanskrit:	*Shani*
Mongol:	*Sanchir*
Chinese:	*Tǔxīng*
Arabic:	*Zuhal*
Maori:	*Pareārau*
Nahuatl:	*Tzitzimicītlalli*
Swahili:	*Zohali*

Mars  780 days

Mars	
Latin:	*Mars*
Greek:	*Pyroeis*
Sanskrit:	*Mangala*
Mongol:	*Angarag*
Chinese:	*Huǒxīng*
Arabic:	*Marrīkh*
Maori:	*Matawhero*
Nahuatl:	*Chichilcitlalpol*
Swahili:	*Mirihi*

Jupiter  398 days

Jupiter	
Latin:	*Iuppiter*
Greek:	*Phaethon*
Sanskrit:	*Brihaspati*
Mongol:	*Barhasbadi*
Chinese:	*Muxing* or *Suixing*
Arabic:	*Mushtarī*
Maori:	*Hine-i-tīweka* or *Pareārau*
Nahuatl:	*Huēyitzitzimicītlalli*
Swahili:	*Jupita*

Mercury  116 days

Mercury	
Latin:	*Mercurius*
Greek:	*Stilbon***
Sanskrit:	*Budha*
Mongol:	*Bud*
Chinese:	*Shuǐxīng*
Arabic:	*Atarad*
Maori:	*Whiro*
Nahuatl:	*Payīnalcītlalli*
Swahili:	*Zebaki*

Venus  584 days

Venus	
Latin:	*Venus*
Greek:	*Phosphoros* or *Hesperos**
Sanskrit:	*Shukra*
Mongol:	*Sugar*
Chinese:	*Jīnxīng*
Arabic:	*Zuhra*
Maori:	*Kōpū (Tāwera* or *Meremere-tū-ahiahi*)*
Nahuatl:	*Citlalpol (Tlahuizcalpantecuhtli* or *Xólotl*)*
Swahili:	*Zuhura*

 Synodic period (the time it takes for a planet to orbit the Sun and return to its original position)

* Known by different names when viewed as a morning star or an evening star.

** Also called *Apollo* as a morning star and *Hermes* as an evening star.

*** The Inuit did not traditionally name the individual planets.

Heavenly bodies in many cultures

The Sun, the Moon and the five planets that are visible to the naked eye may well be the reason why such symbolic importance is attached to the number 7 in many cultures (days in the week, wonders of the world, deadly sins). Appearing as simple dots of light with no details when seen with the naked eye, these celestial bodies remained an enigma up until the 17th century, when their mass was first deduced from their motion and then calculated. Telescopes showed that their surfaces were discs, and allowed details such as Jupiter's Great Red Spot to be seen. The rings of Saturn were discovered, as were the moons of Mars, Jupiter and Saturn. The Greeks and Romans named the planets after their gods, while the Chinese named them after the traditional elements: earth, fire, water, metal and wood.

Celestial phenomena

Throughout human history, the regular paths of objects moving through the heavens and the unchanging nature of the starry sky contrasted with brief phenomena such as rainbows, the northern and southern lights and other celestial events, including eclipses and comets.

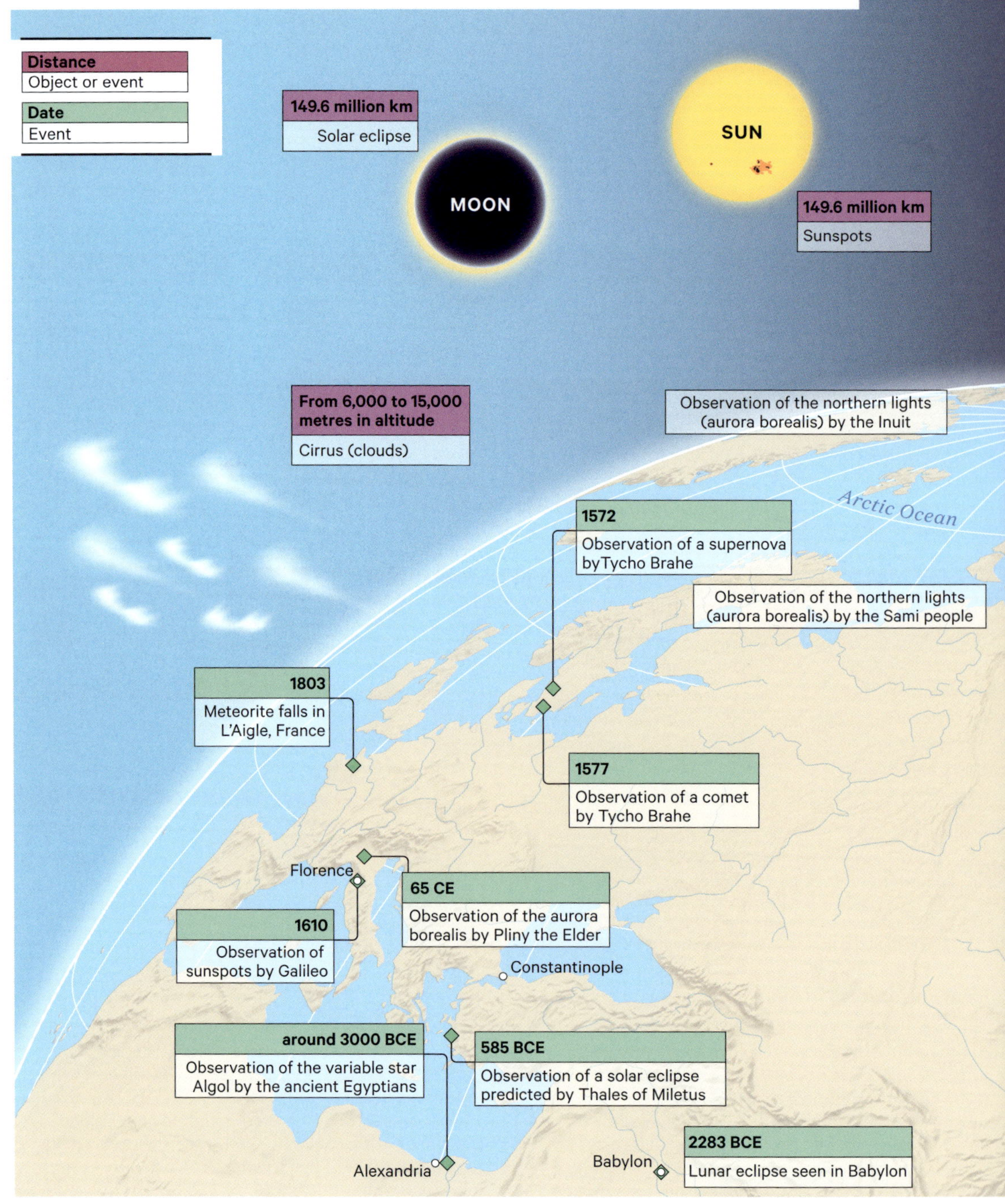

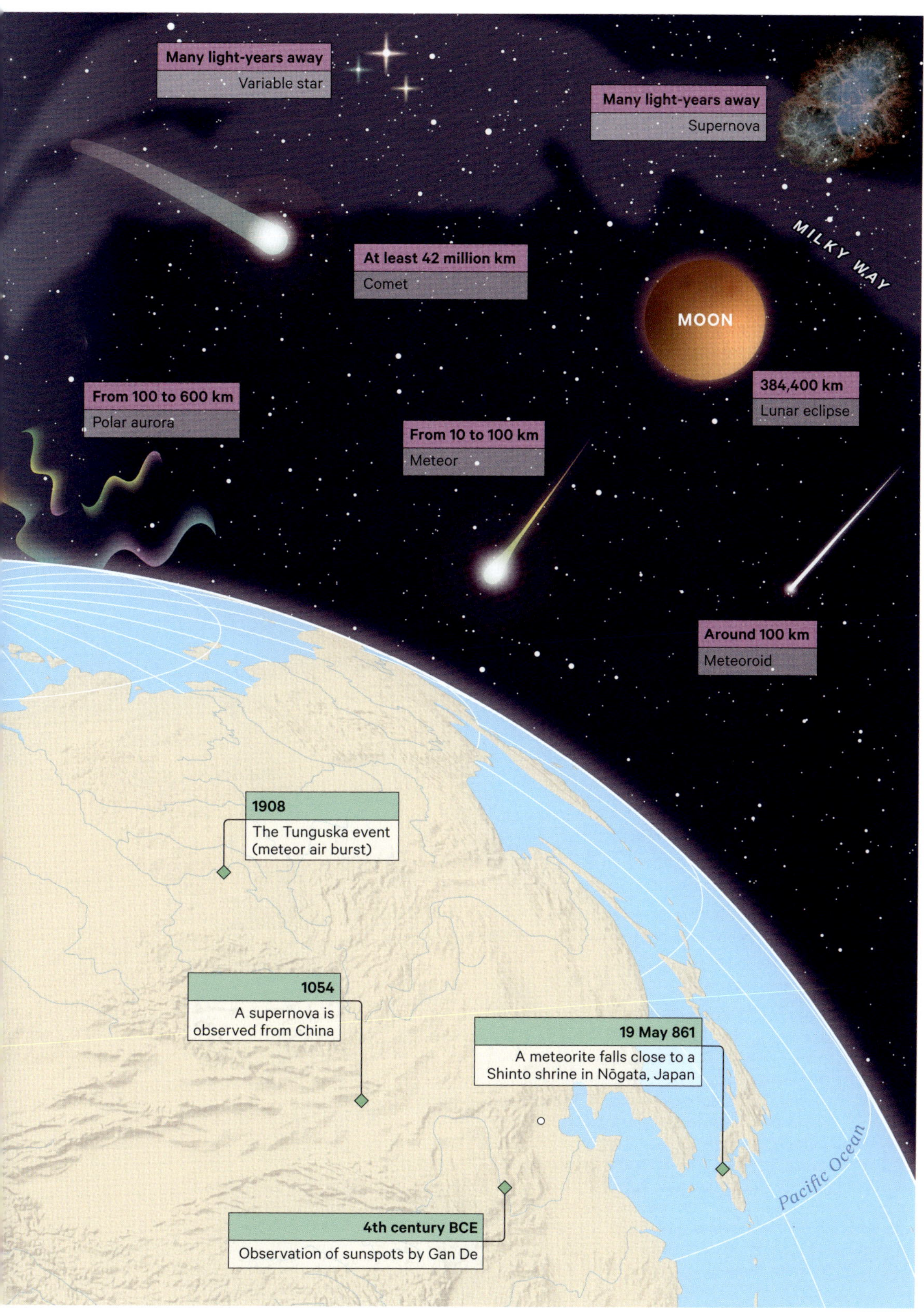
Many light-years away
Variable star
Many light-years away
Supernova
MILKY WAY
At least 42 million km
Comet
MOON
From 100 to 600 km
Polar aurora
From 10 to 100 km
Meteor
384,400 km
Lunar eclipse
Around 100 km
Meteoroid
1908
The Tunguska event
(meteor air burst)
1054
A supernova is
observed from China
19 May 861
A meteorite falls close to a
Shinto shrine in Nōgata, Japan
4th century BCE
Observation of sunspots by Gan De
Pacific Ocean

Humans and gods

When humans first looked up at the sky and the stars, they were faced with a vastness that was so much greater than them and raised profound questions. Astrology, divination, philosophy and religion sought answers to these questions. Where do humans come from? How was the world made? What happens after death? What should we base our morality on? The same universal questions also gave rise to science.

Widespread creation myths* (oral tradition)

△ Myth that humans originated from the ground

▽ Myth that humans originated from the ocean

✦ Myth that humans originated from the sky

* A single point may refer to several sources inside a small geographical area.

Places of emergence of major contemporary faiths

(j) Judaism

(c) Christianity

(i) Islam

(h) Hinduism

(b) Buddhism

(s) Shinto

(z) Zoroastrianism

(c) Confucianism

(t) Taoism

(s) Sikhism

(j) Jainism

Greco-Roman polytheism, source of many astronomical names

Reaching for the heavens

Every continent has its own myths that seek to explain how humans came into existence. Some say that we came from the sky, others from the Earth, or even from the sea. These myths are often found in the stories and foundational texts of religions; however, access to the spiritual is not reserved for their believers alone. With varying degrees of explicitness, many religions incorporated the sky into their beliefs. It was the home of the gods in Greek and Roman culture. It is where the souls of the dead go, to be rewarded for their life on Earth. It communicates with humans to guide or punish them. An image of stability, eternity and infinity, the sky is the epitome of unachievable perfection and creative omnipotence. It goes beyond anything our minds can conceive of, a concept that is called transcendence.

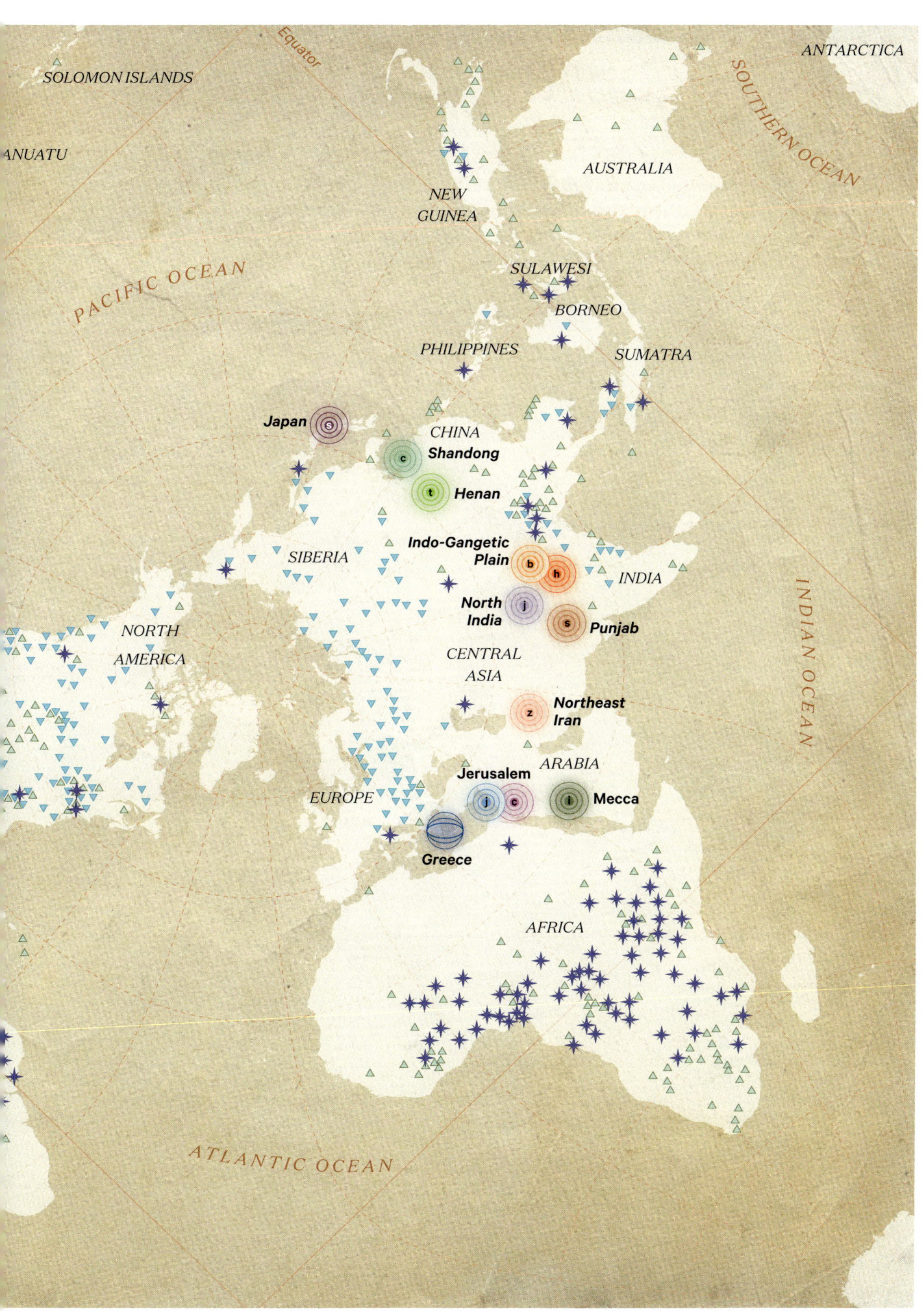

ANTARCTICA
SOUTHERN OCEAN
SOLOMON ISLANDS
Equator
VANUATU
AUSTRALIA
NEW GUINEA
PACIFIC OCEAN
SULAWESI
BORNEO
PHILIPPINES
SUMATRA
INDIAN OCEAN
Japan
CHINA
Shandong
Henan
Indo-Gangetic Plain
SIBERIA
INDIA
North India
Punjab
NORTH AMERICA
CENTRAL ASIA
Northeast Iran
ARABIA
Jerusalem
Mecca
EUROPE
Greece
AFRICA
ATLANTIC OCEAN

Knowledge of the sky is power: 4,000 BCE to the present day

For forty centuries, understanding the sky has been a source of power for those who study the heavens, as well as for rulers. Observing and predicting changes in the sky, giving structure to space and time based on celestial events, creating shared points of reference, or even – in the 20th century – going to space are just a few aspects of this power.

Observing and predicting

The regularity of celestial phenomena means that they can be predicted, which is useful for agriculture as well as science. Irregular events, such as eclipses or comets, must not be allowed to catch those in power off guard.

Structuring time and space

Measuring time meant that those in power could set the dates of ceremonies and events using a calendar. The unchanging nature of the sky also allowed them to organize space: the layout of towns, religious buildings, or the movements of soldiers.

Creating universal reference points

Arithmetic and geometry were developed. Units of measurement for time and size, set out by those in power, made trade relationships easier.

Entering space

Technological advances have allowed humans to explore the Solar System and beyond, but have also opened new possibilities for surveillance and warfare.

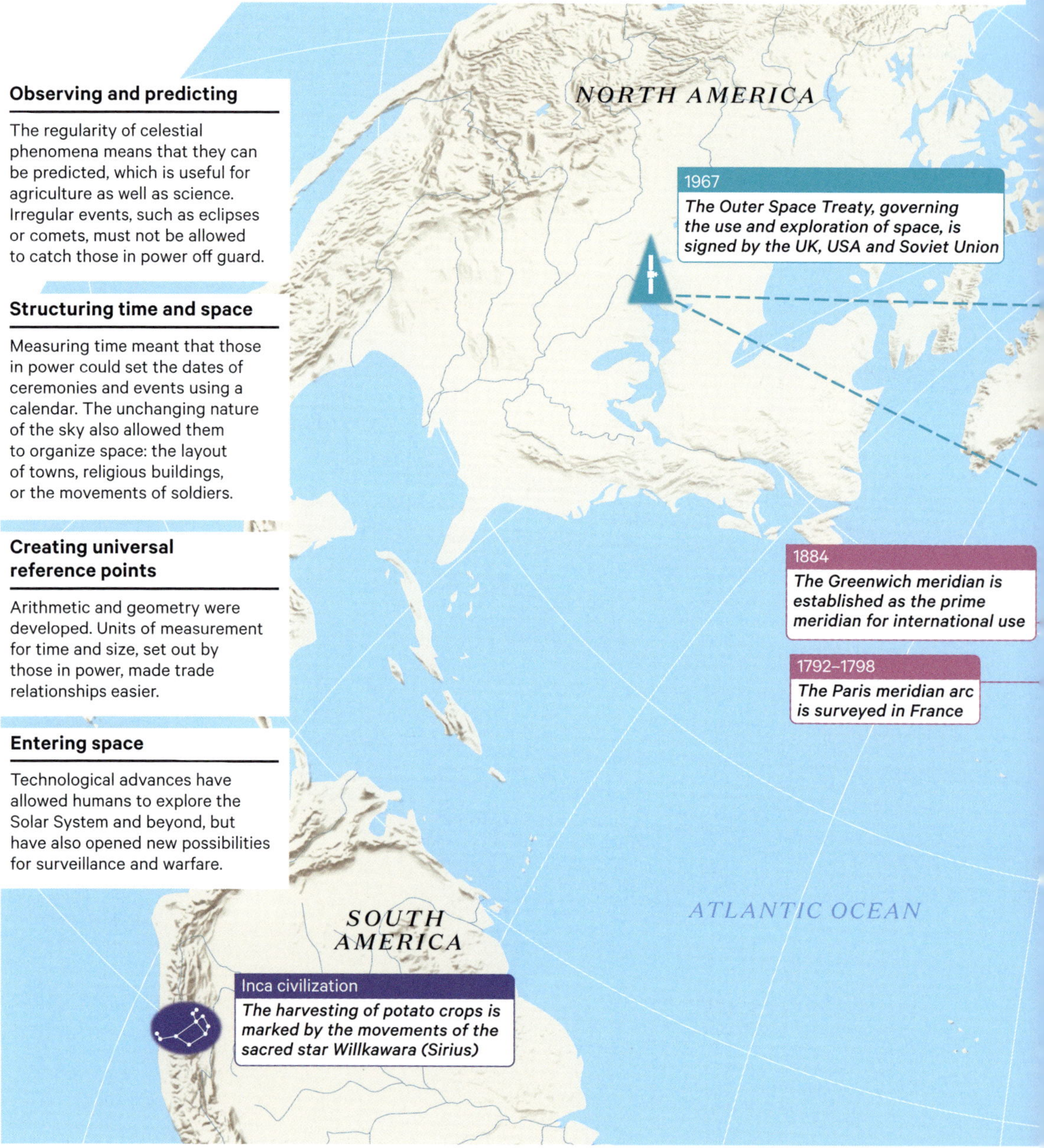

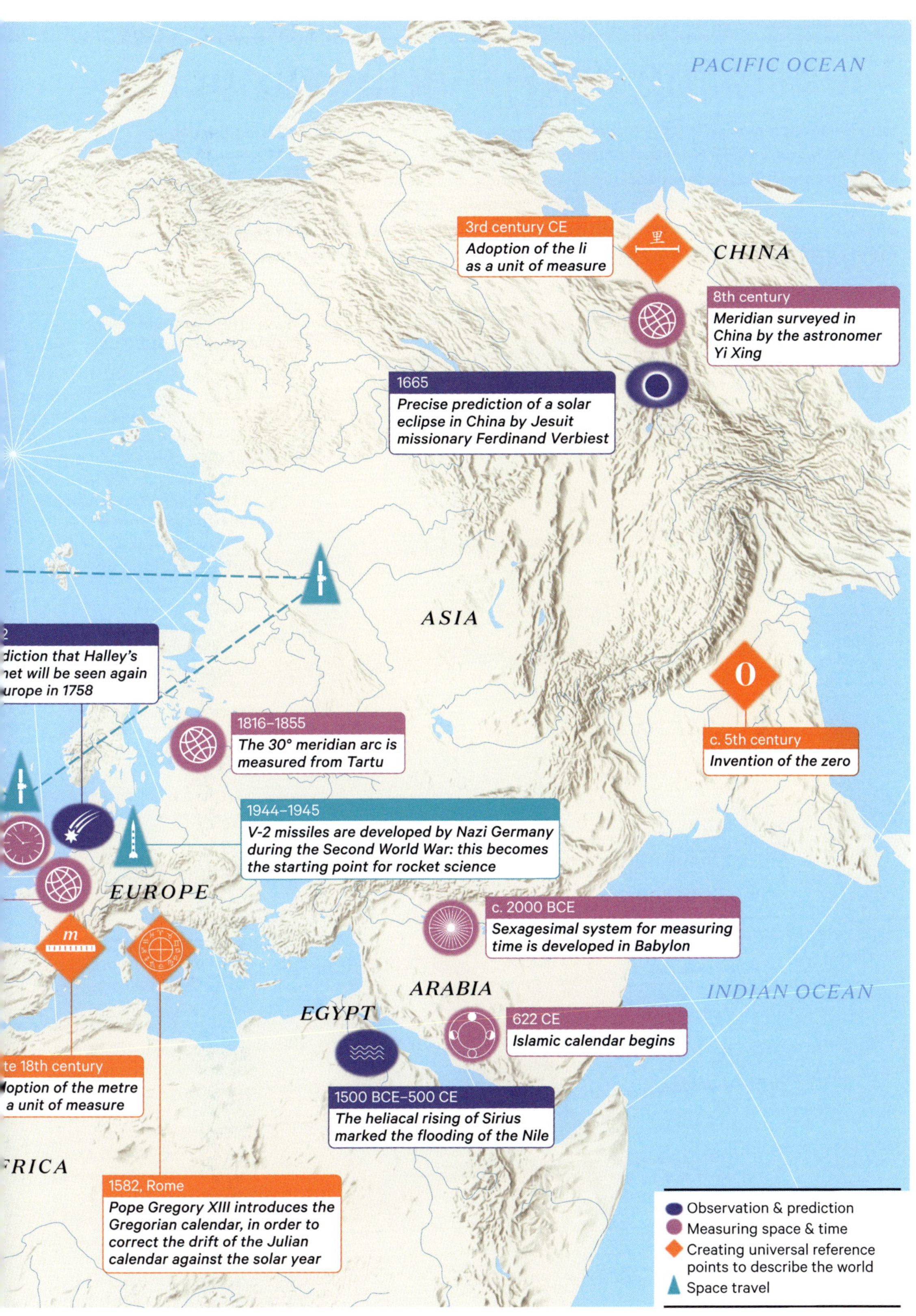

PACIFIC OCEAN
CHINA
里
3rd century CE
Adoption of the li as a unit of measure
8th century
Meridian surveyed in China by the astronomer Yi Xing
1665
Precise prediction of a solar eclipse in China by Jesuit missionary Ferdinand Verbiest
ASIA
diction that Halley's
net will be seen again
urope in 1758
1816–1855
The 30° meridian arc is measured from Tartu
0
c. 5th century
Invention of the zero
1944–1945
V-2 missiles are developed by Nazi Germany during the Second World War: this becomes the starting point for rocket science
EUROPE
m
c. 2000 BCE
Sexagesimal system for measuring time is developed in Babylon
ARABIA
INDIAN OCEAN
EGYPT
622 CE
Islamic calendar begins
te 18th century
loption of the metre
a unit of measure
1500 BCE–500 CE
The heliacal rising of Sirius marked the flooding of the Nile
FRICA
1582, Rome
Pope Gregory XIII introduces the Gregorian calendar, in order to correct the drift of the Julian calendar against the solar year
Observation & prediction
Measuring space & time
Creating universal reference points to describe the world
Space travel

Measuring and dividing space

Many societies looked to the sky to understand where they were on Earth. The rising, setting and midday sun, and the stars that are aligned with the Earth's axis of rotation (Polaris, the Southern Cross) allowed them to determine directions, the cardinal points. These markers were used for travelling and drawing up maps. The realization that when you move north or south, the height of the stars varies, made it possible to measure latitude. Understanding and organizing space in the sky and on Earth therefore went hand in hand.

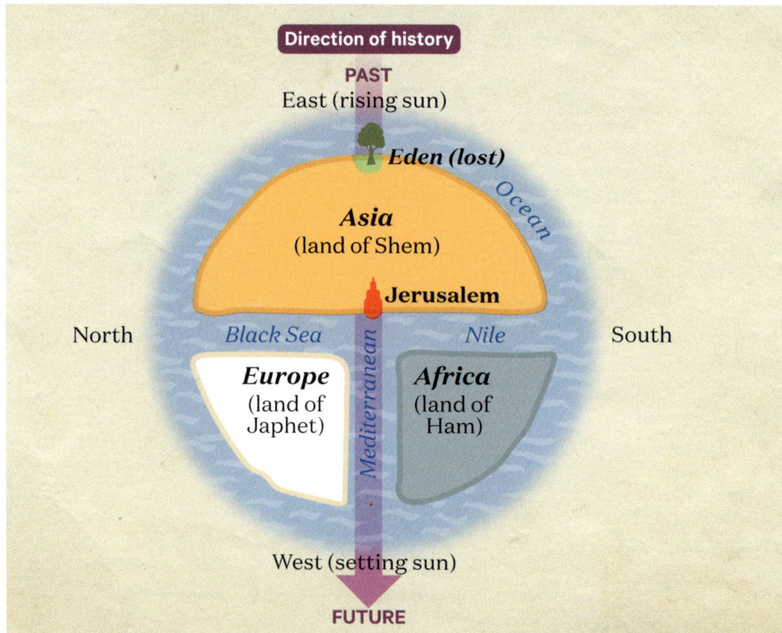

The medieval *mappa mundi*

In medieval thought (Isidore of Seville, 7th century), the known world was divided between the three sons of Noah. Their territories had ancient names: Asia and Europe (the two sides of the Aegean Sea, where the sun rises and sets) and Africa (Roman Tunisia). These three parts of the world were surrounded by the ocean and became the sources of our current continents. This way of organizing space is related to how time was organized: history moved in the same direction as the Sun. The Garden of Eden was in the east and Jerusalem was at the centre, just as time was divided into before and after the birth of Christ.

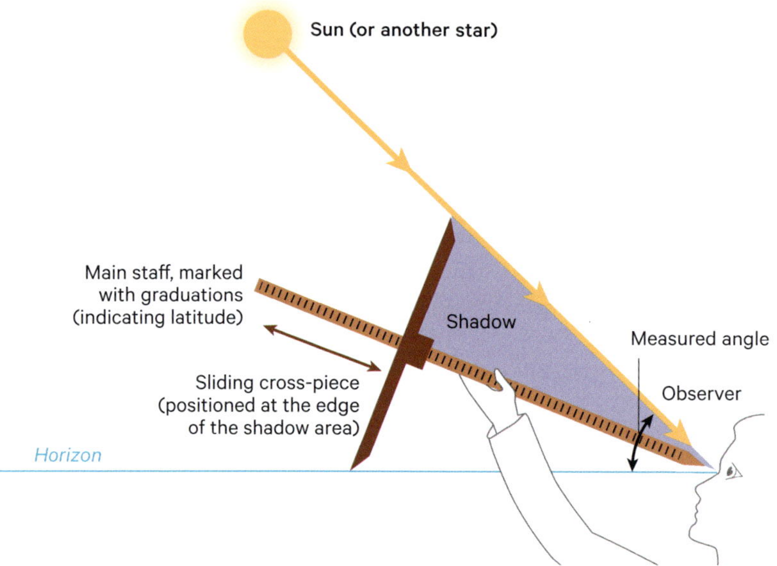

Jacob's staff

The Jacob's staff, or cross-staff, was used by sailors to measure angles and distances, drawing on simple principles of trigonometry. It was used by navigators from the 16th to the 18th century, when it was replaced by the octant, and was also used by astronomers and surveyors. It was made from a long stick marked with graduations (the main staff) with another stick that slid up and down (the cross-piece). The observer trained it on a fixed point, often the horizon, and aligned the Sun or another star with the cross-piece. The graduations marked on the main staff showed the angle.

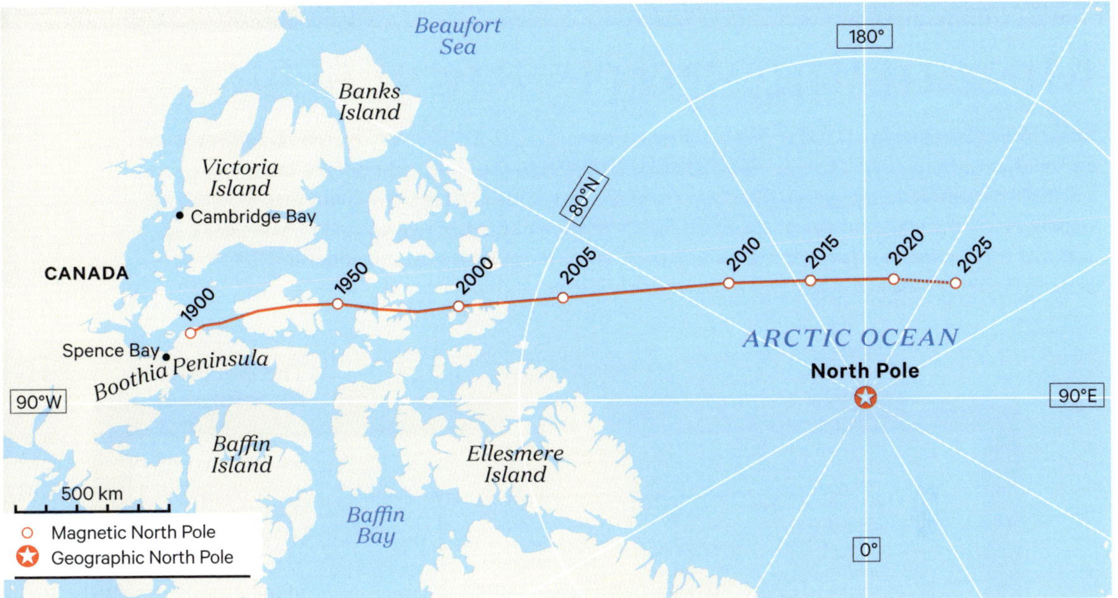

▲ The moving magnetic poles

A compass points to magnetic north, not geographic north. Magnetic north is the place where the Earth's magnetic field, which protects it from cosmic rays and solar wind, points vertically down. Referring to it as the magnetic 'north' pole is pure convention, due to its proximity to geographic north. It moves position rather quickly, depending on the movements of the Earth's iron core, from 15 km a year in 1980 to 55 km a year since 1990, from Canada towards Siberia. What's more, the Earth's magnetic field regularly flips and magnetic 'north' shifts to near the geographic South Pole.

▼ The compass rose

Nautical maps known as 'portolan charts' were used on great European voyages from the time compasses first appeared in the West in the 12th century up until the 18th century. These maps are organized around a web-like grid of rhumb lines that radiate out from compass roses or 'windroses', leading in the major navigational directions. This type of cartography is the origin of the way maps are traditionally oriented towards north. As seen in the example below, the starlike shape of the compass rose is often topped with a fleur de lys to mark north.

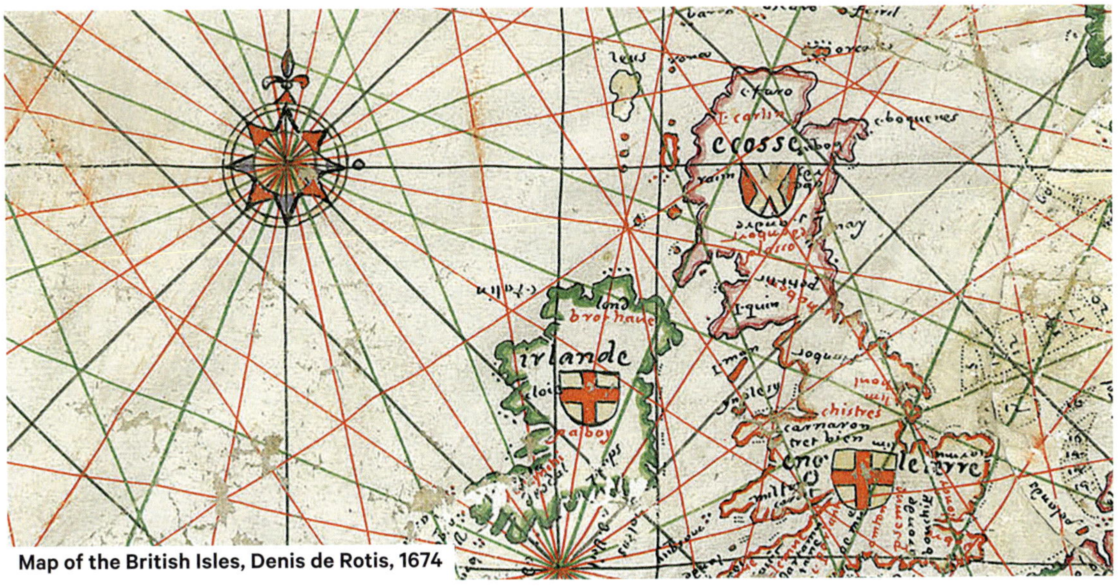

Map of the British Isles, Denis de Rotis, 1674

Organizing space

Many examples of both sacred and vernacular architecture are deliberately constructed to face the sun or align with the axis of other celestial bodies. Christian churches usually face the rising sun, as a symbol of Christ's resurrection. The corners of the pyramids of Giza are aligned with the cardinal points, as are the longhouses built by the Iroquois of North America. However, interpretations of many of these monuments remain disputed, such as the megaliths of Stonehenge.

NORTH GATE
WEST GATE
EAST GATE
SOUTH GATE
Chamber to the left of the Dark Hall
1st moon of winter
玄堂個左
Chamber to the right of the Zongzhang Temple
3rd moon of autumn
総章個右
Dark Hall (Xuantang)
2nd moon of winter
玄堂大廟
Chamber to the left of the Qingyang Temple
1st moon of spring
青陽個左
Chamber to the right of the Dark Hall
3rd moon of winter
玄堂個右
Zongzhang Temple
2nd moon of autumn
総章大廟
Great Hall of the Palace
大廟大至
Qingyang Temple
2nd moon of spring
青陽大廟
Chamber to the left of the Zongzhang Temple
1st moon of autumn
総章個左
Chamber to the right of the Bright Hall
3rd moon of summer
明堂個右
Bright Hall (Mingtang)
2nd moon of summer
明堂大廟
Chamber to the left of the Bright Hall
1st moon of summer
明堂個左
Chamber to the right of the Qingyang Temple
3rd moon of spring
青陽個右
10 m

The 'Divine Palace' of China

The diagram above shows the ground-floor layout of the Wanxiang Shen'gong, or 'Divine Palace of Myriad Phenomena'. This three-storey wooden building, measuring 98 metres in height, was the largest building in the world in the 7th century. Built by Wu Zetian, a female emperor of the Tang dynasty, in 687 in Luoyang, China, it was destroyed by war in 762. Aligned with the celestial axis of the surrounding palace complex, its layout reflects the Earth and its seasons, with the emperor holding court in different rooms depending on the time of year. The Chinese year is made up of twelve lunar months, each lasting twenty-nine or thirty days, sometimes complemented by a thirteenth. The careful orientation of the building reflected the mandate of the emperor and her advisors, whose decisions had to reconcile earth and sky and assure harmony for the people, according to the Taoist conception of the universe. Similar palace buildings existed before this one – the concept can be traced back to 1000 BCE – but none were quite as impressive.

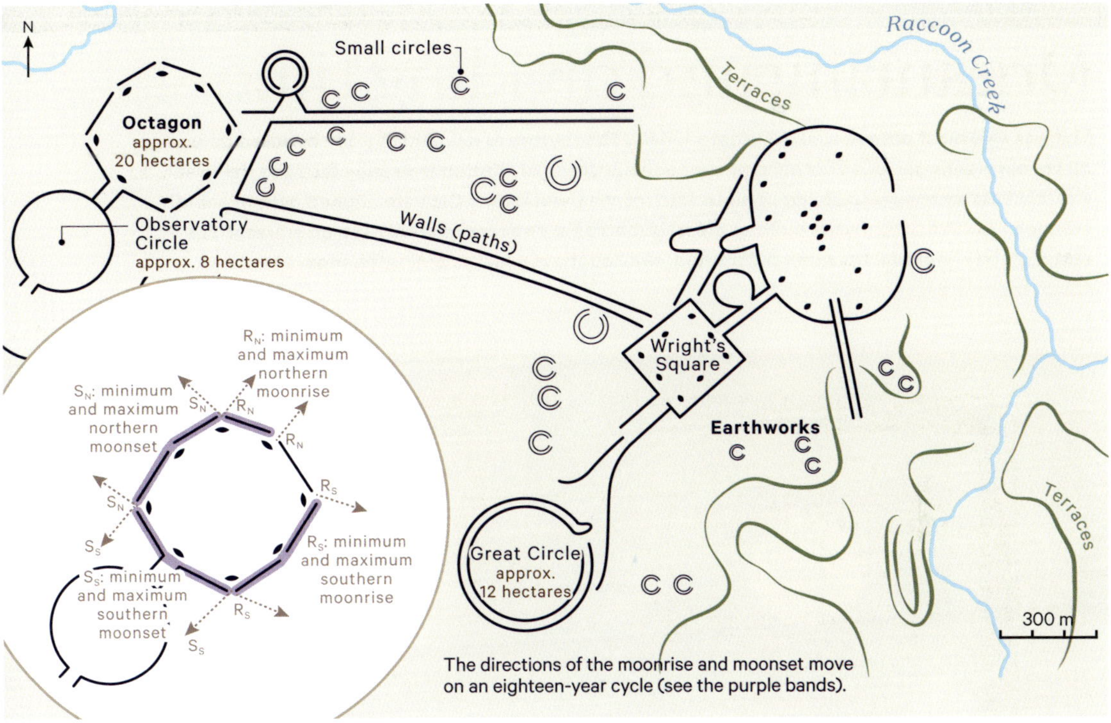

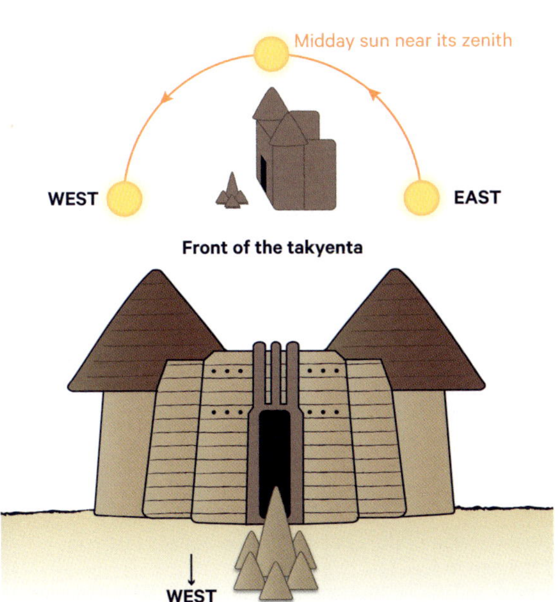

The directions of the moonrise and moonset move on an eighteen-year cycle (see the purple bands).

The Newark Earthworks

This pre-Columbian archaeological site in Newark, Ohio, was built by the Hopewell culture (a subgroup of the Mound Builders) between 100 BCE and 400 CE. Less than 10% of the site has been preserved since the arrival of European colonizers. The surviving area (83 of the original 1,200 hectares) includes three major earthwork complexes: the Great Circle, the Octagon and Wright's Square. These were used as sites for ceremonies, social gatherings and trade. The Octagon complex was a lunar observatory, created to track the movements of the Moon, according to the saros, a cycle of 223 lunar months (just over 18 years), at the end of which the Moon, Earth and Sun return to the same positions.

Tower-houses of the Batammariba

Koutammakou is a region in Togo and Benin that is home to the Batammariba or Tammari people, who build traditional earthen tower-houses known as 'takyenta'. The Batammariba seek to follow the example of Kuiye, the sun god and creator, and build their mud-brick homes over two levels to honour this god. The main façade always faces west. The celestial village where Kuiye lives is believed to be in the western sky and, when the sun sets in the evening, Kuiye's rays are welcomed into the house. Every evening, because at low latitudes the apparent movements of the Sun do not change as much with the seasons, the sanctuaries of ancestors beneath the lower room are lit with natural light. This direct contact allows the spirits of the ancestors to speak to the god and intercede on their families' behalf.

Measuring and dividing time

Humans, like most primates, are diurnal animals. The rhythm of day and night is essential to us. All known human societies noted changes to the position of the Sun over the course of the year and came up with ways to divide up time: into months, weeks and days then hours, minutes and seconds. However, the rhythms of the celestial bodies are not completely regular, which led to complicated compromises when determining the length of years and months. Hours and their subdivisions are less complicated.

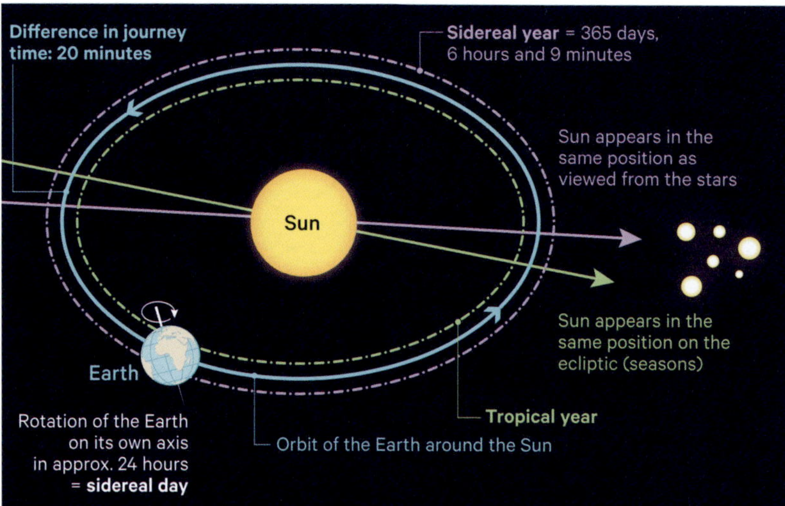

The solar year

The Sun is used to define a year. The Earth takes around 365.25 days to complete one orbit around the Sun, using unmoving distant stars as a reference point. This is known as a sidereal year. It lasts around 20 minutes longer than a tropical year, which is the time it takes for the Sun, as viewed from the Earth, to return to the same position on the ecliptic. The tropical year is the time it takes to complete a full cycle of astronomical seasons.

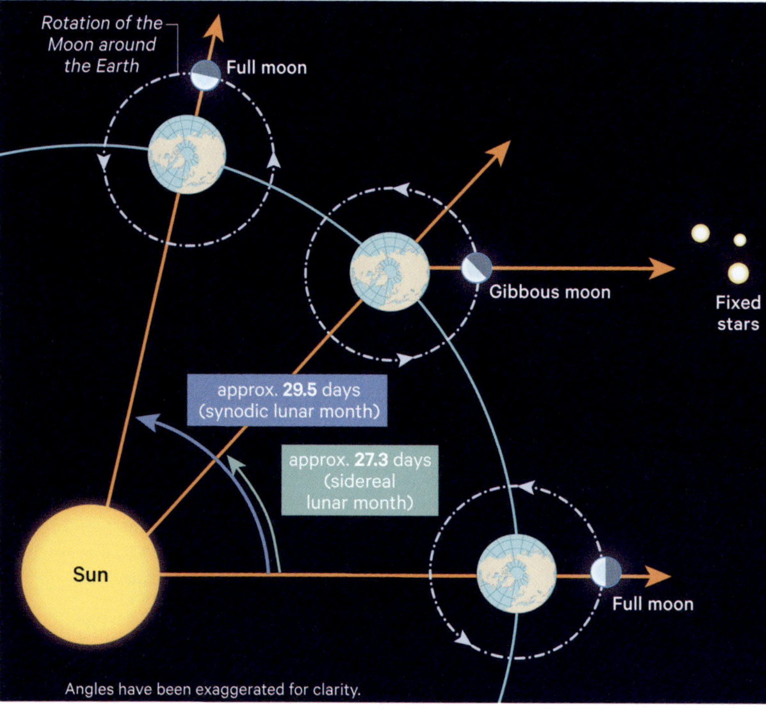

The lunar month

The length of time between two identical phases of the Moon (two full moons, for example) is 29.5 Earth days. That is therefore the length of a lunar day, the time it takes for the Moon to complete one rotation around its axis with respect to the Sun. The time it takes for the Moon to complete one rotation relative to the fixed stars is 27.3 Earth days. The difference between these two periods (synodic and sidereal) is due to the fact that, while the Moon is orbiting around the Earth, the Earth itself is moving relative to the Sun. The synodic period is used as the basis for defining a month.

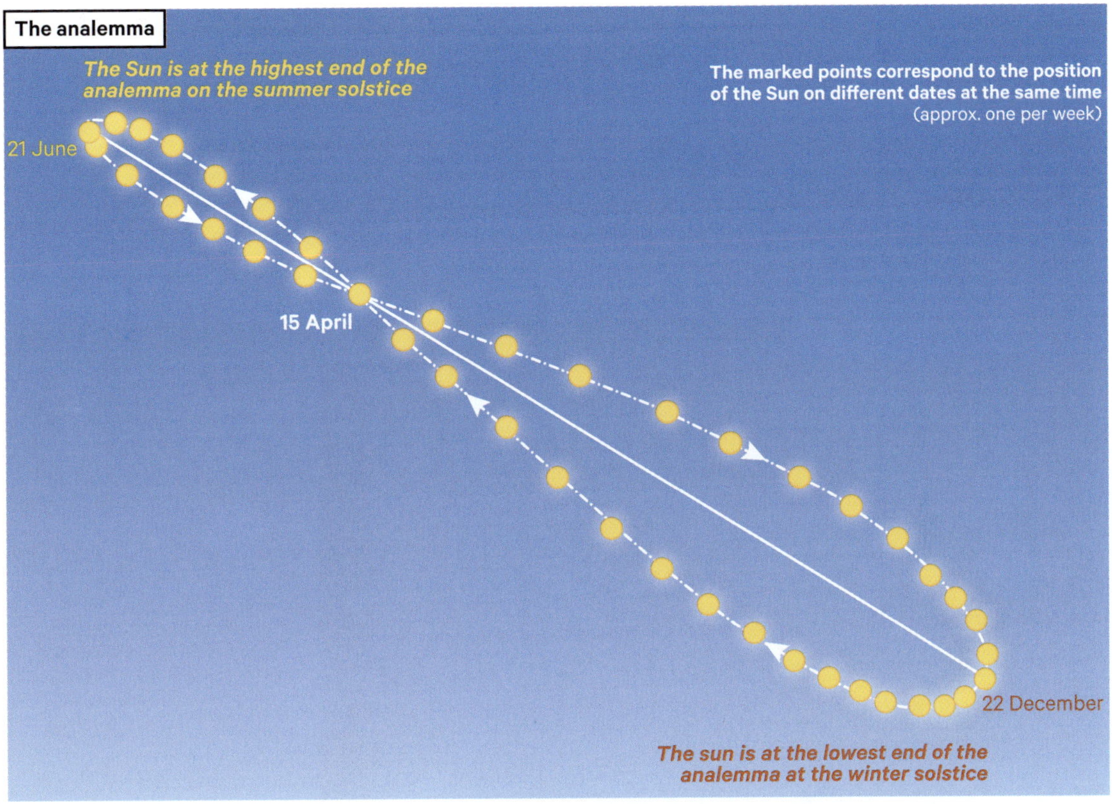

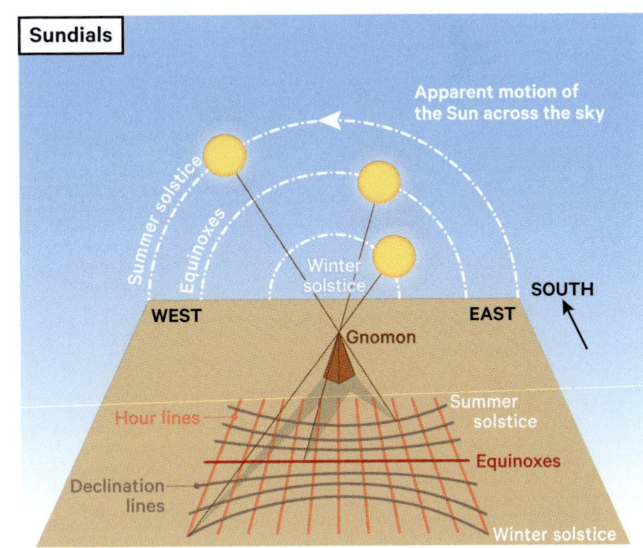

Hours and seconds

Since the time of the Sumerian civilization, the length of a day, which is determined by the Earth's rotation, has been divided into 24 hours. A sundial was one of the first objects used by humans to measure the passing of time. The time of day (the hour) is indicated by the length and direction of the shadow cast by the gnomon. An analemma is a diagram showing the position of the Sun in the sky at the same time and from the same place over the course of a year. It can be used to correct the time shown on a sundial.

The definition of a second is derived from that of an hour. It is a sixtieth of a minute, which is a sixtieth of an hour. The way of measuring a second has changed over time, as shown by the timeline below.

The evolving definition of a second

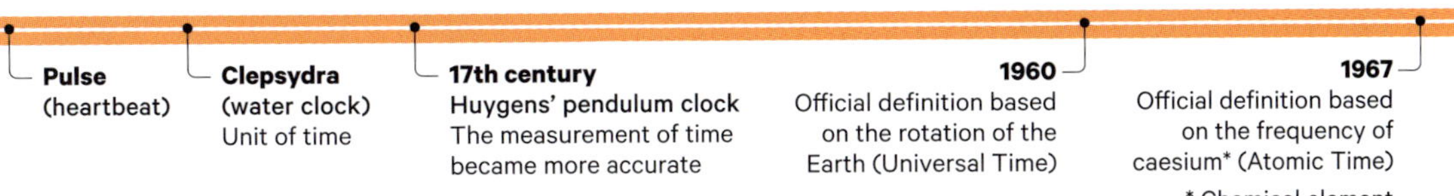

Calendars

Western societies have imposed a linear concept of time on much of the rest of the world. Most ancient societies, however, lived according to cycles of shorter duration. The choice of where to begin counting calendar years is purely cultural. Using the supposed date of Christ's birth, as proposed in the 6th century by Dionysius Exiguus, made it possible to assign numbers to the years. It should be noted that the concept of zero did not exist in the West at that time.

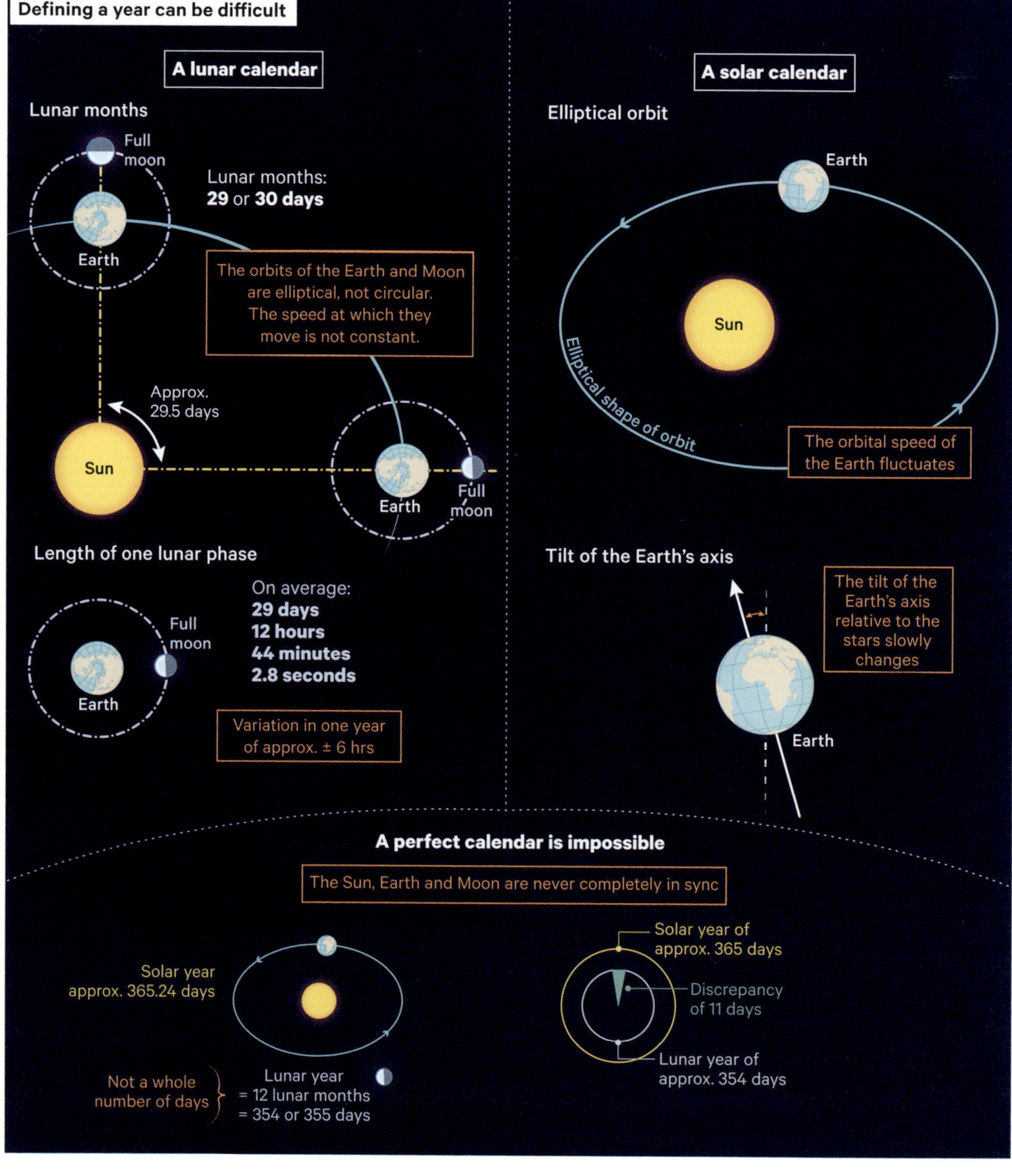

1 January 2025 in different calendars

Calendars in use around the world

Gregorian • 1 January 2025
(starting point: the birth of Christ)

Julian • 19 December 2024
(starting point: the birth of Christ)

Hijri (Islamic) • 1 Rajab 1446
(starting point: the Hijra)

Hebrew • 1 Tevet 5785
(starting point: creation of the world, according to the Torah)

Chinese • 2nd day of the 12th month of the year Jia Chen
(starting point: start of the reign of the Yellow Emperor)

Julian day (used in astronomy) • Day number 2,460,677
(starting point: 1 January in the year 4713 BCE)

Calendars no longer in use

Babylonian • 20th day of the 9th month of the year 2773
(starting point: beginning of the reign of the Babylonian king Nabonassar, on 26 February 747 BCE)

Ancient Rome • 14th day before Kalends of January 2778
(starting point: the foundation of the city of Rome)

If astronauts were to spend several months on the surface of Mars, what would their calendar be like?
Answer on page 225

The date of Easter

←——— **PASSOVER (PESSACH)** ———→
26 March   26 April

Begins on the 15th day of the first month (Nisan); lasts 7 days in Israel and 8 days in the diaspora

| March | April | May | June |

←——— **EASTER (CATHOLIC** ———→   ←——— **ASCENSION** ———→
22 March **& PROTESTANT)** 25 April   30 April   3 June

The Sunday following the first full moon after the spring equinox

40 days after Easter

←——— **PENTECOST** ———→   *50 days after Easter*
10 May   13 June

The date of Ramadan

It takes almost 40 years for Ramadan to fall roughly on the same dates again

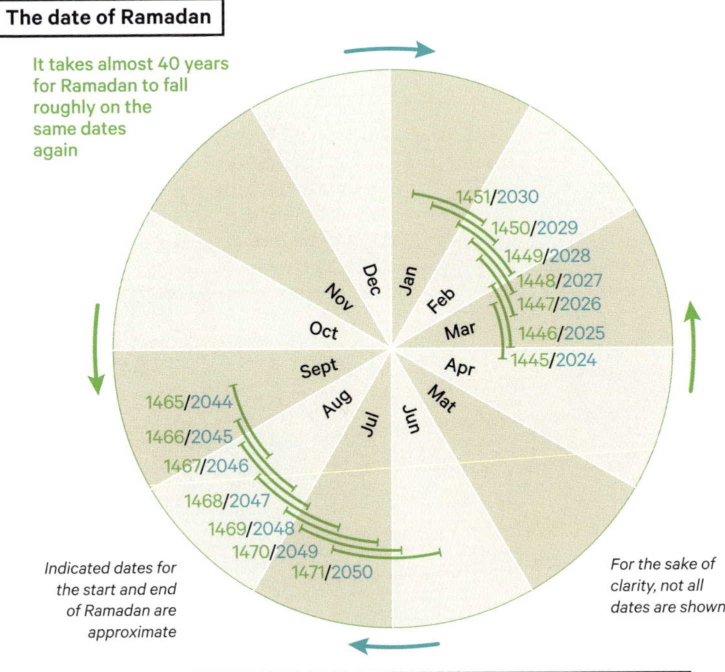

Indicated dates for the start and end of Ramadan are approximate

For the sake of clarity, not all dates are shown

Hijri (Islamic) calendar: 354 or 355 days a year
1445 Year   ⊢—⊣ Dates of Ramadan (the 8th month)
Gregorian calendar: 365 or 366 days a year
2024 Year   → Movement of time in the course of a year
Oct. Month   → Shift in the dates of Ramadan (≈ 10 or 11 days a year)

A multitude of calendars

Different civilizations invented their own calendars, sometimes based on the Moon and sometimes based on the Sun or on other celestial bodies. Introduced by Pope Gregory XIII in 1582, the Gregorian calendar is a solar calendar that includes leap years. While it is the main international calendar in use today, other calendars are also still used, for religious reasons (the Julian calendar by Orthodox Christians, the Hijri calendar by Muslims) or cultural reasons (the Chinese calendar). Others serve a technical purpose, like the calendar of Julian Days. Used to date astronomical events, this daily calendar makes calculations involving dates easier because it does not rely on complex calendar cycles, such as the uneven lengths of months or leap years.

Astrology and the zodiac

Every society has tried to predict the future, but not every society has used the sky to do it. Drawing on Mesopotamian traditions, the Greeks created the zodiac and Western astrology. Thousands of miles away in Asia, China developed a different system of astrology based on a twelve-year cycle.

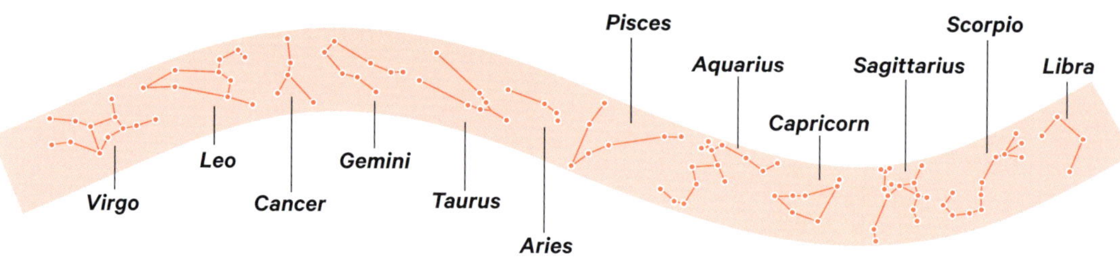

From Mesopotamia to the Western zodiac

In the Babylonian world, the gods were thought to speak to humans through the changing sky. Over the course of its cycle, the Moon moves through twelve different constellations. In winter, these form an arch, while in summer, they form a bowl shape. When combined, they form an S on its side, a belt of sky divided into twelve sections, within which the Moon, Sun and planets move.

This was used to draw up horoscopes, which originally applied to the community as a whole. After the conquests of Alexander the Great, the Greek world came into contact with this way of interpreting the stars, particularly through Eudoxus of Cnidus (406–355 BCE), who introduced the twelve zodiac signs. Horoscopes later developed into personalized predictions.

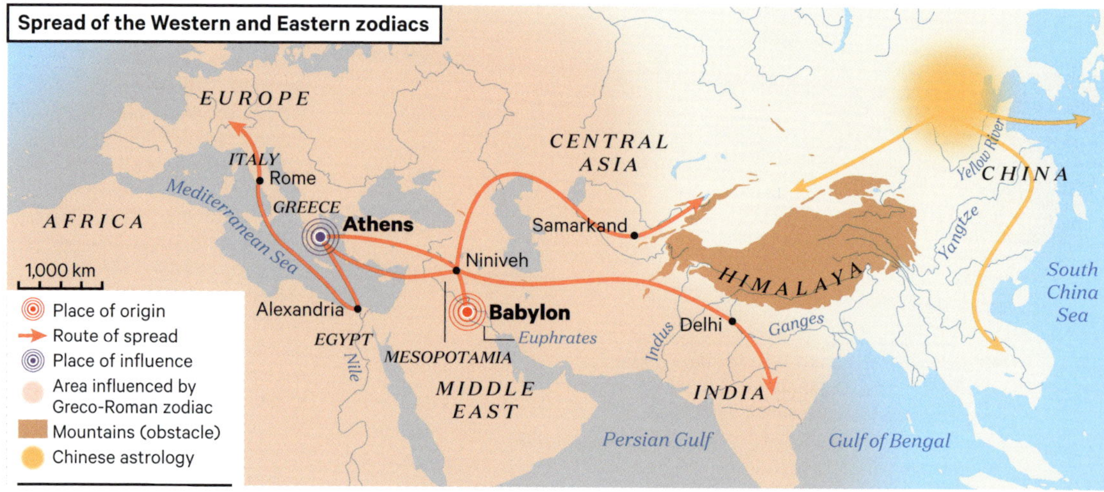

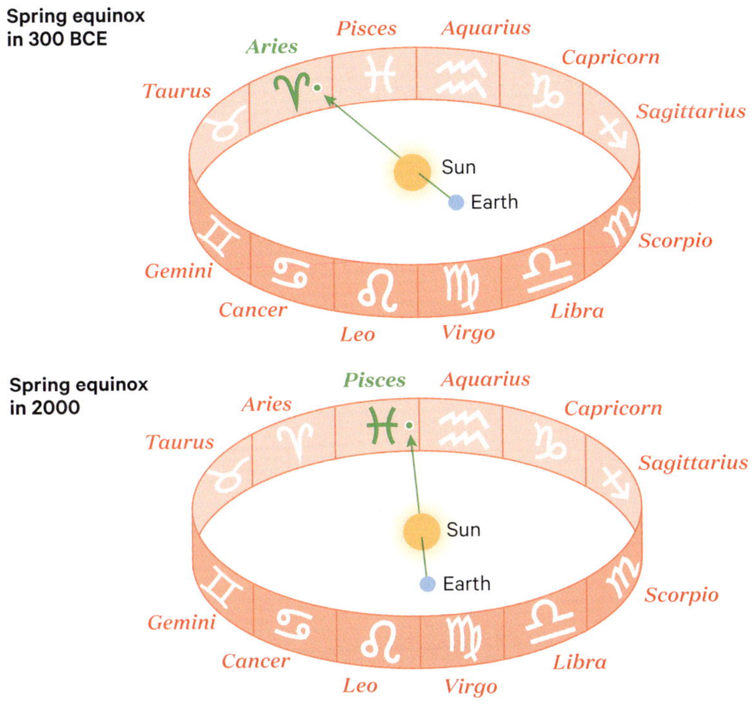

Astronomy takes over from astrology

Astrological predictions, based on the idea that celestial bodies have a direct effect on human lives, have no basis in science. This can be seen this from the position of the vernal point – the point at which the circle of the ecliptic crosses the circle of the equator – and the astrological role of the constellation Aries. The vernal point was situated in Aries during the time of Hipparchus, but has moved west and is now found in Pisces, and will be in Aquarius next century. This effect is called the precession of the equinoxes.

Zodiac belt

Vernal point (position of the Sun seen from the Earth at the spring equinox)

The human zodiac

In the 2nd century, Claudius Ptolemy wrote a major work of astronomy called the *Almagest*, and also the equally influential *Tetrabiblos*. In it, he set out a theory of astrology that involved a series of connections between the human world and the universe, the microcosm and the macrocosm. It sees the cosmos as an immense body whose planets and constellations have anatomical and psychological equivalents in humans. The planets that are situated in the corresponding sign at the moment when an event takes place, or when an individual is born, imbue them with personality traits associated with those planets (the wisdom of Jupiter, the impulsiveness of Mars, the conservatism of Saturn). This allows astrologers to assign predispositions to a person or characteristics to a specific moment in time. The signs of the zodiac are each associated with a specific part of the body.

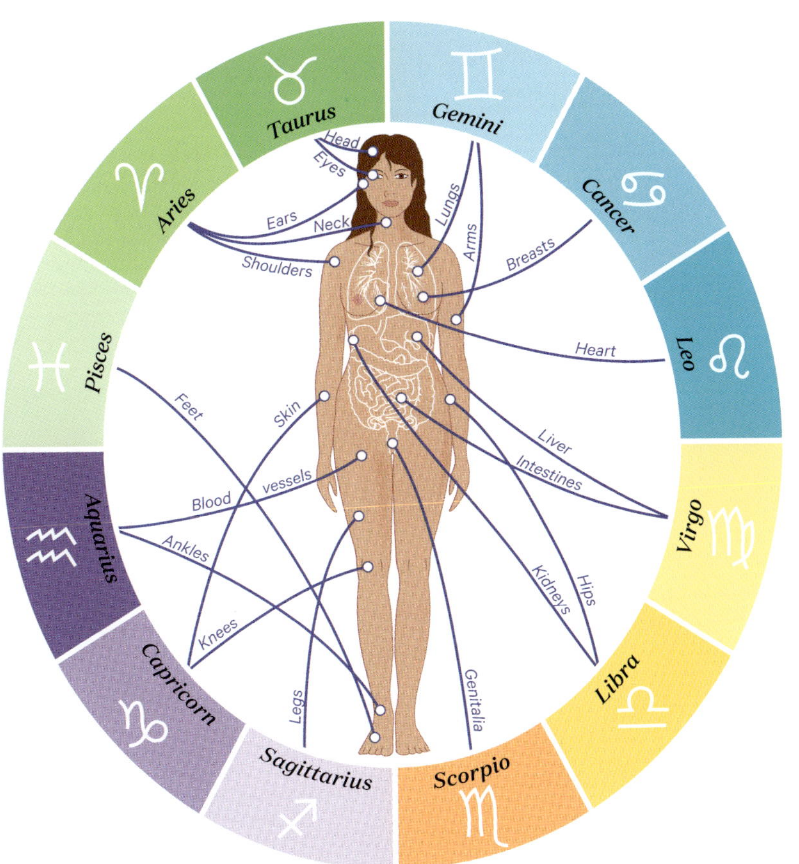

Indigenous peoples of the Americas

All indigenous American cultures placed great religious and societal importance on the stars. Their understanding of astronomy, particularly that of the Maya, was often as advanced as that in ancient Europe. Their agricultural life was governed by the rising of celestial bodies. They also used the stars to guide them on journeys; Polaris was used in this way by for the First Nations of modern-day Canada.

The Aztec sun stone

This *cuauhxicalli* (sacrificial stone) of 1479 is sometimes incorrectly called an 'Aztec calendar'. In fact, it illustrates the Mexica people's concept of the cosmos, which is based on eight concentric circles. In the centre, the first circle is the face of the Sun. The second contains the four past eras, while the Sun's two clawed hands each hold a heart to anchor it to the universe in the current fifth era. The third circle corresponds to the 20 days of the Mexica month; the Mexica year was made up of 18 months. The next circles are dedicated to Mars, Saturn, Jupiter and Venus. On the outer ring, two serpents mark the boundaries of the visible world.

From the centre outwards:

- 1st ring: the Sun (and its rays extending into outer rings)
- 2nd ring: the four previous 'suns' or eras
- 3rd ring: the 20 days of the month
- 4th ring: Venus
- 5th, 6th and 7th ring: Mars, Jupiter and Saturn
- 8th ring: The Milky Way, the edge of the visible world

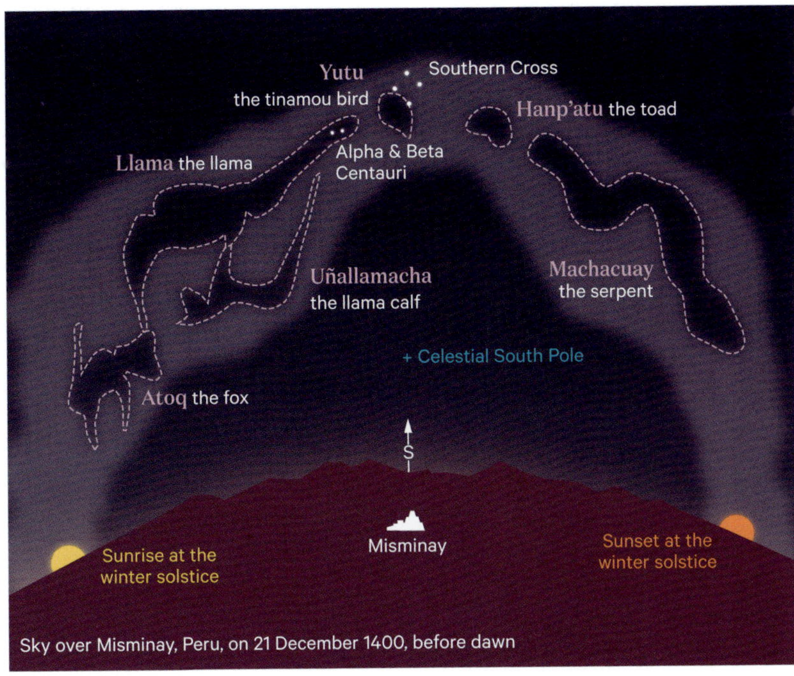

Sky over Misminay, Peru, on 21 December 1400, before dawn

The dark constellations of the Inca

Within the bright band of the Milky Way lie darker patches. The Inca people saw these 'constellations' as animals. Llama and Uñallamacha, representing a mother llama and her suckling calf, are the largest dark constellations. The stars Alpha and Beta Centauri form the llama's eyes, which are easy to spot when the constellations rise in November. The rising of the Toad (Hanp'atu) signified that it was time to plant crops, as the Inca believed that the more toads were croaking, the more likely it was to rain. The Coalsack Nebula, between the Southern Cross and the southern celestial pole, was named Yutu and represented a partridge-like bird called a tinamou.

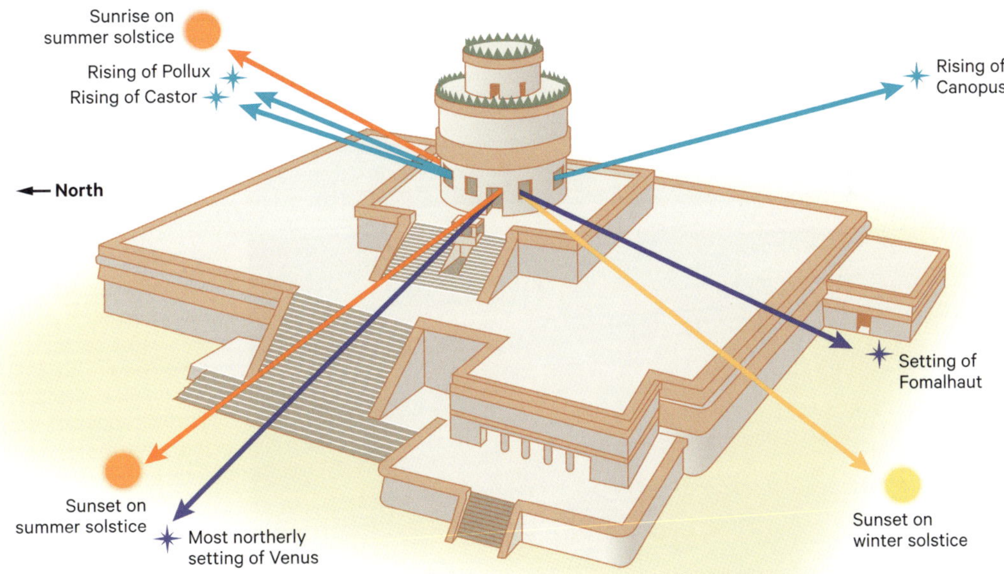

The observatory of Chichén Itzá

This Mayan observatory, which dates to 906 CE and is known as 'the snail' (*el caracol*) because of its spiral staircase, was used to observe the stars and planets. Around twenty different astronomical events that were important to the Maya and their solar calendar (eclipses, equinoxes, solstices), including the rising and setting of various stars and planets, can be viewed from the centre of the building through openings in the walls. The Maya not only studied the Sun and the Moon, but were also experts in the movements of Venus, which was associated with the god Kukulkan (the feathered serpent). It is possible that the constellations they considered most important formed a zodiac. The planet Venus, which has the direction of its rising and setting marked on the monument, played a role in all aspects of Mayan life, from indicating when to plant corn and the start of the rainy season to war, death and religious affairs.

Aboriginal Australians

The sky is very important in Indigeneous Australian cultures, playing a central role in ceremonial and artistic practices. The movements of different communities followed calendars determined by the rising and setting of stars.

The Celestial Emu

Many Aboriginal Australian cultures place great importance on the Milky Way, which is visualized as a river full of fish and water lilies. The 'water lilies' are the areas that do not have many stars, hidden by large quantities of interstellar dust. The region that, in Western astronomy, stretches from the Coalsack Nebula to the constellation of Scorpio forms a figure known as the Celestial Emu. The emu is the largest native bird in Australia. The motif of the constellation is seen in some rock carvings and in contemporary art, as in this painting inspired by the traditions of the Gidhabal people of Queensland.

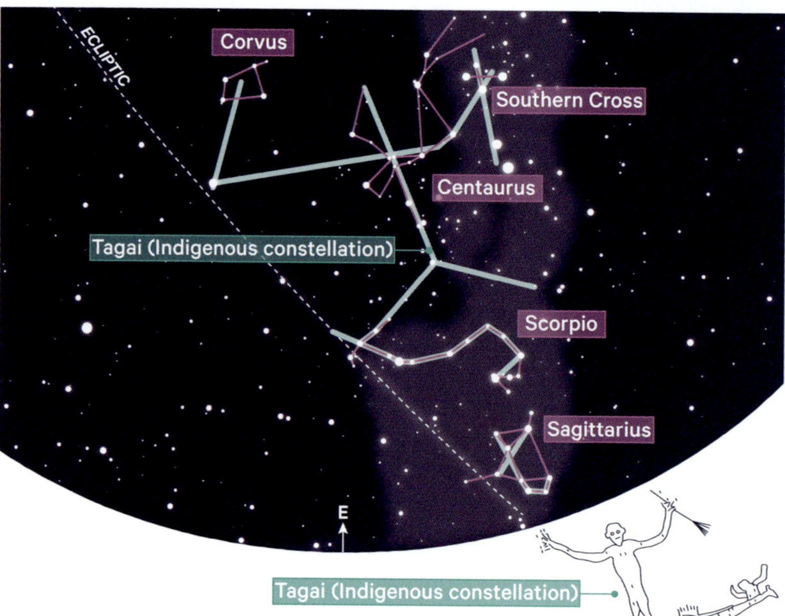

Sky over Melbourne on 13 March 2024
at 2 am (local time)

Tagai: a hero in the sky

To the community of Torres Strait Islanders on the island of Mabuiag, a combination of several Western constellations represents the mythical hero Tagai, his canoe and his crew. In the myth, the sailors ate all of their provisions before they set off, so, in order to punish them, Tagai threw them into the sky. They became the Western constellations of Orion and the Pleiades. Tagai's body is Centaurus, holding a spear in his left hand (the Southern Cross) and a piece of fruit in his right hand (Corvus). Scorpio is his boat and Sagittarius is its anchor. Tagai guides travellers around the islands of the Torres Strait.

Sub-Saharan Africa

Relatively little is known about the astronomical knowledge of many ancient African civilizations. Surviving verbal accounts and artistic representations are sometimes hard to interpret with certainty, but manuscripts in Arabic and Swahili record the construction of buildings along specific orientations.

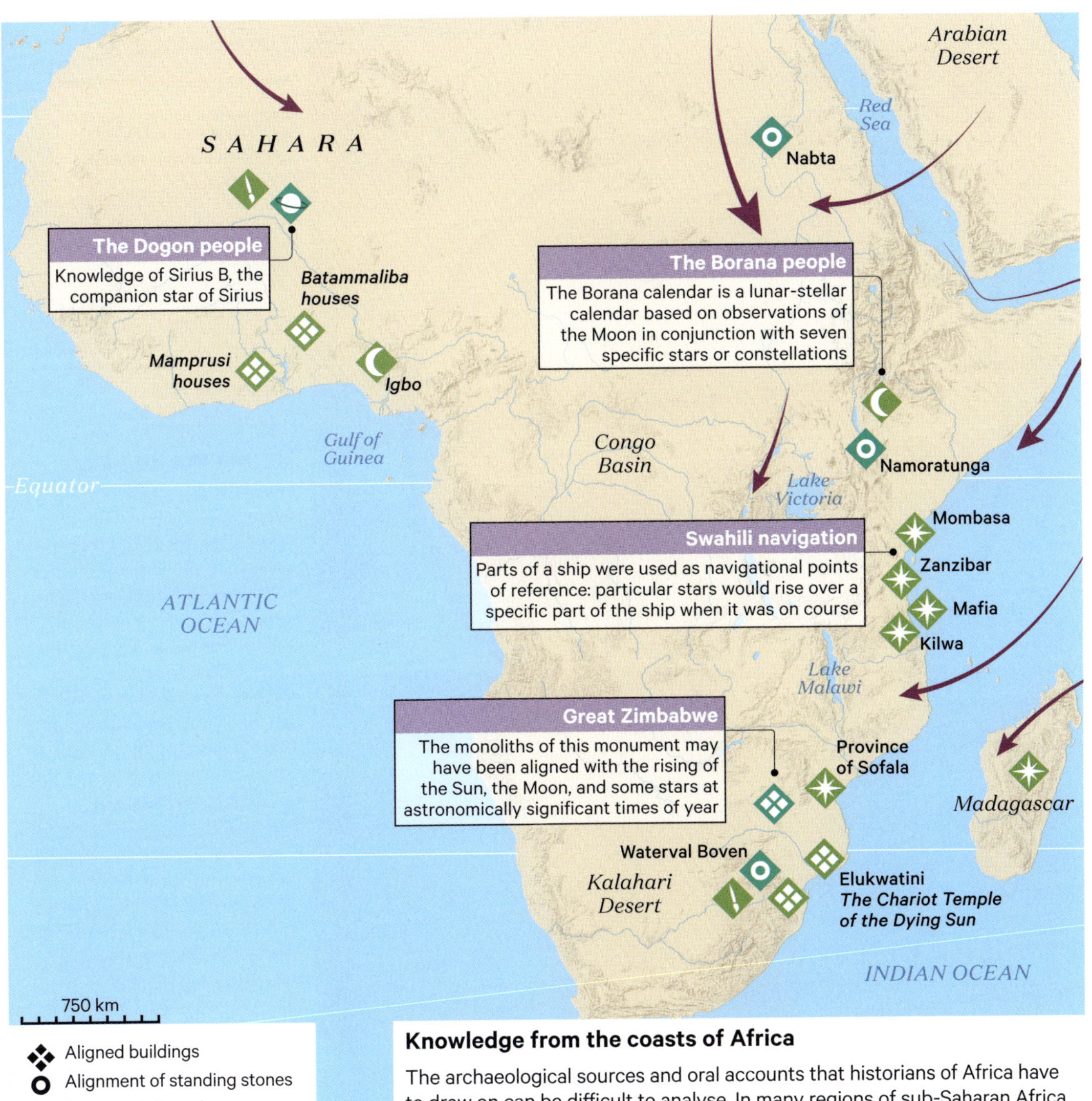

Knowledge from the coasts of Africa

The archaeological sources and oral accounts that historians of Africa have to draw on can be difficult to analyse. In many regions of sub-Saharan Africa, buildings orientated according to astronomical reference points still survive. The ruins of Great Zimbabwe, an imperial centre from the 13th to the 15th century, seem to have been laid out in a way that takes into account the solstices and Orion, but this view is not accepted by all. The calendars that have been found are very varied, but most are lunisolar. The knowledge of the Dogon people, in particular recognizing Sirius as a binary star, was recorded by French anthropologist Marcel Griaule, but has since been called into question. Swahili navigation systems used the stars as reference points.

The peoples of Polynesia

Peerless navigators, the Polynesians could read the heavens like an inverted map. They looked to the stars to pinpoint the geographical position of their canoes, using their hands as a way of measuring them. Aside from the stars, they used a number of different techniques to find their way by sea: the position of islands, the trajectory of flocks of birds, cloud formations and different kinds of waves.

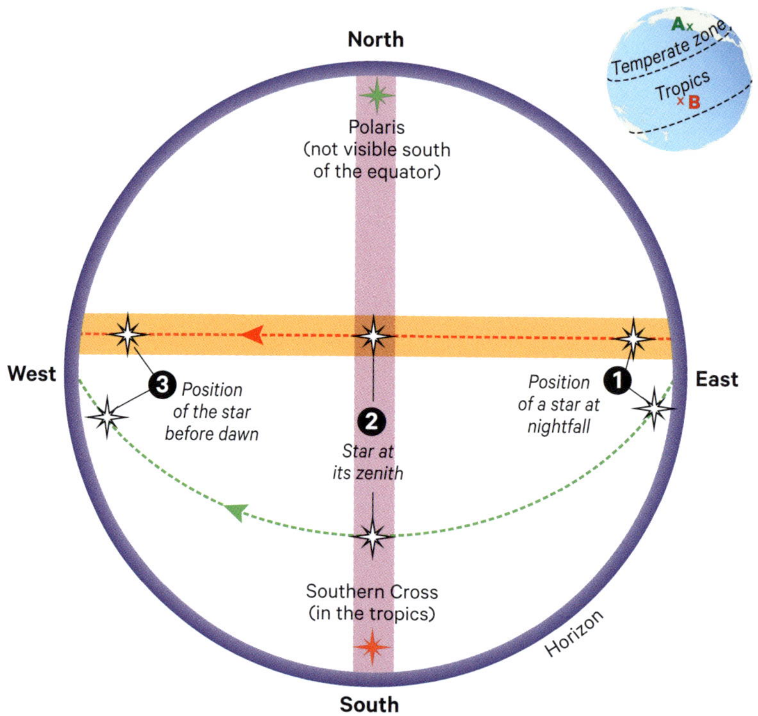

The *rua* and the *pou*

The apparent movements of the stars are very different for Polynesians and Europeans. At mid-latitudes in the northern hemisphere, the sky seems to be laid out in a circle around Polaris, the North Star. In the intertropical zone, the stars seem to move in a more vertical fashion, rising then descending. The Polynesians therefore developed navigation techniques based on this particularity. The stars emerge in alignment one after another, rise, reach their peak and descend again in the opposite direction, tracing a trajectory across the skies. They form a *rua* (a star path). These star paths are used to navigate in the east–west direction. The *pou* (sky pillars) are used to navigate in the north–south direction. The northern *pou* is formed by Polaris, when it is visible, then Dubhe (Alpha Ursae Majoris) takes over.

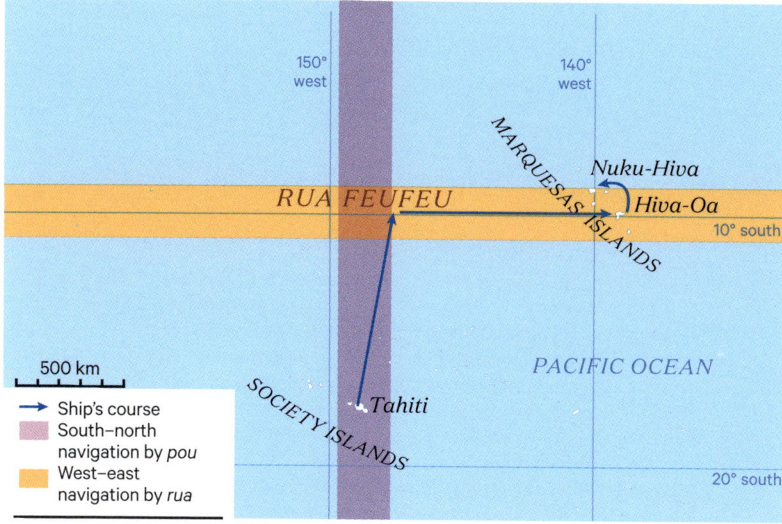

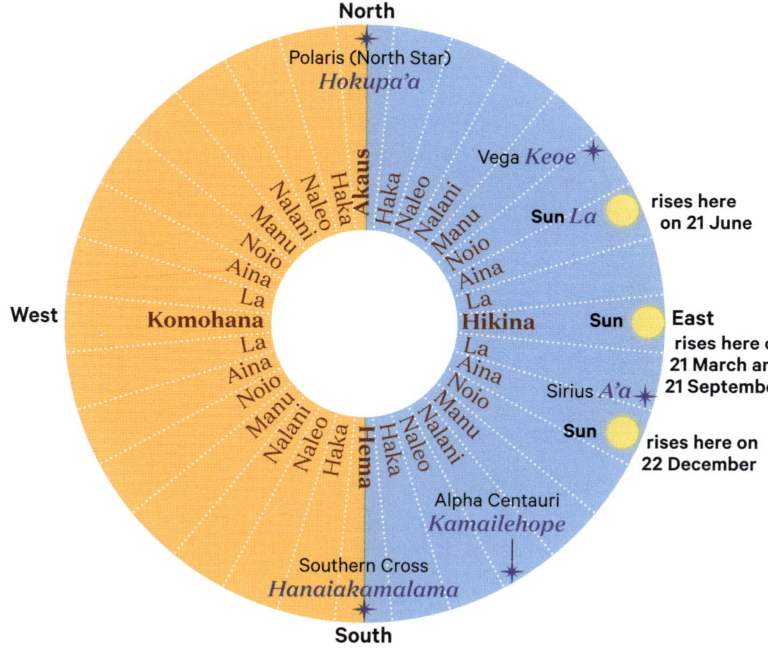

Houses in the sky

In traditional Polynesian culture, the stars are said to rise and set in the same 'house'. These reference points are used in navigation to determine the course followed by canoes. The 'compass' shown here has been recreated by the Hawaiian navigator Nainoa Thompson. Traditionally, Polynesians memorized this information but did not store it in written form.

Sky zones

Noio House name (the compass is divided into 32 houses)

Zone where stars rise

Zone where stars set

Stars

Sun

Other notable stars

La Polynesian star names

A comet carved in stone on Easter Island

Easter Island, or Rapa Nui, one of the most isolated inhabited places in the world, is famous for its Moai, a collection of stone statues around 4 metres high, which are positioned around the island's coast, facing inland. Mystery also surrounds the many rock carvings found on the island. The one reproduced below, which is etched into the wall of the Ana O Keke cave, shows the comet C/1743-X1 (known as the Great Comet of 1744), which would have been visible from Rapa Nui in March and April 1744. Venus is also depicted, and would have been visible at the same time. The other motifs in the carving have not been definitively identified.

A **The Great Comet of 1744** (C/1743-X1), also observed in Switzerland from December 1743 to April 1744

B **Venus** (the Great Comet passed close to Venus in April 1744)

C Possibly **Halley's Comet**?

D Unknown

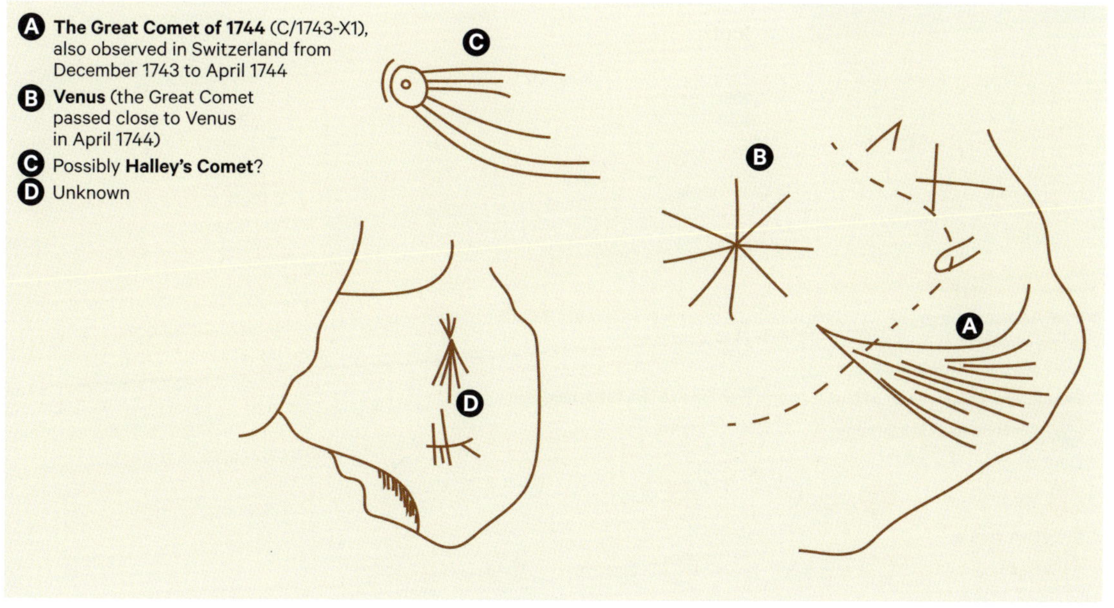

From Persia to India

From the 2nd millennium BCE onwards, the people of Persia were able to draw on the astronomical knowledge of Mesopotamia, and later Greece. The Hindu world of the Sanskrit language was born on the Indo-Gangetic plain. The study of the sky, arithmetic and algebra all developed there. From the 8th century onwards, these societies and their relationship with the skies were reshaped by Islam.

The background map reflects the geopolitical situation in the year 200 CE

Trade routes:
— land   — sea

People who made major scientific contributions to the following fields

Astronomical knowledge

Mathematics

Arithmetics   Algebra   Geometry

Religious faiths

Jainism   Buddhism   Hinduism   Zoroastrianism

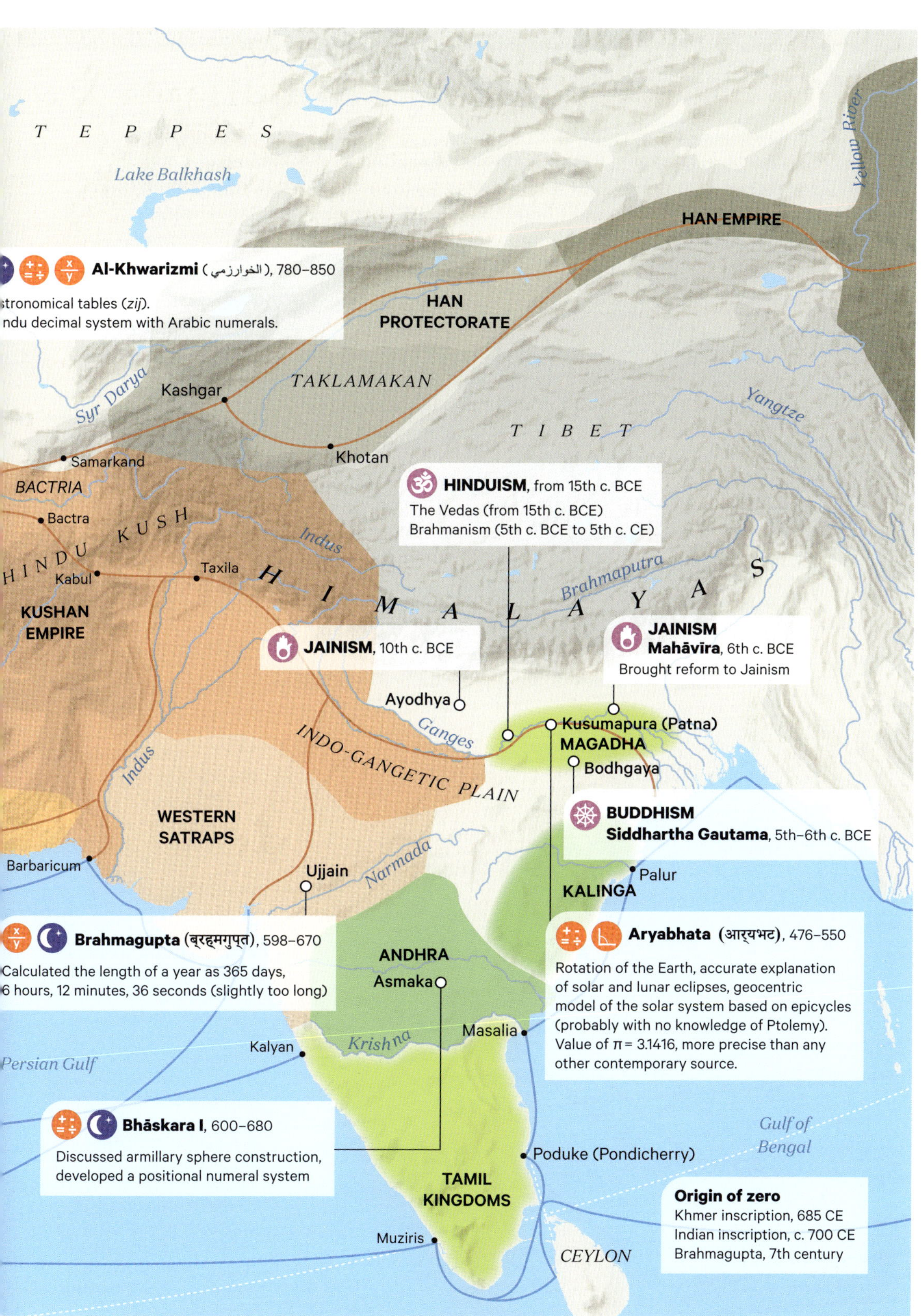
T E P P E S
Lake Balkhash
HAN EMPIRE
Yellow River
Al-Khwarizmi (الخوارزمي), 780–850
Astronomical tables (zij).
Hindu decimal system with Arabic numerals.
HAN PROTECTORATE
TAKLAMAKAN
Syr Darya
Kashgar
Yangtze
Khotan
T I B E T
Samarkand
BACTRIA
Bactra
HINDUISM, from 15th c. BCE
The Vedas (from 15th c. BCE)
Brahmanism (5th c. BCE to 5th c. CE)
H I N D U   K U S H
Indus
Kabul
Taxila
H I M A L A Y A S
Brahmaputra
KUSHAN EMPIRE
JAINISM, 10th c. BCE
JAINISM
Mahāvīra, 6th c. BCE
Brought reform to Jainism
Ayodhya
Ganges
Kusumapura (Patna)
MAGADHA
Indus
INDO-GANGETIC PLAIN
Bodhgaya
BUDDHISM
Siddhartha Gautama, 5th–6th c. BCE
WESTERN SATRAPS
Barbaricum
Ujjain
Narmada
Palur
KALINGA
Brahmagupta (ब्रह्मगुप्त), 598–670
Calculated the length of a year as 365 days,
6 hours, 12 minutes, 36 seconds (slightly too long)
Aryabhata (आर्यभट), 476–550
Rotation of the Earth, accurate explanation
of solar and lunar eclipses, geocentric
model of the solar system based on epicycles
(probably with no knowledge of Ptolemy).
Value of π = 3.1416, more precise than any
other contemporary source.
ANDHRA
Asmaka
Masalia
Krishna
Kalyan
Persian Gulf
Gulf of Bengal
Bhāskara I, 600–680
Discussed armillary sphere construction,
developed a positional numeral system
Poduke (Pondicherry)
TAMIL KINGDOMS
Origin of zero
Khmer inscription, 685 CE
Indian inscription, c. 700 CE
Brahmagupta, 7th century
Muziris
CEYLON

The Eastern Mediterranean

Across the Eastern Mediterranean, ideas about the sky developed independently in Mesopotamia, Egypt, and later Greece. Over time, trade and interactions between these civilizations, accelerated by the conquests of Alexander (334–323 BCE), led to intellectual cross-pollination. However, on the edges of the Mediterranean world, independent conceptions of the sky endured, for example on the Germanic plains.

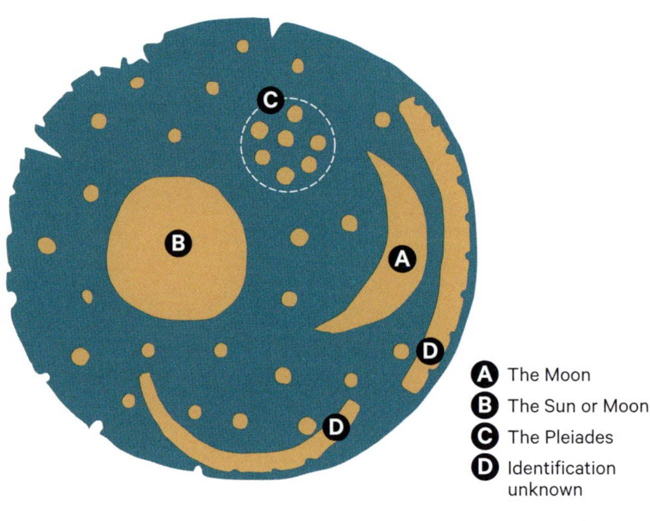

The Nebra sky disc

Discovered in 1999 by unlicensed treasure-hunters, around 180 km southwest of Berlin, this bronze disc, measuring 32 cm in diameter and embellished with gold leaf, dates to 1600 BCE. It is the oldest known depiction of the sky. It features the crescent moon and a round object (the full moon or the Sun?), as well as the Pleiades. The 82° arc on the right was matched by a symmetrical one on the left, which has been lost. This may represent the distance between the points on the horizon where the Sun rises and sets at the solstices.

The temple of Apollo at Delphi

Inside this Panhellenic sanctuary, the temple of Apollo (4th century BCE) was home to the Pythia, or Oracle, except during the three months of winter when the god lived with the Hyperboreans. The temple was built facing northeast–southwest, as determined by the setting of the star Beta Lupi at 45° southwest on 28 July, announcing the start of the Pythian Games. Greek temples were laid out so that the rising Sun would light the interior during major festivals. In Delphi, the ridge line of Mount Parnassus had to be taken into account. For two days in the year, the Sun's rays appear at an angle, then pass between two columns and through a window believed to strike the adyton, the room where the oracle lived.

Sky on 20 July 1340 BCE, over Gizeh, around 4.30 am

The skies of ancient Egypt

Nut, the Egyptian goddess of the sky, was a personification of the heavens. Her toes rest on the rising sun and her fingers on the setting sun. She is supported by the god Shu, who stands on Geb, the Earth. Nut and Geb's father is Shu, the air, while their mother, Tefnut, is the goddess of water. Nut and Geb are the parents of Osiris, Set, Isis and Nephthys. Nut is also the mother of Ra, the Sun, whom she swallows every evening and gives birth to again every morning.

The time of year when Sirius is once again visible on the western horizon at dawn is its heliacal rising. Personified by the goddess Sopdet, represented by a small dog, its appearance marked the beginning of summer and anticipated the flooding of the Nile.

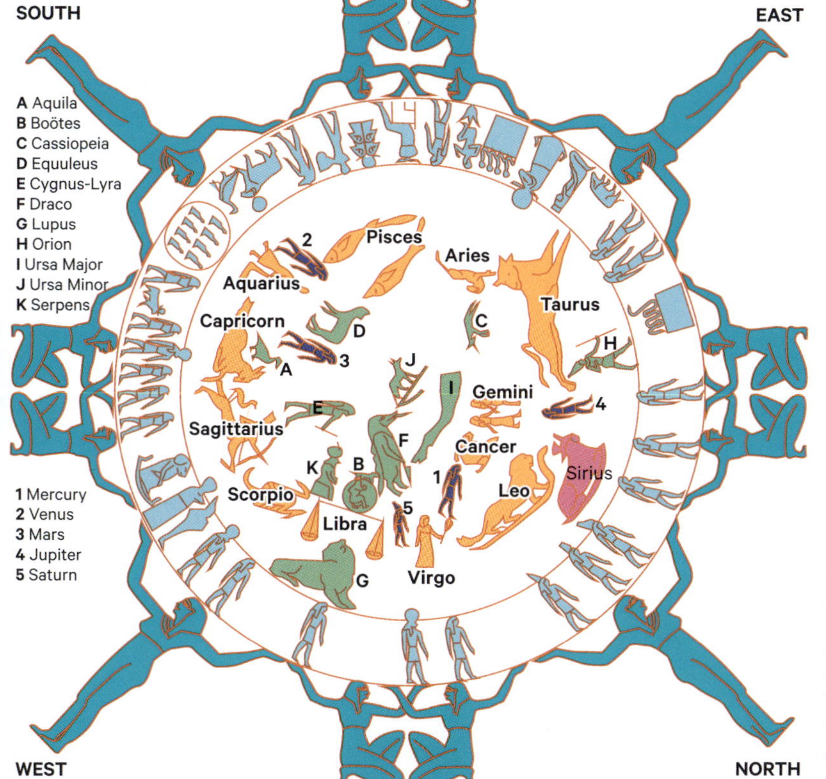

The Dendera zodiac

This monumental Egyptian stone bas-relief (c. 50 BCE), now in the collection of the Louvre, Paris, depicts a syncretic concept of the sky. It shows the twelve constellations of the zodiac (shown in orange) plus the five planets that can be seen with the naked eye (dark blue), eleven other constellations (green) and the star Sirius (pink). The figures on the outer ring represent the 36 decans, small groups of stars that divide the ecliptic into 36 parts. The Egyptians divided the night into 12 hours and the year into 360 days.

The image is a planisphere, depicting the sky and stars as a flat projection. It shows how Egyptian and Greek concepts of the sky merged after the conquests of Alexander. The Egyptians integrated the Babylonian zodiac (reinterpreted by the Greeks) into their calendar.

2

The sky over Europe and Asia (3000 BCE to 1600)

Until the 17th century, humans had to rely on the naked eye to observe the stars and planets. Many civilizations sought to understand the sky. In the fertile crescent of Mesopotamia, the demand for astrological predictions led to observations becoming more precise. Around the Mediterranean, where the skies were very clear, the sixth century BCE saw the beginnings of a science that brought together philosophical and metaphysical reflections with geometry and ways of measuring objects in the sky. Through Aristotle in Athens and Ptolemy in Alexandria, this philosophy eventually led to the development of a science of the skies, which gradually spread across Europe and Asia. In the 16th century, the slow maturing of this science, facilitated by international trade, led to the intellectual flourishing of the European Renaissance. However, on the eastern side of Asia, the Chinese world was developing its own understanding of the sky and mostly remained separate from this process.

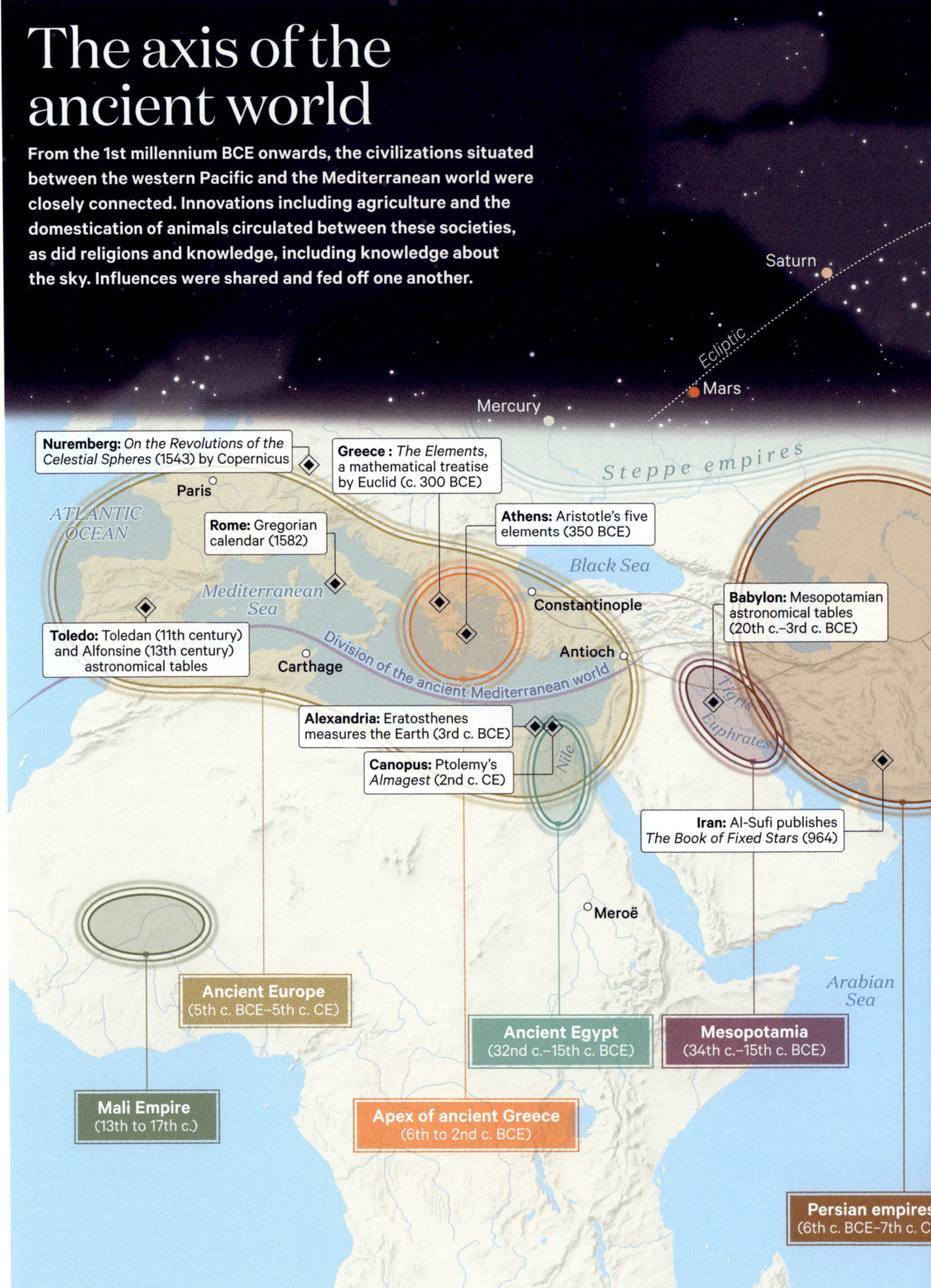

The axis of the ancient world

From the 1st millennium BCE onwards, the civilizations situated between the western Pacific and the Mediterranean world were closely connected. Innovations including agriculture and the domestication of animals circulated between these societies, as did religions and knowledge, including knowledge about the sky. Influences were shared and fed off one another.

Saturn
Ecliptic
Mars
Mercury

Nuremberg: On the Revolutions of the Celestial Spheres (1543) by Copernicus
Greece : The Elements, a mathematical treatise by Euclid (c. 300 BCE)
Steppe empires
Paris
ATLANTIC OCEAN
Rome: Gregorian calendar (1582)
Athens: Aristotle's five elements (350 BCE)
Black Sea
Mediterranean Sea
Babylon: Mesopotamian astronomical tables (20th c.–3rd c. BCE)
Constantinople
Toledo: Toledan (11th century) and Alfonsine (13th century) astronomical tables
Antioch
Carthage
Division of the ancient Mediterranean world
Tigris
Euphrates
Alexandria: Eratosthenes measures the Earth (3rd c. BCE)
Nile
Canopus: Ptolemy's Almagest (2nd c. CE)
Iran: Al-Sufi publishes The Book of Fixed Stars (964)
Meroë
Arabian Sea
Ancient Europe (5th c. BCE–5th c. CE)
Ancient Egypt (32nd c.–15th c. BCE)
Mesopotamia (34th c.–15th c. BCE)
Mali Empire (13th to 17th c.)
Apex of ancient Greece (6th to 2nd c. BCE)
Persian empires (6th c. BCE–7th c. CE)

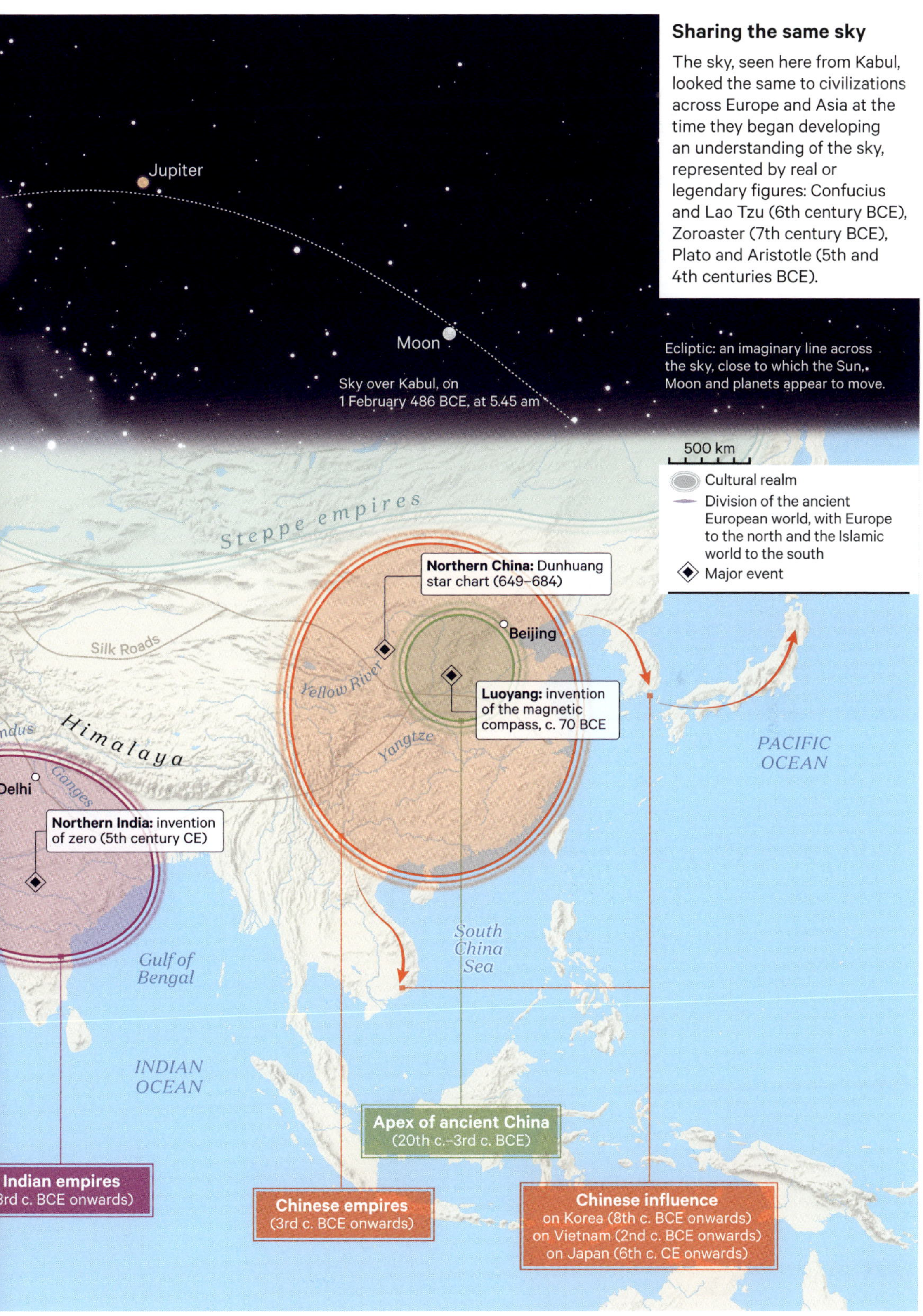
Jupiter
Moon
Sky over Kabul, on
1 February 486 BCE, at 5.45 am
Sharing the same sky
The sky, seen here from Kabul,
looked the same to civilizations
across Europe and Asia at the
time they began developing
an understanding of the sky,
represented by real or
legendary figures: Confucius
and Lao Tzu (6th century BCE),
Zoroaster (7th century BCE),
Plato and Aristotle (5th and
4th centuries BCE).
Ecliptic: an imaginary line across
the sky, close to which the Sun,
Moon and planets appear to move.
500 km
Cultural realm
Division of the ancient
European world, with Europe
to the north and the Islamic
world to the south
Major event
Steppe empires
Northern China: Dunhuang
star chart (649–684)
Beijing
Silk Roads
Yellow River
Luoyang: invention
of the magnetic
compass, c. 70 BCE
Himalaya
Yangtze
Indus
Ganges
Delhi
PACIFIC
OCEAN
Northern India: invention
of zero (5th century CE)
Gulf of
Bengal
South
China
Sea
INDIAN
OCEAN
Apex of ancient China
(20th c.–3rd c. BCE)
Indian empires
(3rd c. BCE onwards)
Chinese empires
(3rd c. BCE onwards)
Chinese influence
on Korea (8th c. BCE onwards)
on Vietnam (2nd c. BCE onwards)
on Japan (6th c. CE onwards)

The legacy of Mesopotamia

Over the course of two millennia, a detailed science of astrology developed in Mesopotamia. It was based on a series of ancient recorded observations and aided by the invention of arithmetic. At the same time, however, it served the needs of the state. Divination was used to interpret celestial phenomena as auspicious or ominous predictions from the gods to those in power. This knowledge was passed on to the Greeks after the conquests of Alexander.

Sky over Jerusalem on 22 February 6 CE at 6 pm

The Star of Bethlehem: a planetary conjunction?

Every twenty years or so, the planets Jupiter and Saturn draw close to one another in their journey across the sky. This conjunction, which was studied for divinatory purposes, is even more spectacular when Mars is also present, as was the case over Jerusalem in 6 BCE. In 1614, Johannes Kepler calculated this conjunction of Jupiter and Saturn, and theorized that it may have been the star seen by the Wise Men of the New Testament, thus suggesting a date for the birth of Christ.

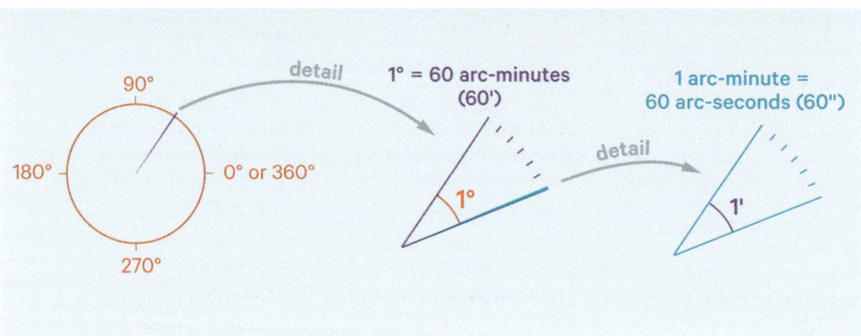

Divisions of a circle, divisions of time

The Babylonians divided the circle into 360 degrees, then these degrees into arc-minutes and arc-seconds. The decision to divide each degree into 60, chosen because 60 is easy to divide by 2, 3, 4, 5 and multiples of these numbers, is still a convention today.

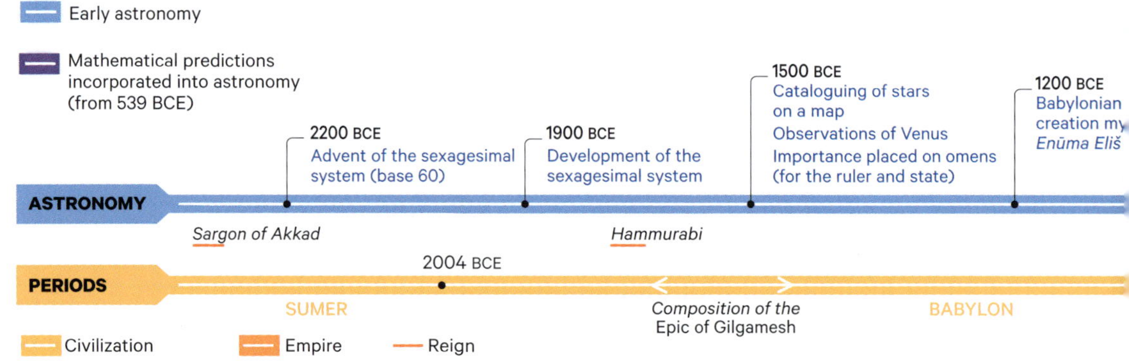

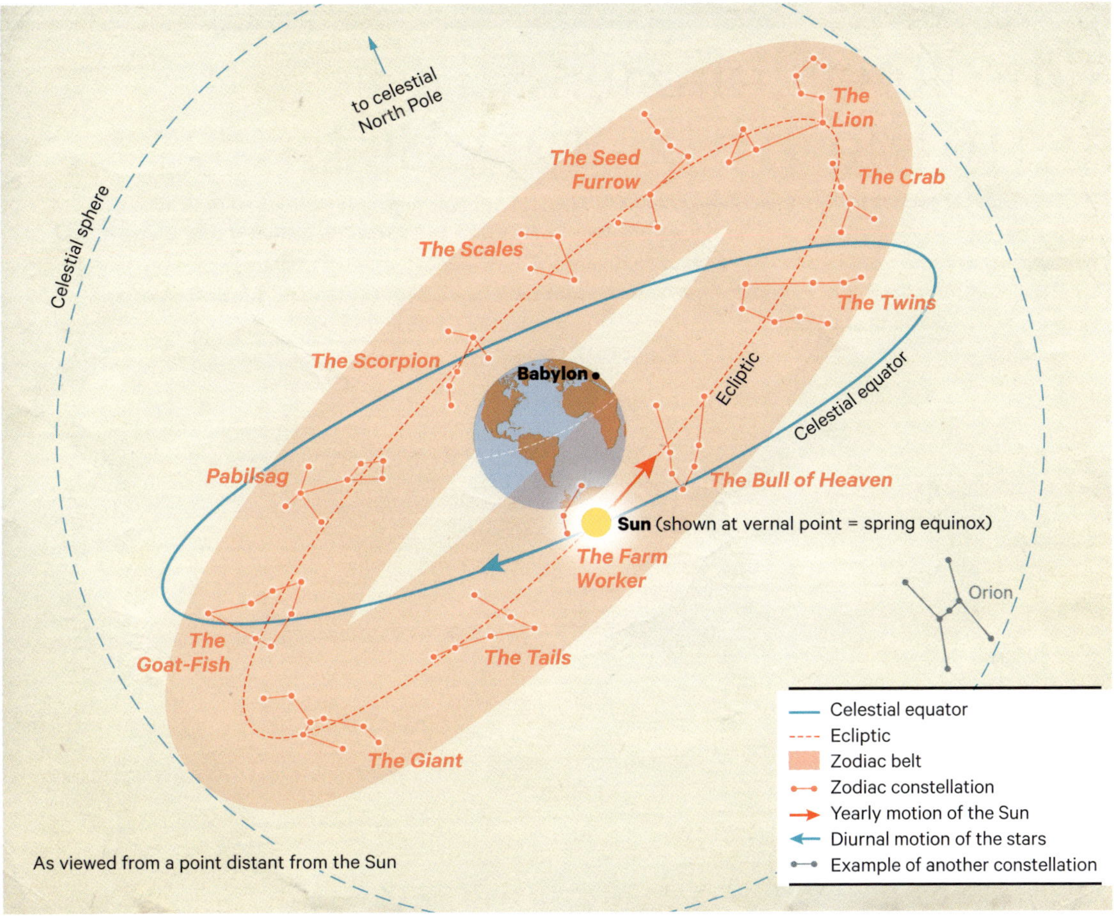

The Babylonian zodiac

The positions of the Sun, Moon and planets change in relation to the stars. Knowing their position in the sky, both now and in the future, was important for divination, calendars and predicting eclipses. Over time, these celestial bodies move within a narrow belt of sky, tilted at an angle to the celestial equator. The Babylonians noted their positions relative to bright stars that form constellations within this belt. It is divided into twelve sections, each one measuring 360°/12 = 30°, which creates a useful correlation with the twelve months of the year – the time it takes for the Sun to travel the 360° of the ecliptic. Adopted by the ancient Greeks, this belt was called the 'zodiac', in reference to the 'little animals' (*zōdion* in Greek) after which some of the constellations were named. These names are still used by astronomers today.

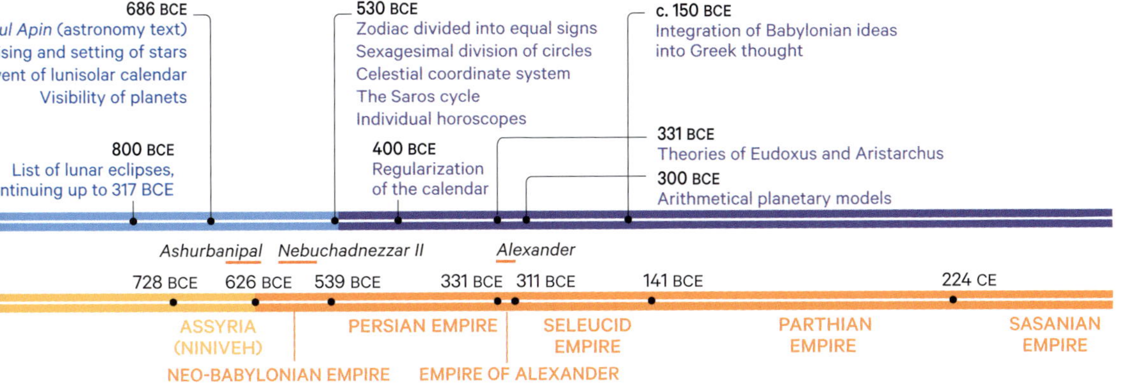

Greek trailblazers: up to the 4th century BCE

A philosophical conception of the universe and the place of humans within it developed in Greece over the course of four centuries, culminating in major works by Plato and Aristotle, which would shape the centuries that followed. Despite only having rudimentary instruments with which to measure the sky over the Mediterranean, a number of schools of thought developed hypotheses and suggested initial responses to questions about the universe, matter and celestial bodies.

A concept of the universe is established

After Homer and Hesiod's poetic descriptions of the skies, philosophical and scientific thought developed hand in hand. Philosophy, with its emphasis on reason and proof, was constantly asking questions about why things existed and why observed phenomena took place, especially in the sky. It reached its peak with Plato, then Aristotle, whose ideas remained highly influential up until the 16th century. The scientific tradition seeks to answer questions by formulating hypotheses, then testing the results: ideas such as 'does the Earth move?', 'Is space infinite?', 'Are numbers infinite?' and so on. The birth of geometry (etymologically, 'measuring the Earth') drew connections between these questions and ways of measuring the sky, which were used in sundials. Circles and spheres were shapes suggested by celestial bodies, which were then explored in detail using geometry.

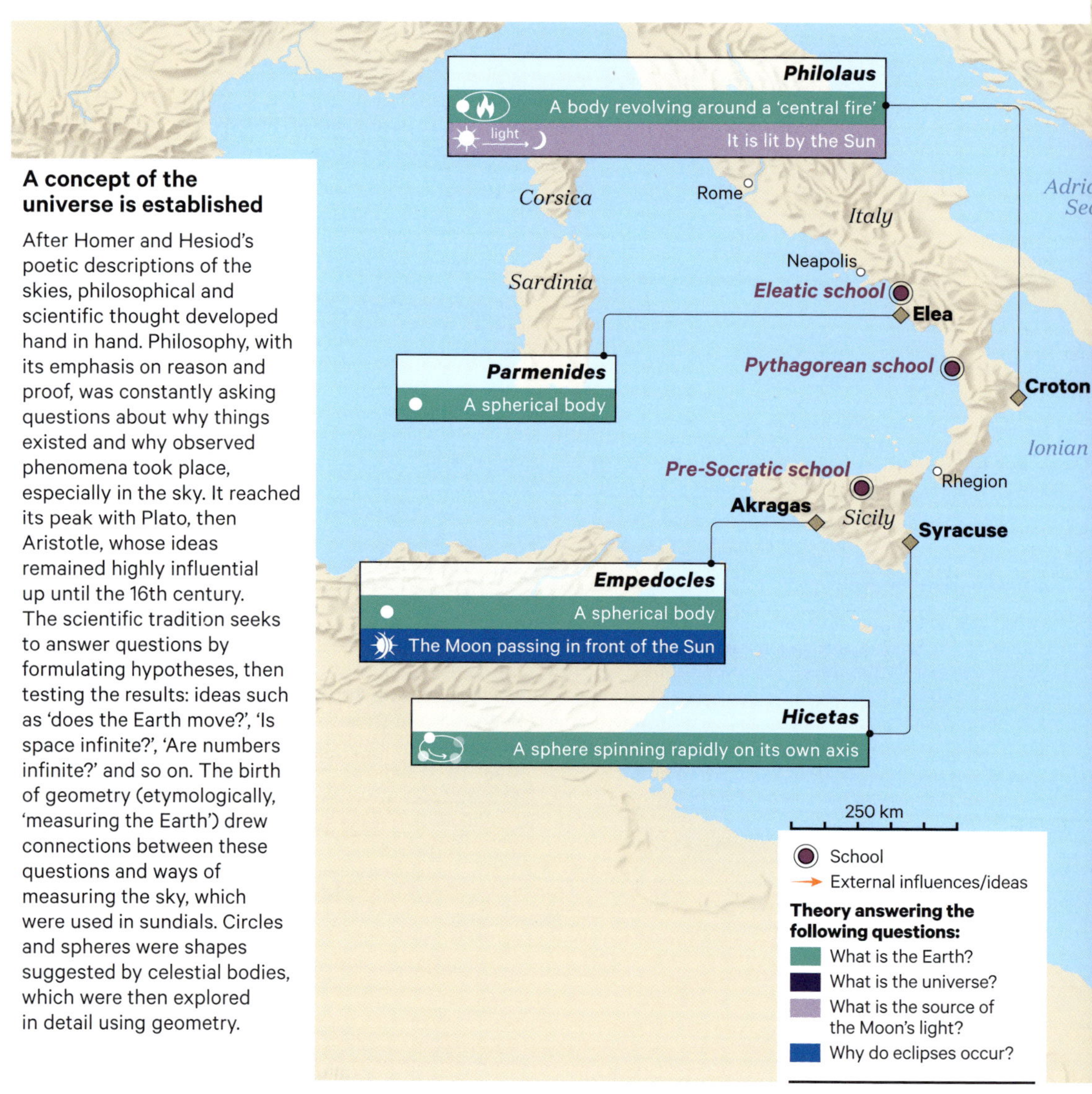

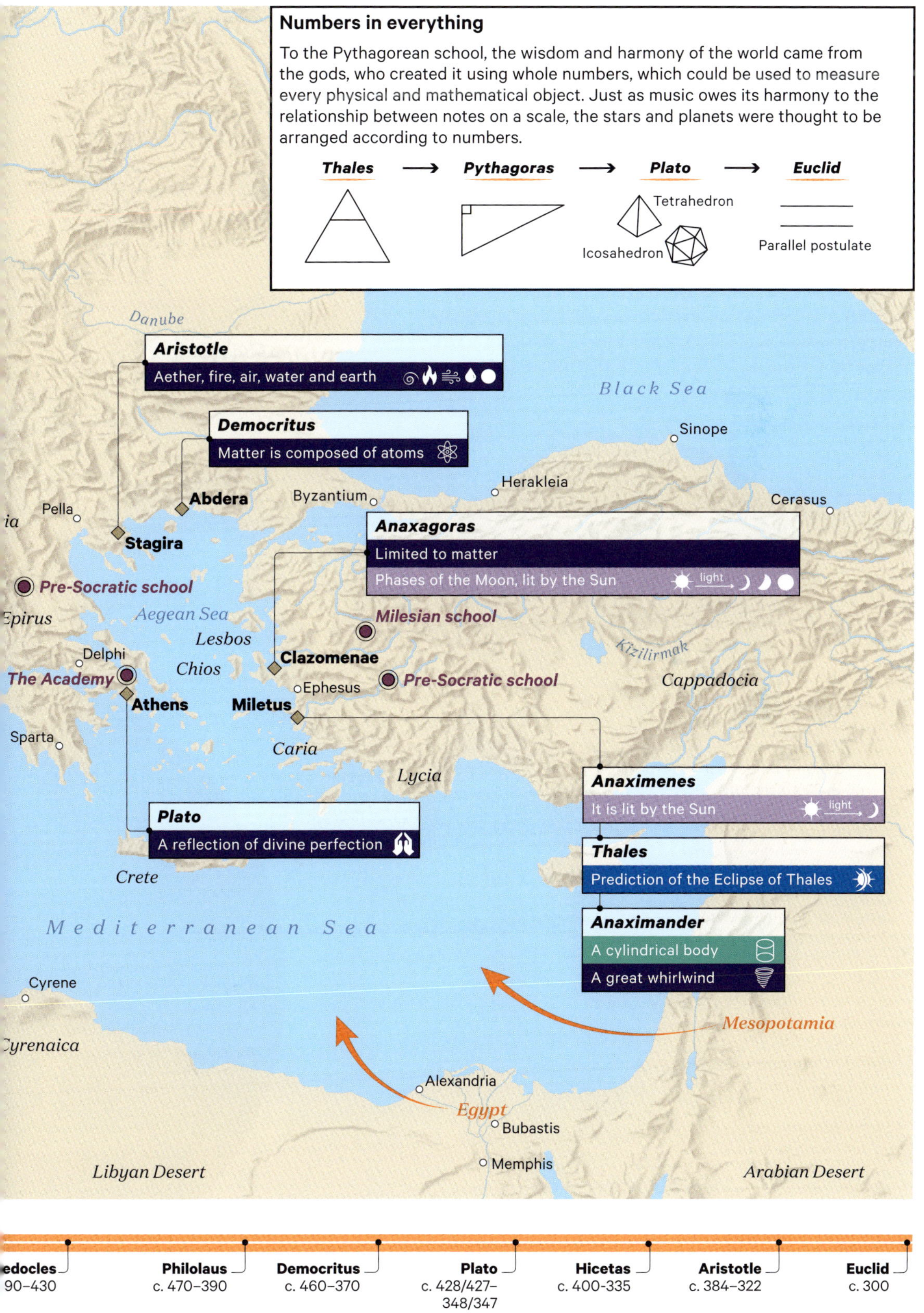

Numbers in everything
To the Pythagorean school, the wisdom and harmony of the world came from the gods, who created it using whole numbers, which could be used to measure every physical and mathematical object. Just as music owes its harmony to the relationship between notes on a scale, the stars and planets were thought to be arranged according to numbers.
Thales
Pythagoras
Plato
Euclid
Tetrahedron
Icosahedron
Parallel postulate
Danube
Aristotle
Aether, fire, air, water and earth
Black Sea
Democritus
Matter is composed of atoms
Sinope
Herakleia
Byzantium
Cerasus
Abdera
Anaxagoras
Limited to matter
Phases of the Moon, lit by the Sun
light
Pella
ia
Stagira
Pre-Socratic school
Milesian school
Epirus
Aegean Sea
Lesbos
Kizilirmak
Delphi
Chios
Clazomenae
Cappadocia
The Academy
Ephesus
Pre-Socratic school
Athens
Miletus
Sparta
Caria
Lycia
Anaximenes
It is lit by the Sun
light
Plato
A reflection of divine perfection
Thales
Prediction of the Eclipse of Thales
Crete
Anaximander
A cylindrical body
A great whirlwind
Mediterranean Sea
Mesopotamia
Cyrene
Cyrenaica
Alexandria
Egypt
Bubastis
Memphis
Libyan Desert
Arabian Desert
edocles
90–430
Philolaus
c. 470–390
Democritus
c. 460–370
Plato
c. 428/427–348/347
Hicetas
c. 400–335
Aristotle
c. 384–322
Euclid
c. 300

Circles, spheres and geometry

The idea of a celestial sphere is suggested by the movement of stars; the idea of a circle traced at a constant speed is suggested by the daily and annual movement of the Sun. It was Thales of Miletus who discovered that spheres and circles had a role in describing the world from a geometric perspective. Measuring angles comes from observing shadows and using a sundial. The Earth, Moon and Sun were described as spheres. Euclid created a mathematical framework that would remain in use for the twenty-three centuries that followed.

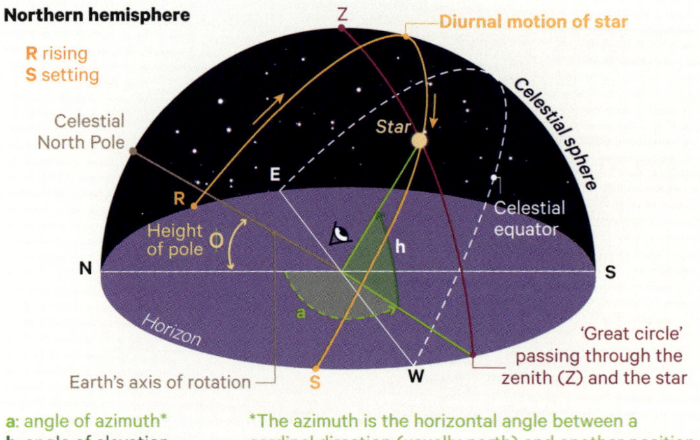

a: angle of azimuth*
h: angle of elevation

*The azimuth is the horizontal angle between a cardinal direction (usually north) and another position.

Pinpointing the position of a star

On the celestial sphere, viewed at a particular time from a particular place on Earth, the location of a star or other body can pinpointed using two angles (altitude h and azimuth a), measured in degrees. A great circle on the sphere, such as the celestial equator, is 360°. The height of the celestial pole is equal to the latitude of the observer.

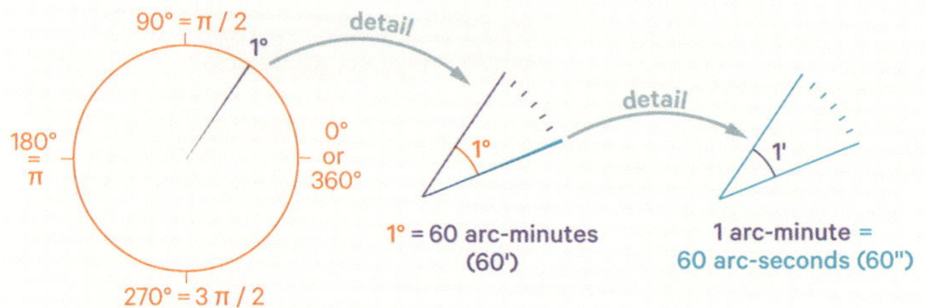

Circles and spheres

A circle is an abstract mathematical concept. For the ancient Greeks, its simplicity was close to divine perfection. The number pi (π) is used to calculate the length of an arc of the circle relative to its radius. A sphere, as described by Autolycus of Pitane and Euclid, is a circle transposed into three-dimensional space.

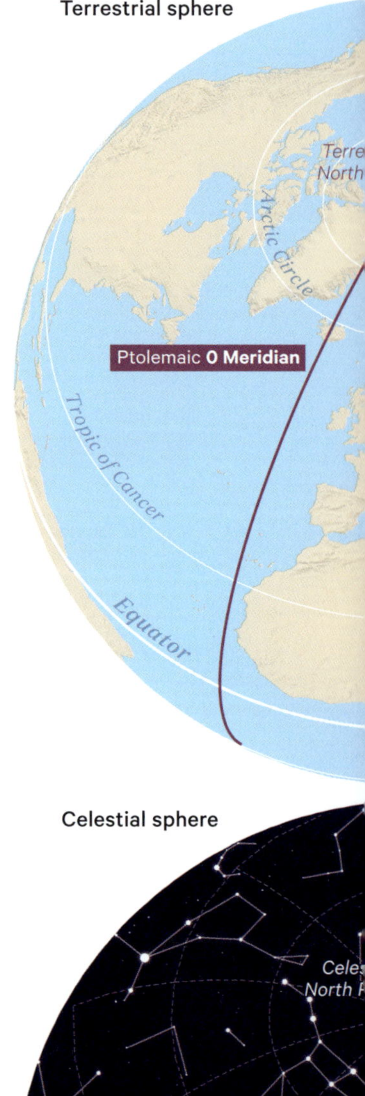

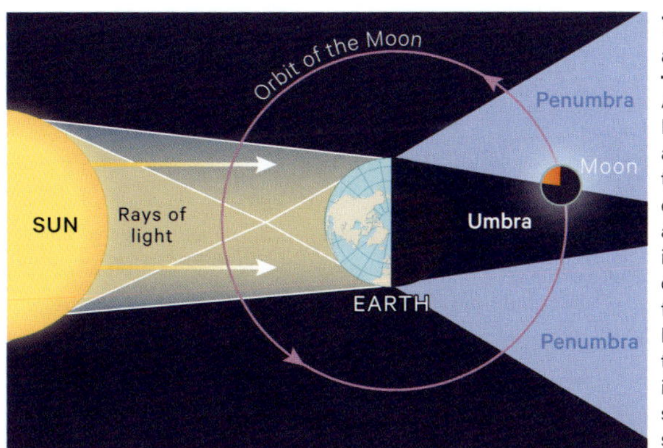

The Earth is a sphere

Aristotle (4th century BCE) offered three arguments to support this hypothesis. Heavy objects all fall towards a centre. When a ship is sailing away from the coast, its hull disappears first, followed by its mast. During a lunar eclipse, the shadow of the Earth is always circular, which suggests that it is a sphere and not a disc.

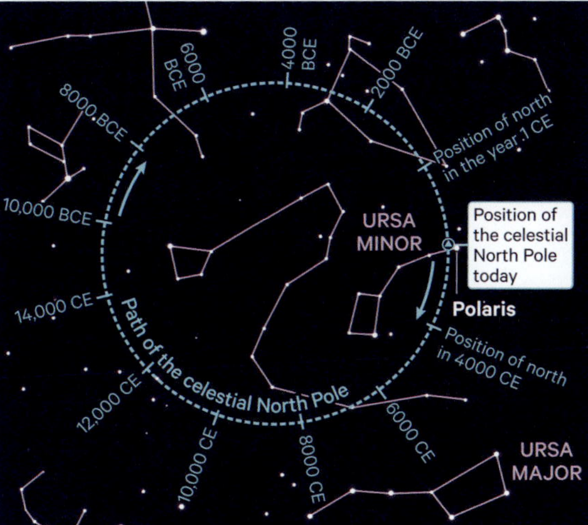

The celestial North Pole moves in a circle

The Earth's north–south axis determines the poles of the celestial sphere. The daily movement of the stars traces circles around these poles. The poles themselves move slowly among the stars, as the Earth's axis traces a circle over the course of 26,000 years: this process is called precession.

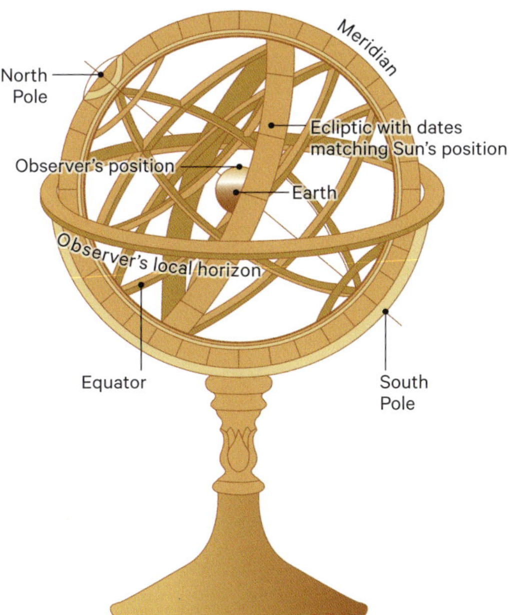

The armillary sphere

From the 4th century BCE onwards, the armillary sphere provided a three-dimensional model that was easier to use than a solid globe and could aid in understanding the movement of the stars and planets. It made it possible to bring multiple elements into alignment, and was also used to measure the positions of celestial bodies, for example by Ptolemy.

The 4th century BCE: a geocentric perspective

The 4th and 3rd centuries BCE brought observations of the movements of the Moon, Sun and planets, while philosophers were seeking to describe how the world was organized. Aristarchus of Samos placed the Sun at the centre of the universe (heliocentrism) with all the celestial bodies in motion around it, with only the Moon orbiting the Earth. However, Eudoxus of Cnidus placed the Earth at the centre (geocentrism).

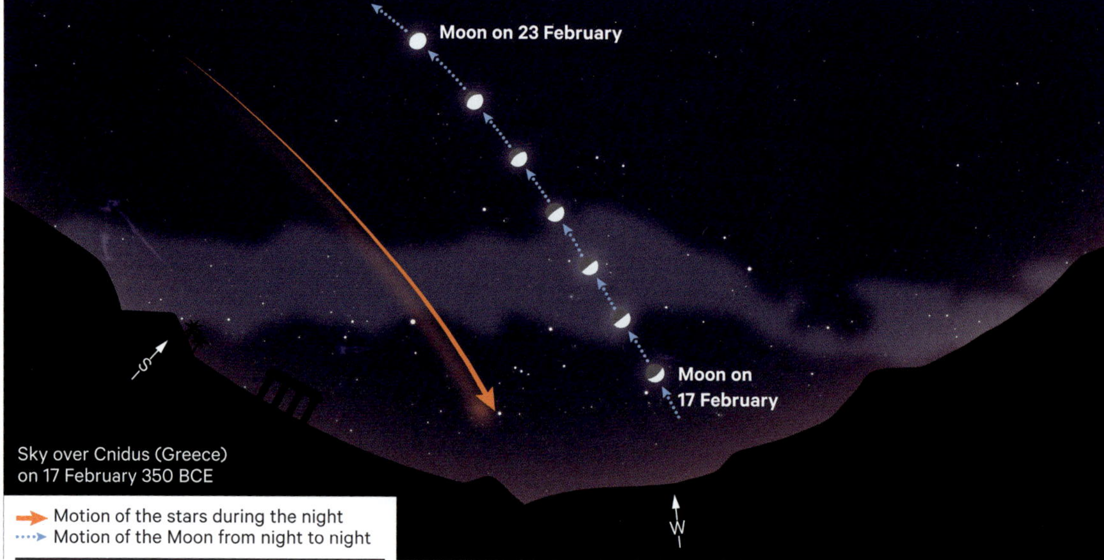

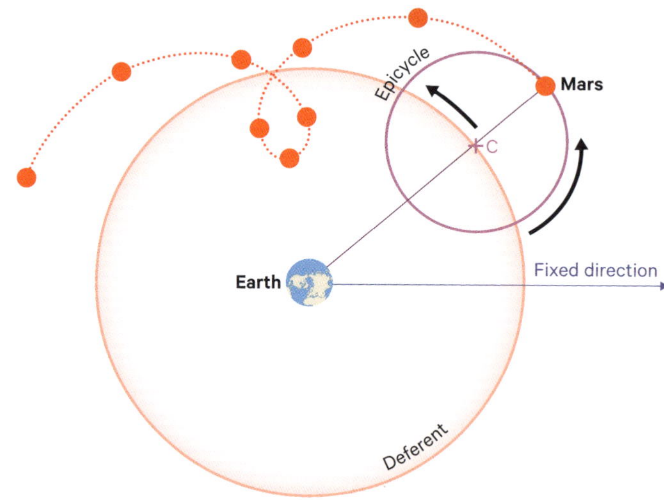

The problem of planetary motion

The Moon constantly changes position relative to the stars and its appearance also changes. As its light comes from the Sun, the idea that it orbits the Earth has been widely accepted since the 6th century BCE. But the irregular motion of the planets raised questions. For example, Mars sometimes seems to stand still or move backwards (retrograde motion), in a pattern that repeats over the years. Apollonius of Perga (c. 200 BCE) proposed a theory that became widely accepted. The motion of a celestial body could best be described using a circle being drawn at a constant speed. In order to reconcile observations of Mars with this idea, he proposed that the observed motion involved two circles: Mars traced a small circle, whose centre followed a large circle centred on a fixed Earth. This idea of epicycles was later applied to models with a rotating Earth.

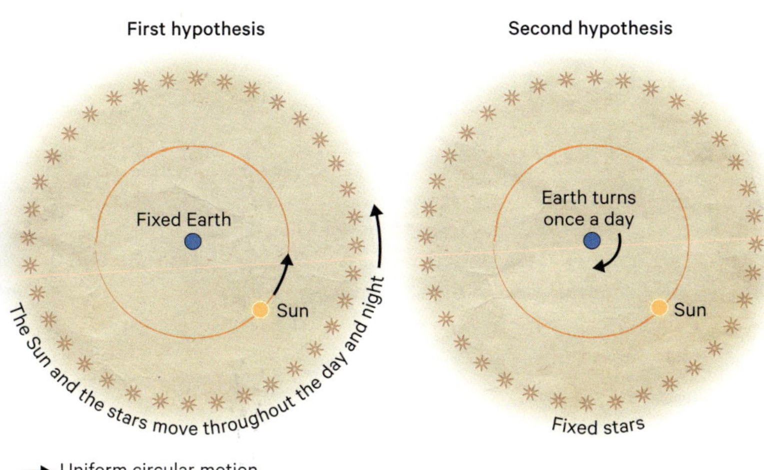

A fixed or turning Earth?

What explanation could there be for the observation that the stars' position in the sky changes slightly every night, and the Sun moves across the sky during the day? Either the distant celestial sphere, which holds the stars, is fixed, and the Earth rotates over the course of 24 hours, or the opposite is true: the Earth is fixed in space and the celestial sphere moves. For Aristarchus of Samos, as quoted by Archimedes (c. 220 BCE), the simplest solution was that the Earth was rotating, not the stars.

Is the Earth or the Sun at the centre?

How could the movement of the Sun over the course of a year be explained? If the Earth is unmoving at the centre of the universe (geocentrism), it followed that the Sun and stars must orbit around it, as proposed by Eudoxus of Cnidus (c. 360 BCE). If the Sun is at the centre, then the Earth is moving in two different ways: rotating once a day and orbiting the Sun once a year (heliocentrism). In around 330 BCE, Aristotle and others put forward a number of arguments against the daily rotation of the Earth, arguing that an object placed on the ground would be moved to the west; what's more, no movement of the Earth could be felt. Experience showed that all objects fell towards the Earth, so it must be the centre of the universe. From that point on, Aristarchus's heliocentrism was forgotten, and geocentrism remained the dominant view up until the Renaissance.

Hipparchus and the movement of the Sun

In the 3rd century BCE, thanks to the conquests of Alexander, Greece came into contact with Babylonian astronomical traditions. In Rhodes, the mathematician and astronomer Hipparchus invented the astrolabe, drew up the first star catalogue, predicted eclipses with his tables of the Moon and the Sun, and developed trigonometry to establish the relationship between angles and chords in a circle.

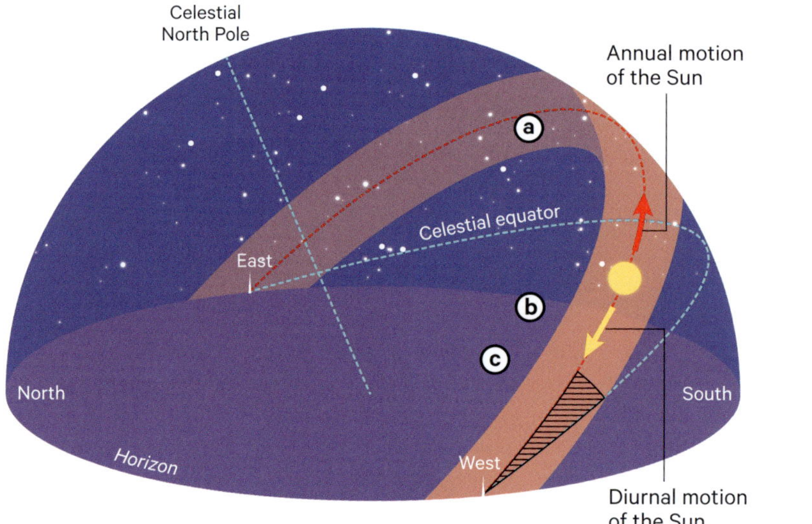

ⓐ ---- **Ecliptic**
Line around the celestial sphere travelled by the Sun in one year. It crosses the celestial equator twice each year, on the spring and autumn equinoxes.

ⓑ ▮ **Zodiac**
Belt on both sides of the ecliptic containing some constellations, through which the Moon and planets appear to move.

ⓒ ◁ **Obliquity**
Angle between the circles of the equator and the ecliptic (23° 26'= axial tilt of the Earth)

How long is a solar year?

To confirm whether the length of the solar tropical year was constant and measured 365.25 days, as was generally accepted, Hipparchus refined the observation of solar shadows at the equinoxes and determined that the length of a solar year was 365.2467 days. As the length of the seasons was already known to be unequal, he came up with more precise definitions: in 130 BCE, spring was six days longer than autumn. If the Sun moved in a geocentric circle at a constant speed on the ecliptic, that would not explain these differences, however. Hipparchus therefore concluded that the circle traced at a constant speed by the Sun was not centred on the Earth, but slightly away from it. This model allowed him to calculate tables of the motion for the Sun, which remained in use for more than a thousand years.

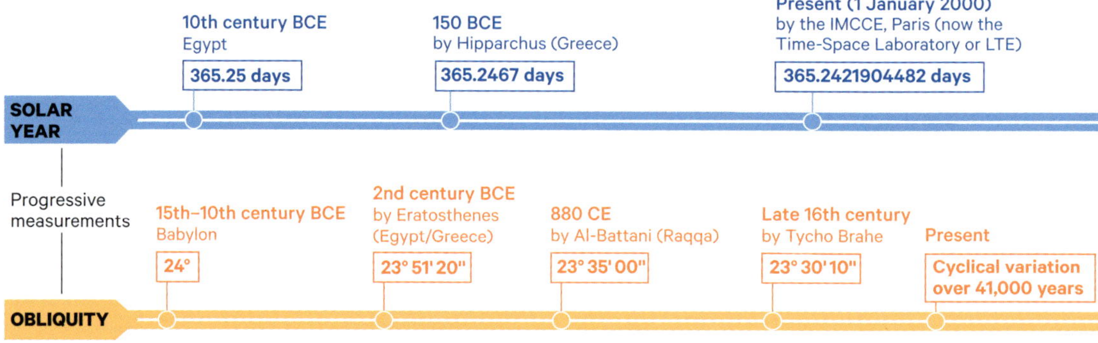

The discovery of the precession of the equinoxes

The precession of the equinoxes

The spring equinox occurs when the Sun, as it travels along the ecliptic, crosses the celestial equator. The intersection of these two circles of the sphere, known as the 'vernal point' (V), marks the start of spring, and the angle between them is known as 'obliquity'. Comparing his observations to those of his predecessors from the previous two centuries, Hipparchus realized that this point was not in the same position relative to the stars around it, and calculated that it must complete a full circle over the course of 26,000 years (known as the 'precession of the equinoxes'). As a result, the celestial North Pole, determined by the orientation of the Earth's axis of rotation, traces a circle among the stars. Polaris, which is known as the North Star today, will no longer be used to find the celestial North Pole in a few hundred years' time.

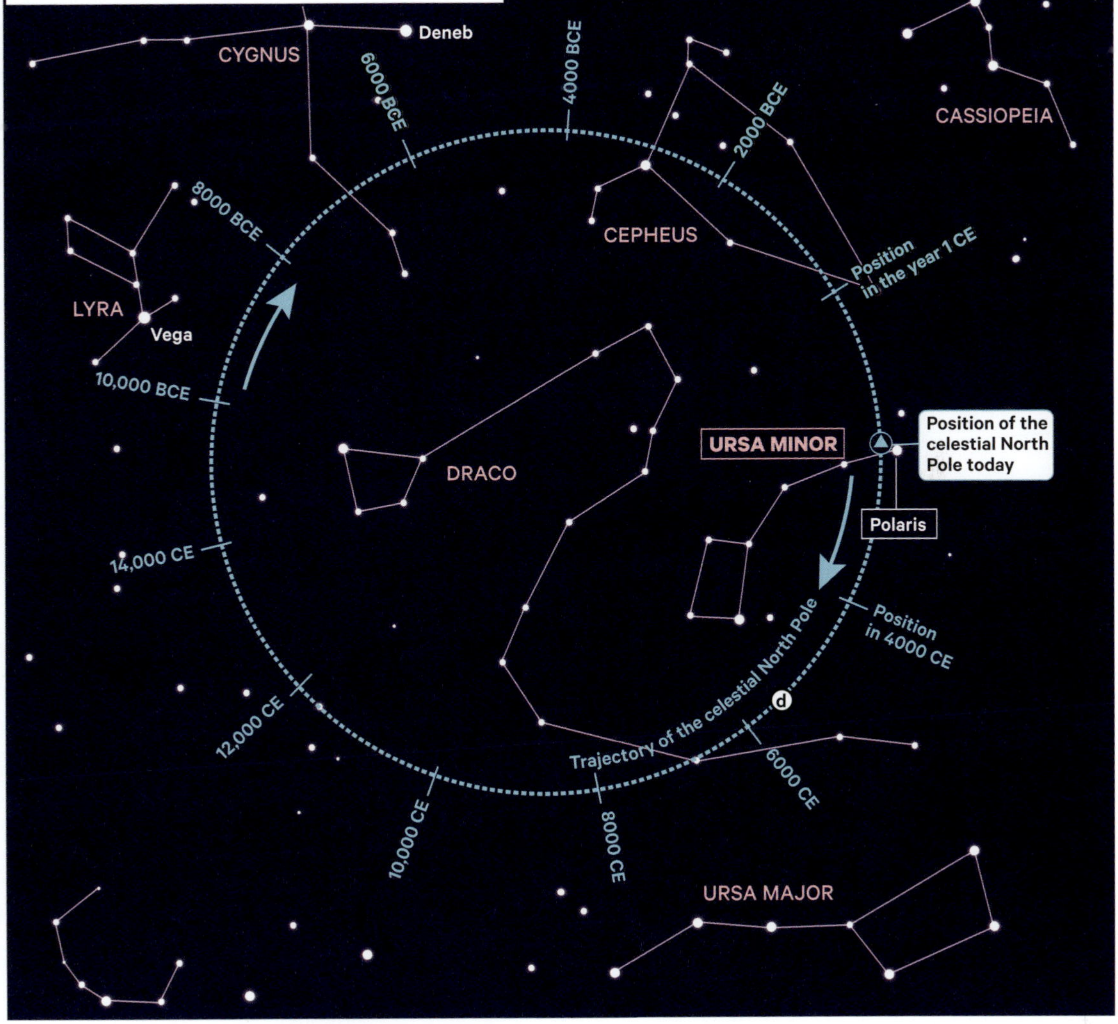

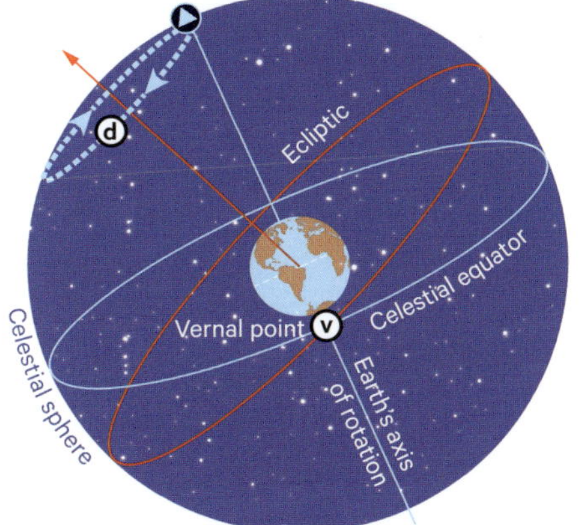

(d) Circular motion of the celestial North Pole over 26,000 years, due to the changing orientation of the Earth's axis in relation to the stars (= precession of the equinoxes, also called axial precession)

Ptolemy and the planets

In Alexandria, the intellectual centre of the Mediterranean world in the 2nd century CE, the astronomer, astrologer and geographer Ptolemy built upon the work of his predecessors. His geocentric model, which was far more precise than those that had come before it and which he set out in his treatise the Almagest, was used for fifteen centuries to draw up calendars, predict eclipses and describe planetary movements, following the principles set out by Aristotle.

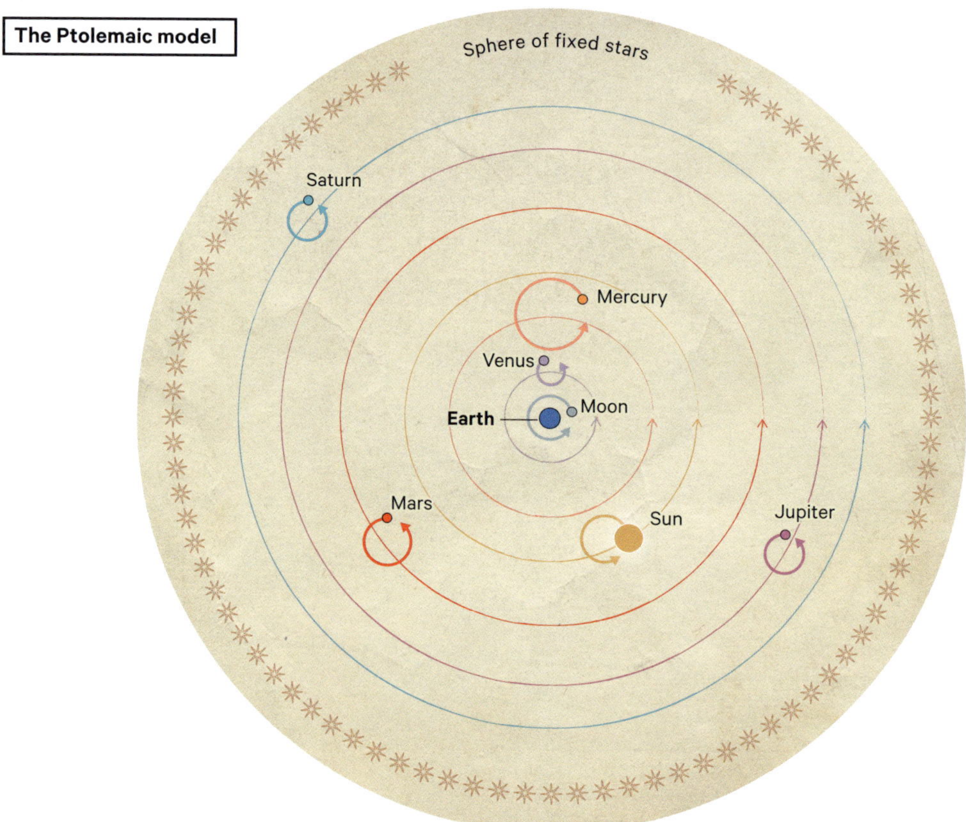

A complex theory of planetary motion

For both Babylonian and Greek scholars, observing the planets over long periods of time meant that they recognized apparent irregularities in their movements. Moving slowly among the stars, in the opposite direction to that of the daily motion of the Sun, they seemed to stop, retrace their path, then set off in their initial direction once again, up until the next retrograde motion (opposite, top). These 'loops' are particularly easy to observe in the case of the planet Mars, because they occur roughly every two years. Ptolemy devised a geometric model to explain these loops. It assigns a number of circles to each of the planets, choosing the positions of their centres and the length of time they take to trace their path according to what best matched his observations. By combining three circles, adjusting their size and these durations, Ptolemy was able to accurately describe the movement of Mars and the other planets. On the plane of the ecliptic (see opposite), each planet traces a small circle (epicycle) at a constant speed. Its centre O_E traces a second circle (known as the 'deferent'), whose centre O_D is close to the Earth. Ptolemy introduced a third circle (the 'equant'), whose centre E is also slightly away from the Earth. Point K moves along it at a constant speed. Ptolemy assigned to each planet a spherical shell containing the planet's orbits (above).

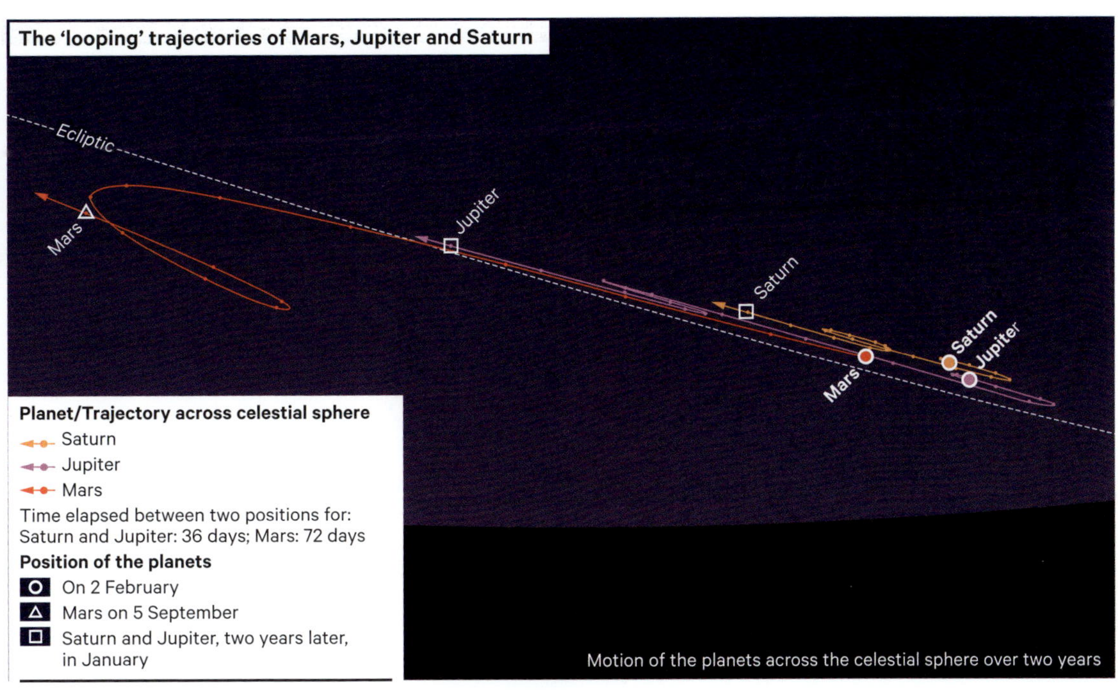

The 'looping' trajectories of Mars, Jupiter and Saturn

Ecliptic
Mars
Jupiter
Saturn
Mars
Saturn
Jupiter
Planet/Trajectory across celestial sphere
Saturn
Jupiter
Mars
Time elapsed between two positions for:
Saturn and Jupiter: 36 days; Mars: 72 days
Position of the planets
On 2 February
Mars on 5 September
Saturn and Jupiter, two years later, in January

Motion of the planets across the celestial sphere over two years

The orbit of Mars according to Ptolemy

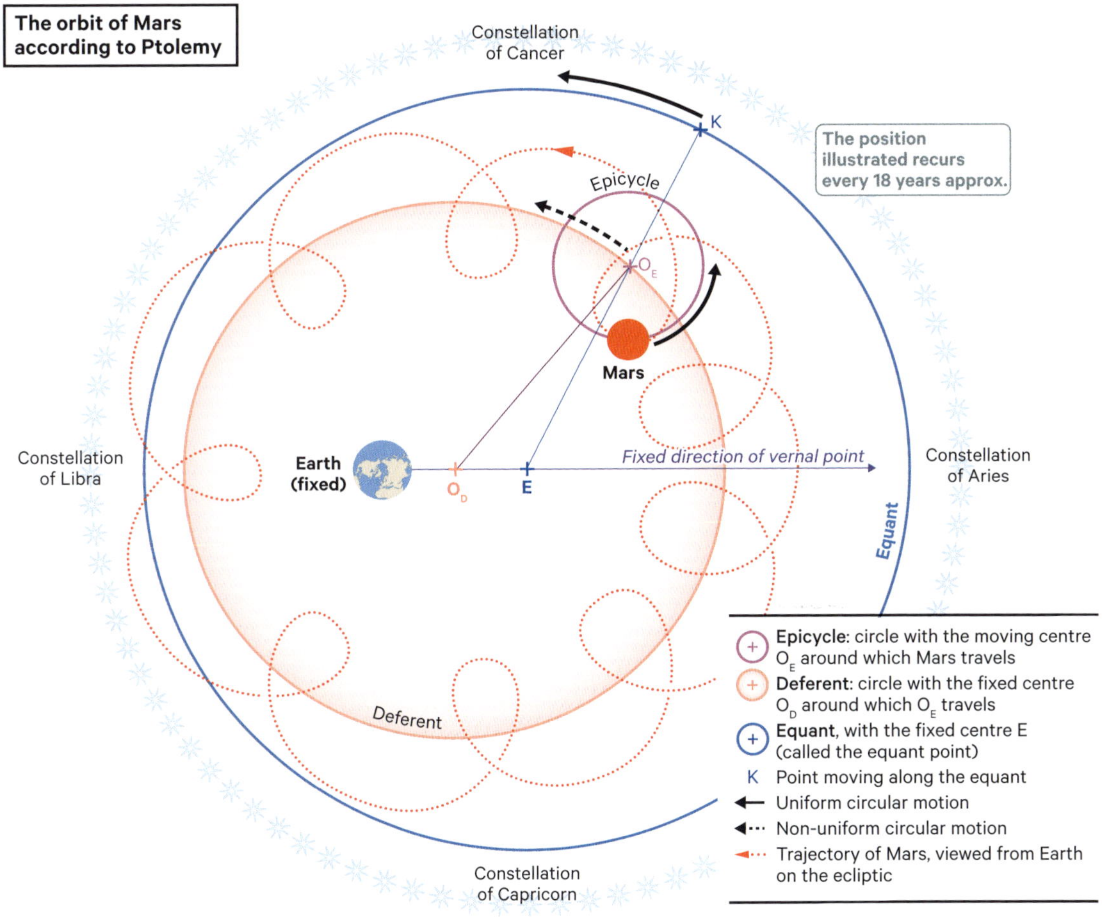

Constellation of Cancer
K
The position illustrated recurs every 18 years approx.
Epicycle
O_E
Mars
Constellation of Libra
Earth (fixed)
O_D
E
Fixed direction of vernal point
Constellation of Aries
Equant
Deferent
Constellation of Capricorn
Epicycle: circle with the moving centre O_E around which Mars travels
Deferent: circle with the fixed centre O_D around which O_E travels
Equant, with the fixed centre E (called the equant point)
K Point moving along the equant
Uniform circular motion
Non-uniform circular motion
Trajectory of Mars, viewed from Earth on the ecliptic

Constellations and star catalogues

Building on the Egyptian and Babylonian constellations, as well as the names of notable stars, Hipparchus and later Ptolemy measured the position of around a thousand stars, each one of which was assigned to a constellation. Ptolemy's measurements, with his armillary sphere, were precise enough to confirm that the positions of the stars relative to one another did not change. These were the first star catalogues.

The *Epic of Creation*

Written in the 12th century BCE, the Babylonian *Epic of Creation* (*Enūma Eliš*) tells the story of the creator god Marduk, who assigns three stars to each of the twelve months of the year and sets them in motion. Three constellations were assigned to each month. They were named after animals and farming tools (see p. 59).

Star catalogues

In Rhodes in 150 BCE, Hipparchus identified 850 stars, noting their coordinates and categorizing them according to their brightness. Three centuries later, Ptolemy listed 48 constellations, 47 of which remain in use today. Looking with the naked eye at an unpolluted sky away from city lights, around 9,000 stars can be seen.

The lost catalogue of Hipparchus

The existence of Hipparchus's star catalogue is contested, because only Ptolemy's has survived. In 2020, a 5th-/6th-century palimpsest parchment was discovered, its original text erased and written over. It contains direct quotes from Hipparchus's catalogue, with the positions of stars specified to the degree, making the measurements more precise than Ptolemy's.

The star Polaris does not appear at the centre of this map of the sky in 150 BCE. This is due to the precession of the equinoxes, a continual shifting of the Earth's axis of rotation on a 26,000-year cycle.

Present constellation whose heritage goes back to the 5th millennium BCE
Present constellation inherited from the Mesopotamians
Present constellation inherited from the Greeks (Ptolemy)
Present constellation of uncertain heritage
Zodiac belt

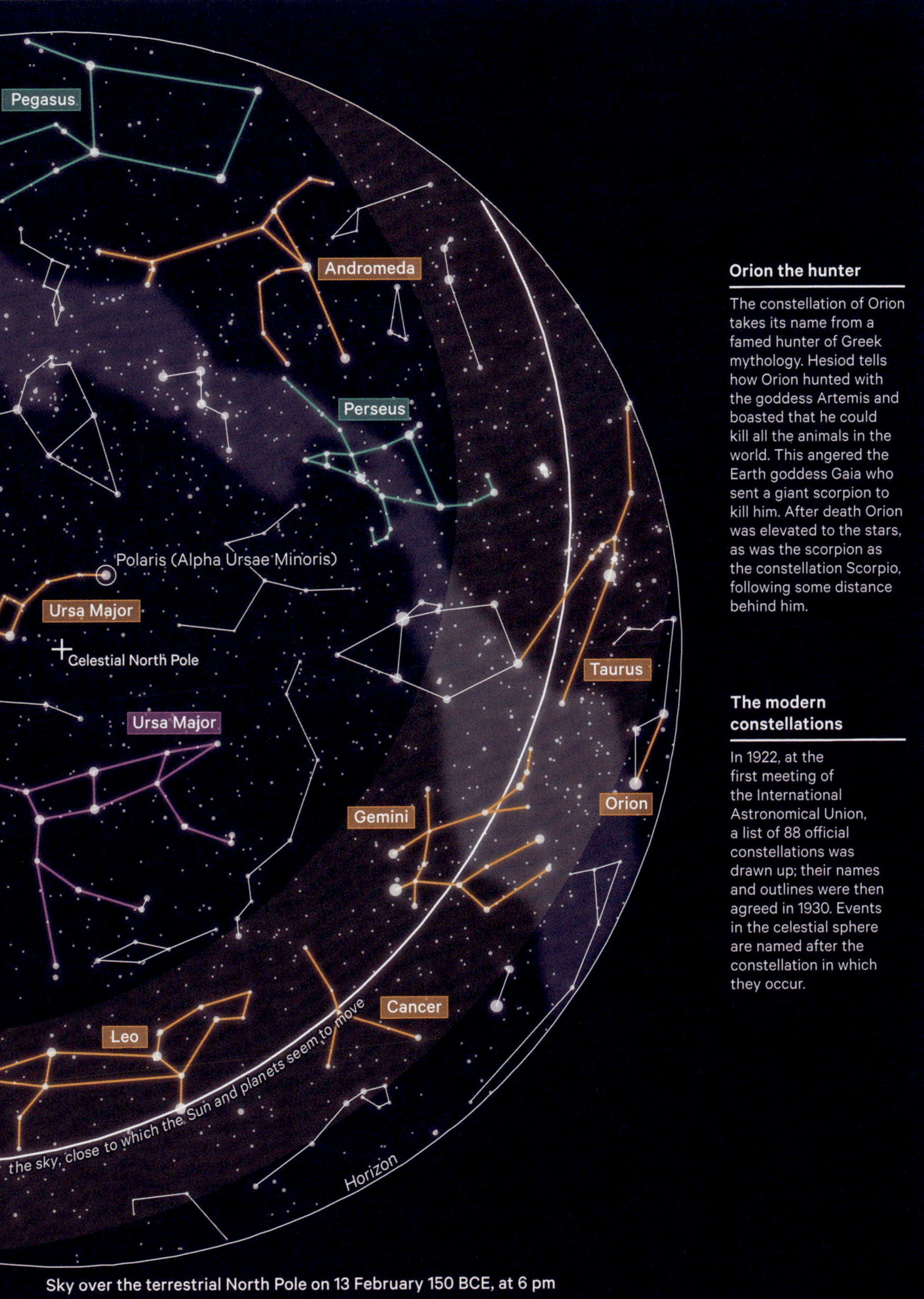

Sky over the terrestrial North Pole on 13 February 150 BCE, at 6 pm

Orion the hunter

The constellation of Orion takes its name from a famed hunter of Greek mythology. Hesiod tells how Orion hunted with the goddess Artemis and boasted that he could kill all the animals in the world. This angered the Earth goddess Gaia who sent a giant scorpion to kill him. After death Orion was elevated to the stars, as was the scorpion as the constellation Scorpio, following some distance behind him.

The modern constellations

In 1922, at the first meeting of the International Astronomical Union, a list of 88 official constellations was drawn up; their names and outlines were then agreed in 1930. Events in the celestial sphere are named after the constellation in which they occur.

Mapping the Earth: 4th century BCE–2nd century CE

As early as the 6th century BCE, the Earth was no longer believed to be flat. Greek philosophers were divided: some thought it was cylindrical, others spherical. Identifying the location of a place on our planet and drawing up a map required a coordinate system. The units that were used to measure the sky (degree, arc-minute, arc-second) were therefore also adopted for the Earth. It became possible to draw maps and use them to determine a location according to a shared system.

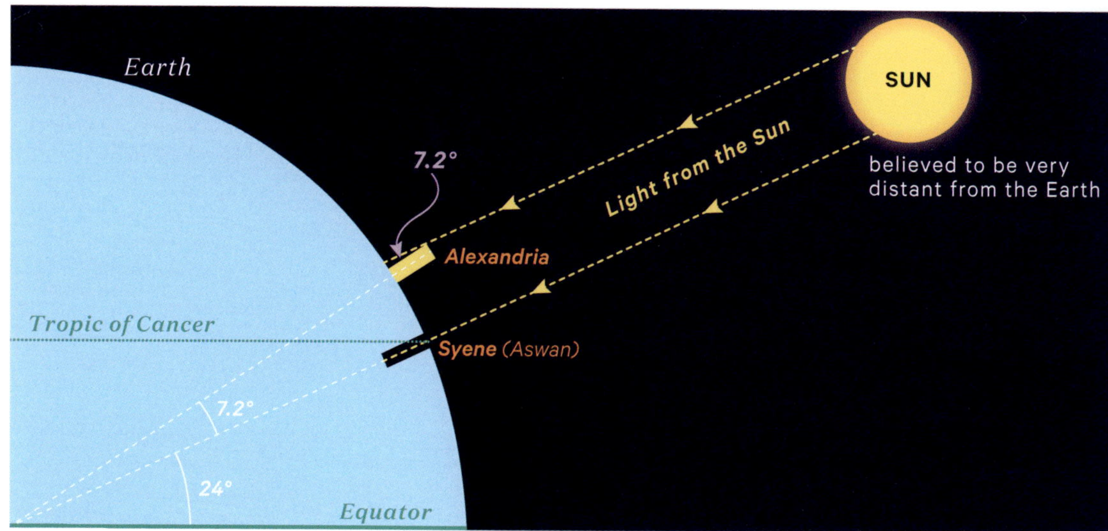

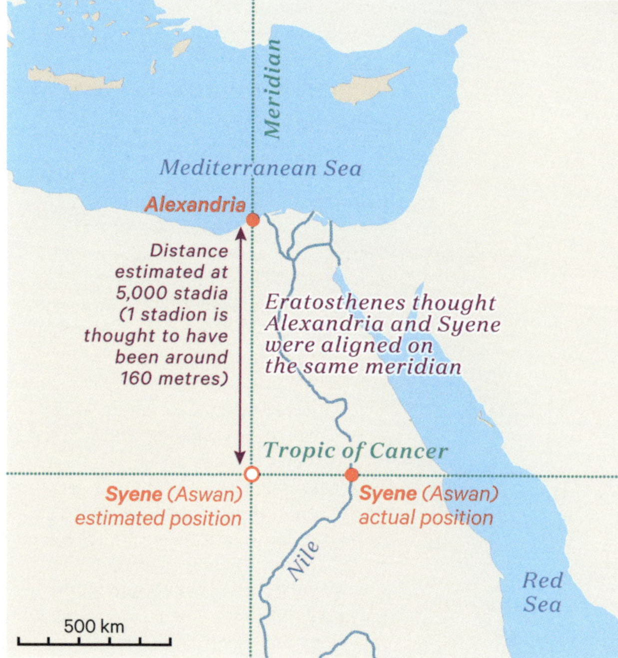

Measuring the size of the Earth

In his treatise *On the Heavens* (350 BCE), Aristotle included arguments in favour of a spherical Earth, along with an estimate of its circumference by unknown mathematicians. In modern units, this estimate is somewhere between 65,000 and 80,000 km. The circumference was calculated more accurately by Eratosthenes of Cyrene, mathematician and chief librarian at the Library of Alexandria from 245 BCE. He noted that at noon on the summer solstice, when the Sun is at its highest point, its rays shine directly into the bottom of the wells in Syene, situated on the Tropic of Cancer. In Alexandria, further north on the same meridian, the shadow cast by a vertical rod is at an angle that measures 1/50 of 360° (7.2°). He concluded that this difference was caused by the Earth's spherical nature. Estimating the distance between the cities at 5,000 stadia (probably the Egyptian stadion, which is 157.5 metres), he concluded that the circumference of the Earth was 50 × 5,000 = 250,000 stadia. This gives a circumference of around 39,400 km, very close to the actual measurement of 40,008 km.

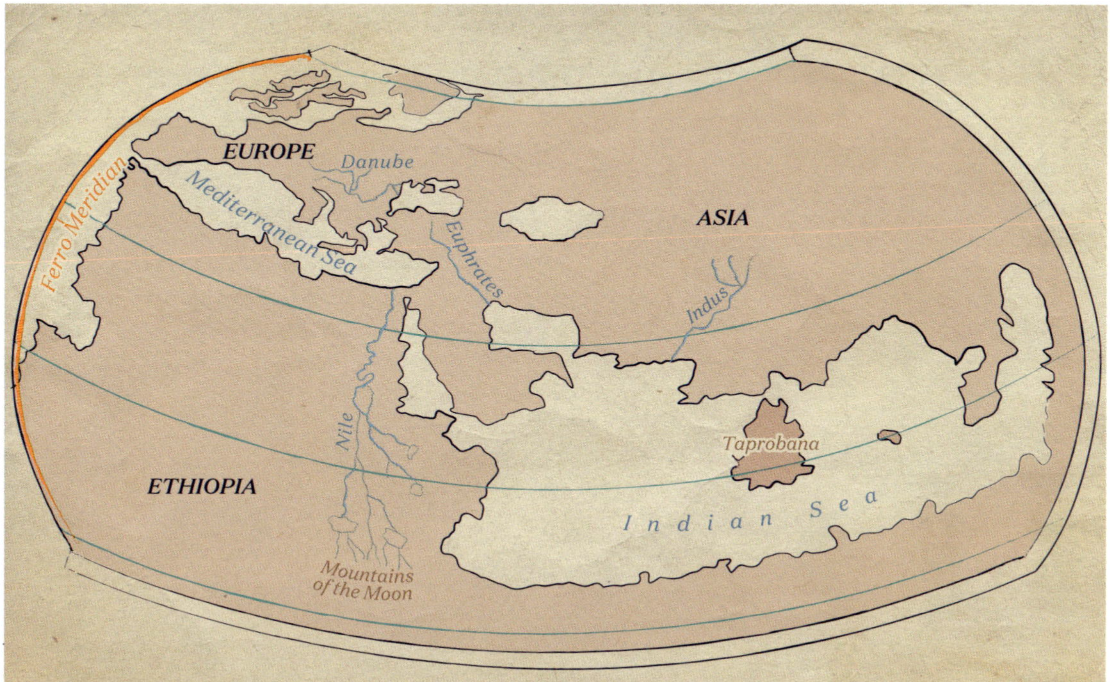

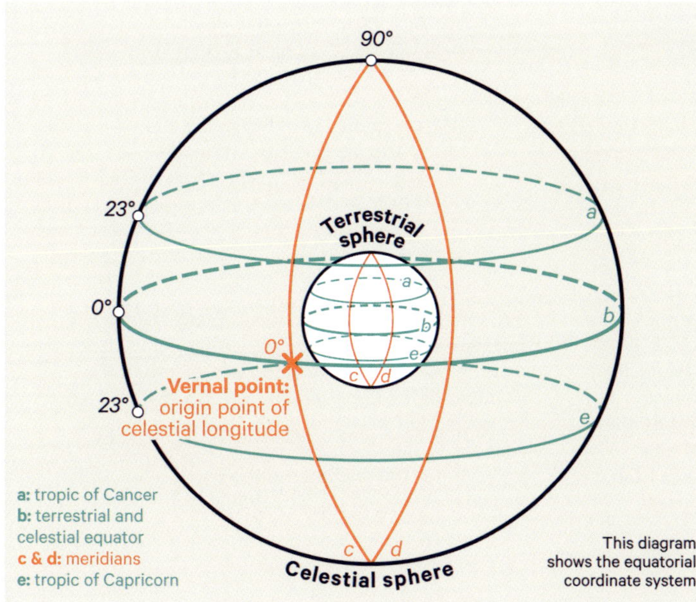

15th-century reconstruction of an ancient map (Ptolemy's interpretation)

The invention of longitude

Observations of the Sun at the equinoxes or the height of the celestial pole give the latitude of a place. Measuring a difference in longitude (east–west) is a different matter. Local time is different in the two places, but how can we know what it is without comparing two clocks? The length of time spent travelling is a very imprecise measurement. To improve the map of the Mediterranean world, Ptolemy used two accounts of the lunar eclipse of 331 BCE, which noted the local time when the eclipse occurred. The difference in time was converted into a difference in longitude and used to adjust the map. He also added a prime meridian at Ferro in the Canaries.

On the Earth as in the heavens

Babylon established the degree and its subdivisions as the units used to measure a circle (360°) and arcs in the sky, and this legacy was inherited by the Greeks. Geometry then established connections between celestial and terrestrial circles. The celestial equator and celestial poles found an equivalent in the equator and poles of the Earth. The tropics of Cancer and Capricorn also had equivalents in the sky. The meridians and parallels on the celestial sphere, which were used to measure the positions of stars, had an equivalent in the terrestrial meridians and parallels, which were used to define a location on Earth. A prime meridian, which Ptolemy placed at Ferro (now El Hierro) in the Canary Islands, was chosen as the starting point for terrestrial longitude. This was later moved to Paris, and then Greenwich, where it still stands.

Measuring distances in space

Before Aristarchus of Samos (280 BCE), how far away the stars and planets were from Earth was a matter of speculation. The Moon, which sometimes eclipsed the Sun, planets and stars, was understood to be relatively close, while the Sun, which was believed to be 'larger than the Peloponnese', was thought to be distant. By combining observations and geometry, Aristarchus managed to calculate some of these distances, just as Ptolemy attempted to do for the planets, a few centuries later.

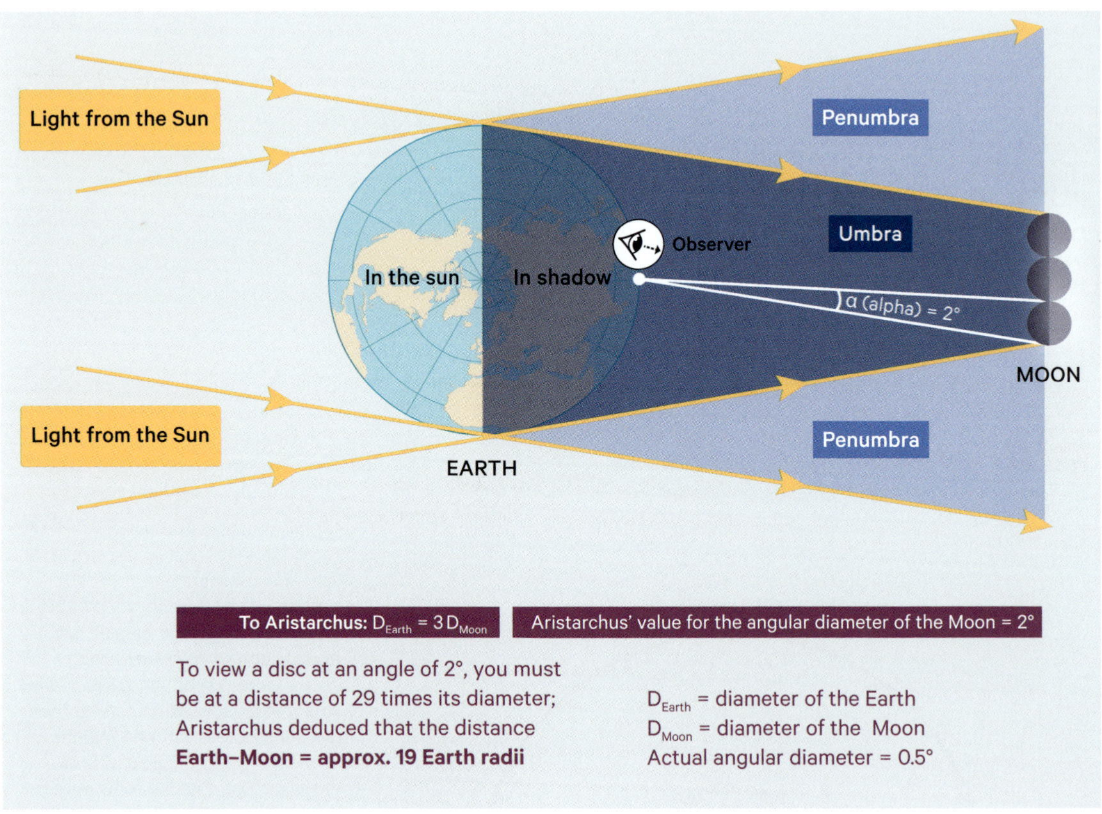

To Aristarchus: $D_{Earth} = 3 D_{Moon}$ Aristarchus' value for the angular diameter of the Moon = 2°

To view a disc at an angle of 2°, you must be at a distance of 29 times its diameter; Aristarchus deduced that the distance **Earth–Moon = approx. 19 Earth radii**

D_{Earth} = diameter of the Earth
D_{Moon} = diameter of the Moon
Actual angular diameter = 0.5°

How far away is the Moon?

Aristarchus used simple reasoning, which he applied to the Moon. Step one: he knew that, when we look at an object with our eyes, the further away it is, the smaller it appears. He worked out this ratio using a shield, whose diameter and distance he knew. Therefore the angle at which we see the Moon in the sky (half a degree) must also be subject to the same ratio. Step two: during a total lunar eclipse, the Moon passes through the Earth's shadow, and Eratosthenes had calculated the size of the Earth. Aristarchus concluded that the diameter of the Moon was a third of the diameter of the Earth, around 4,000 km. Final step: he applied the 'shield ratio' and that gave him the distance to the Moon. He incorrectly estimated it to be ten times the diameter of the Earth, a figure that was later corrected by Hipparchus and Ptolemy.

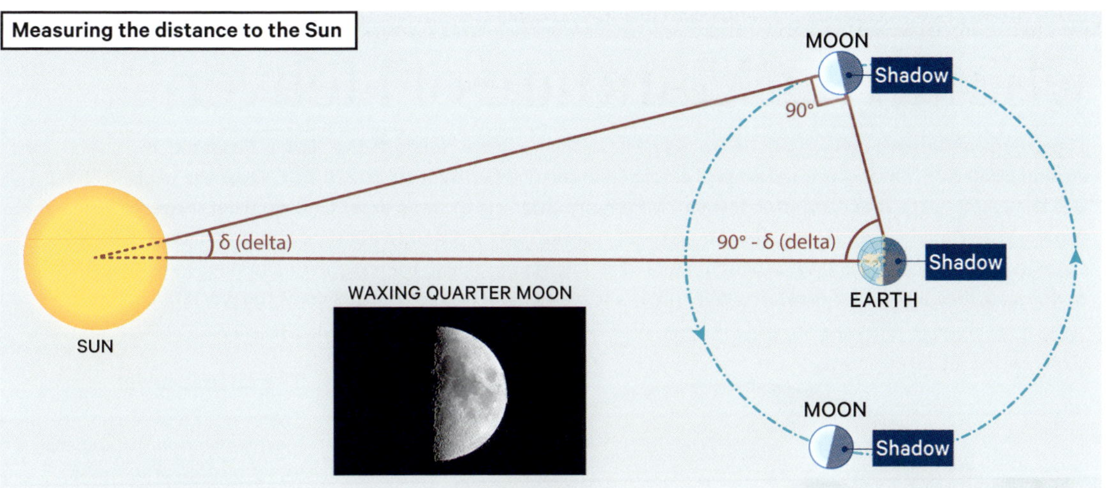

Measuring the distance to the Sun

The Ptolemaic model of nesting celestial spheres

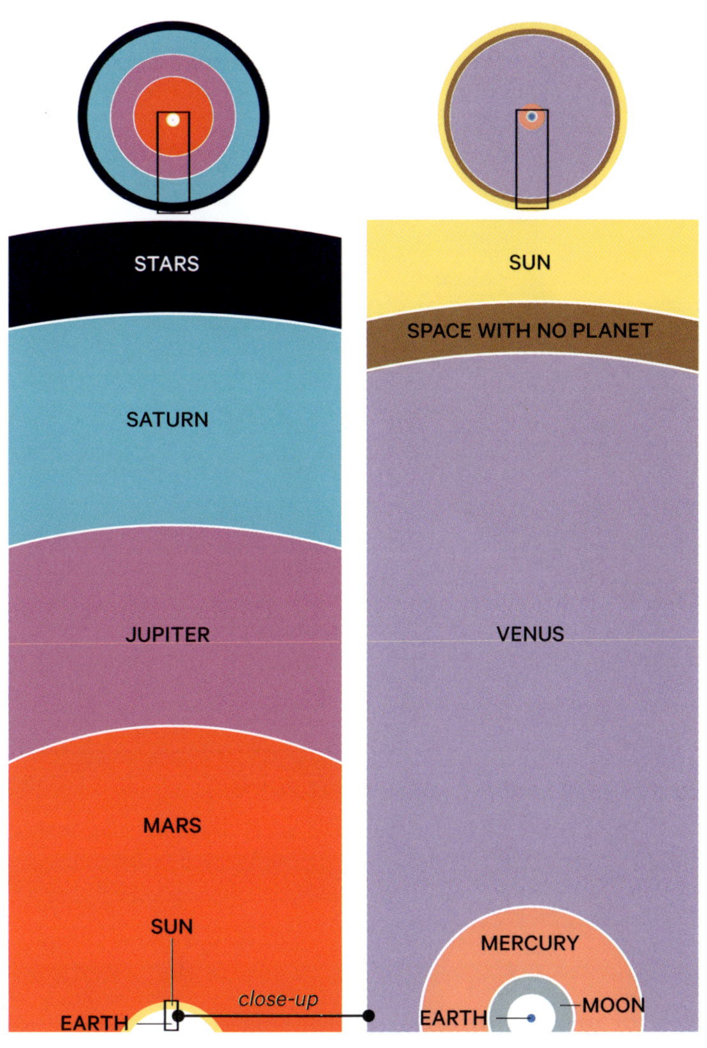

The distance to the Sun

Aristarchus believed that the universe was much larger than his predecessors claimed. Estimating the distance to the Sun for the first time was an impressive feat (see above), despite the possible criticisms of his calculation. He established that when the Moon is exactly in the first or last quarter (quadrature), and therefore divided into two equal parts of shadow and light, the Sun's rays strike it at an angle of exactly 90°. At that moment, for an observer on Earth, the direction to the centre of the Sun and the direction to the centre of the Moon form an angle that differs from 90° by a small value (delta, see above), which Aristarchus estimated to be 3°. This value, when applied to the triangle formed by the Earth, Moon and Sun, suggests that the distance between the Earth and the Sun is between 18 and 20 times larger than the distance from the Earth to the Moon, which he had already measured.

Four and a half centuries later, Ptolemy used this model to estimate the distances to the five visible planets, using the distance to the Moon (see left) as a unit of measurement. He placed Mercury and Venus closer to the Sun; Mars, Jupiter and Saturn further away. He believed the planets were situated on nested spheres of aether, and that neighbouring spheres fitted together like cogs in a machine.

China: the 'Mandate of Heaven'

The origins of Chinese culture can be traced back to the mythical Huangdi, the 'Yellow Emperor', in around 2600 BCE. Later, a unified empire emerged under the Qin dynasty in 220 BCE. Over the next two thousand years, it expanded to form a vast empire that was to some extent cut off from the rest of the world. In Chinese philosophy, the 'Mandate of Heaven' was granted to political rulers and shaped a vision of the sky and the interdependence of all elements of reality. Thinking philosophically and classifying, measuring and recording phenomena were integral to this conception of the world.

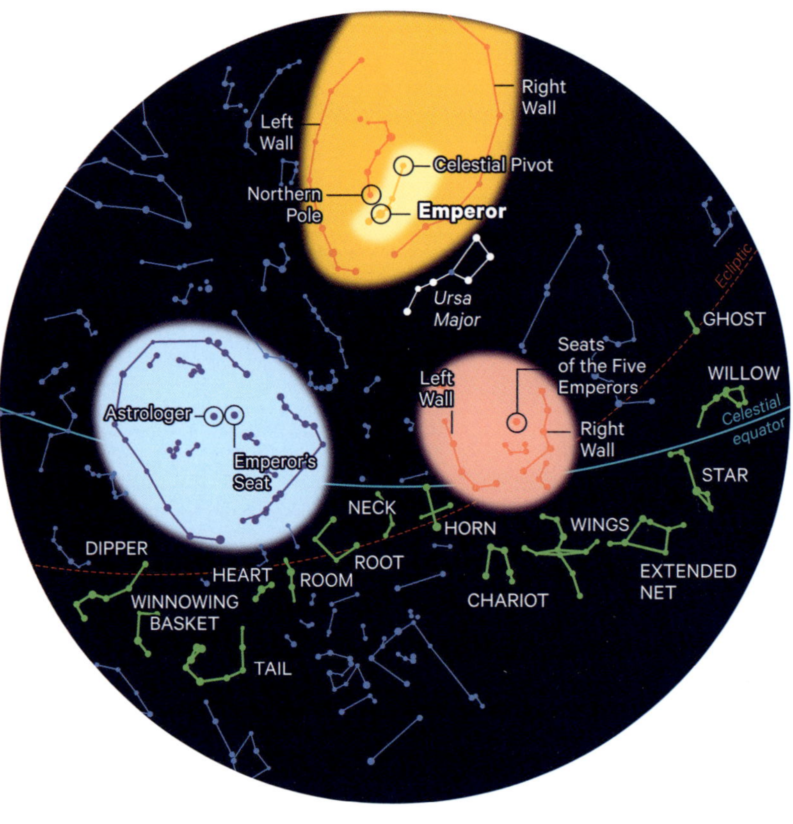

The cosmic order

The structure of society, which revolved around the Emperor, was projected onto the sky. The 'mansions' (positions) occupied by the Moon were used for divination. The position of the stars relative to the celestial equator was noted, although the planets were of little interest. Interpreting the sky meant decoding its intended message for society. Understanding this message was considered vital to avert potential disasters.

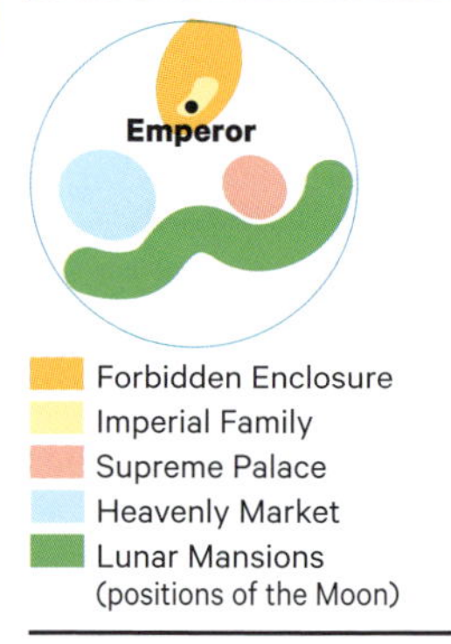

Forbidden Enclosure
Imperial Family
Supreme Palace
Heavenly Market
Lunar Mansions (positions of the Moon)

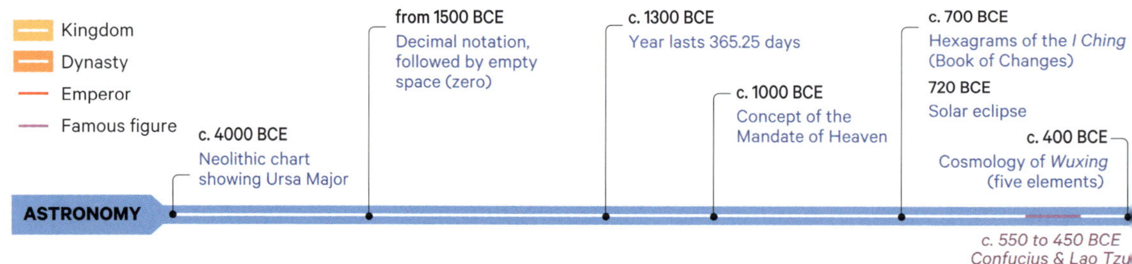

The Chinese calendar cycle

12 Earthly Branches

- Rat (yang)
- Ox (yin)
- Tiger (yang)
- Rabbit (yin)
- Dragon (yang)
- Snake (yin)
- Horse (yang)
- Goat (yin)
- Monkey (yang)
- Rooster (yin)
- Dog (yang)
- Pig (yin)

10 Heavenly Stems

- Wood (yin)
- Wood (yang)
- Fire (yin)
- Fire (yang)
- Earth (yin)
- Earth (yang)
- Metal (yin)
- Metal (yang)
- Water (yin)
- Water (yang)

Yin & yang

→ A dual numbering system

A yang Heavenly Stem can only be linked to a yang Earthly Branch. The same is true of yin.

→ 60 combinations

Sexagenary cycle

1 1984	2 1985	3 1986	4 1987	5 1988	6 1989
7 1990	8 1991	9 1992	10 1993	11 1994	12 1995
13 1996	14 1997	15 1998	16 1999	17 2000	18 2001
19 2002	20 2003	21 2004	22 2005	23 2006	24 2007
25 2008	26 2009	27 2010	28 2011	29 2012	30 2013
31 2014	32 2015	33 2016	34 2017	35 2018	36 2019
37 2020	38 2021	39 2022	40 2023	41 2024	42 2025
43 2026	44 2027	45 2028	46 2029	47 2030	48 2031
49 2032	50 2033	51 2034	52 2035	53 2036	54 2037
55 2038	56 2039	57 2040	58 2041	59 2042	60 2043

→ Example:

From 29 January 2025 **Yin Wood Snake**

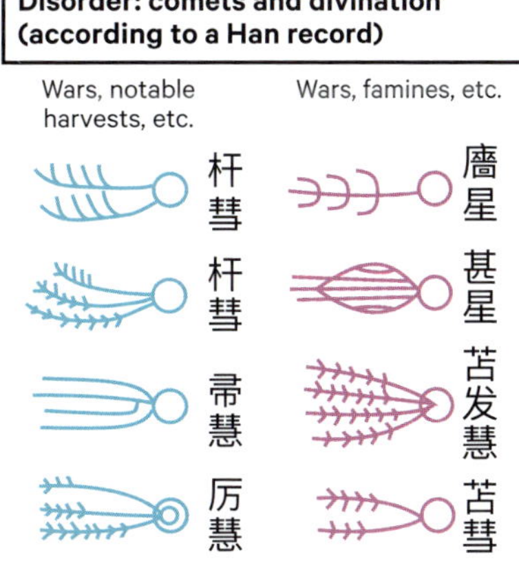

Disorder: comets and divination (according to a Han record)

Heaven's order on Earth

As early as 2000 BCE, turtle shells were used in divination; around 1200 BCE, a writing system was invented that was used to transcribe observations of the sky. A hierarchical, organic and impersonal concept of the universe developed. In infinite space, every phenomenon, whether earthly or heavenly, human or otherwise, was connected: this vision, set out by the moral philosopher Confucius and the cosmology of Lao Tzu, proved an enduring one. The Emperor was the Son of Heaven and received his mandate from it, to establish order on Earth as in the heavens. Calendars, which were regularly improved, and a system of arithmetic, based on a double numbering system using 10 and 12, reflected this heavenly order. Civil servants drew up maps, observed the sky, noted and interpreted phenomena such as eclipses and sunspots, comets and new stars, and recorded measurements. The planets were associated with the five elements (wood, water, fire, earth and metal).

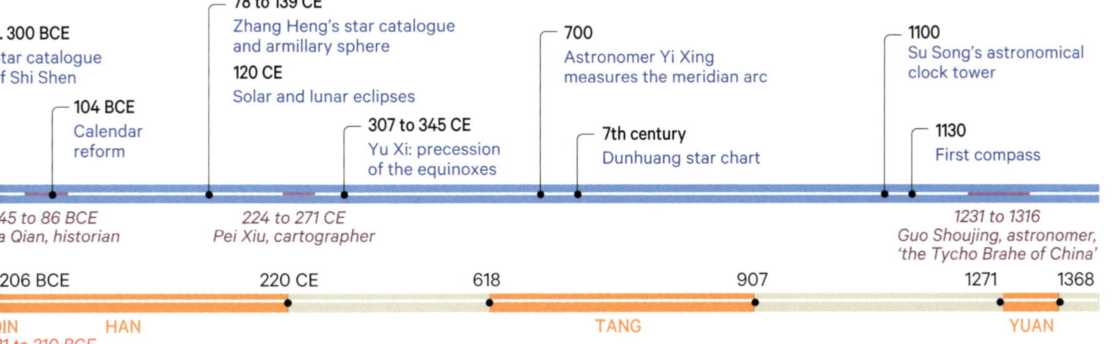

Chinese astronomers and artisans

Based on the cultural tenets established by Confucius and Lao Tzu, China's relationship with the sky, as seen with the naked eye, followed its own path over the course of fifteen centuries, almost totally cut off from the rest of Eurasia. As part of a stable empire, Chinese astronomers, supported by the various dynasties, especially the Tang emperors, inventoried phenomena, measured a meridian, developed arithmetic and invented instruments thanks to the work of skilled craftspeople. In these advances, the Chinese were often ahead of the West. The two concepts of science did not come face to face until the 17th century.

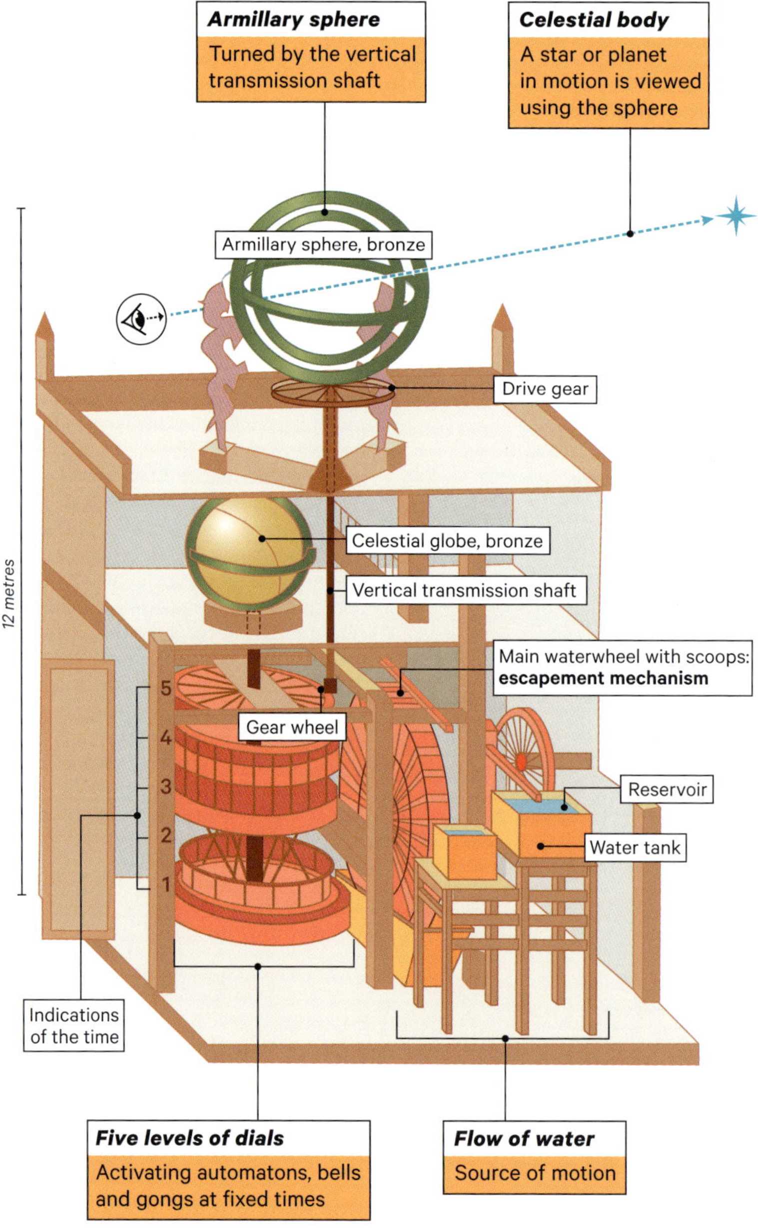

Su Song's astronomical clock (1020–1101)

This may well have been the first mechanical clock in the world. There are a number of descriptions that can be drawn on to recreate it. The armillary sphere at the top was based on older examples (they had been in existence since 150 CE) and it recreates the movement of the stars and planets. Water flows into the lower part (on the right) and drives a large vertical wheel, where the escapement is situated. The machine is built to 'carve up' time into pieces, using small, precise quantities of liquid.

The Buddhist monk Yi Xing first built a clock with a water wheel and chain in 725, but the invention is much improved here, with the clock only losing around three minutes a day. In the West, a clock with a similar level of precision would not be invented until 1560, by Jost Bürgi in Germany. A gear mechanism transmits the movement to the 'column of heaven', the vertical axis that moves the armillary sphere. The left side has five horizontal wheels on the same axis, which is also driven by the large wheel. It communicates the time via 158 wooden figures, which appear with gongs, drums and bells. A celestial globe shows the rising and setting of the Sun.

Magnetism, compasses and 'guest stars'

South-pointing compasses from Han-period China were used as divination tools, and were often decorated with the seven stars that make up Ursa Major. By 1130, compasses made from a magnet fixed to a piece of wood floating on water were used to find south. The fixed geographical north and the shifting magnetic north, marked on compasses, changed according to the era, thus recording the history of the Earth's magnetic field.

The appearance, sometimes in daylight, and subsequent disappearance of novas and supernovas (known as 'guest stars' in Chinese) were seen as harbingers of disorder. Their brightness, position and dates were recorded from 4 BCE onwards; by the 14th century, astronomer Guo Shoujing was aware of around a hundred of these. On 4 July 1054, a bright supernova was documented near the star Tianguan (Zeta Tauri). Traces of it are still visible today. Kepler's Supernova (1604) was also observed.

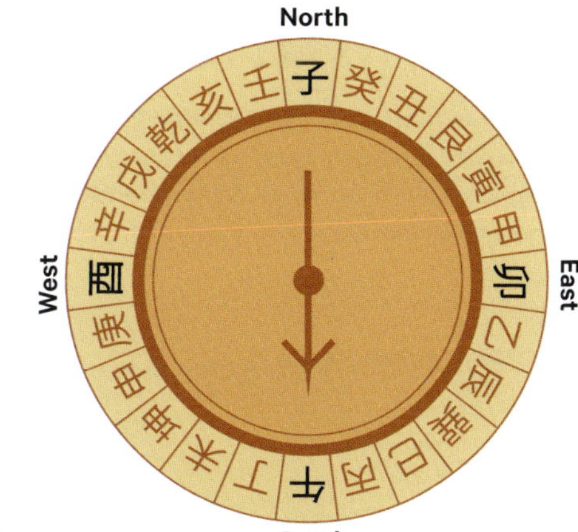

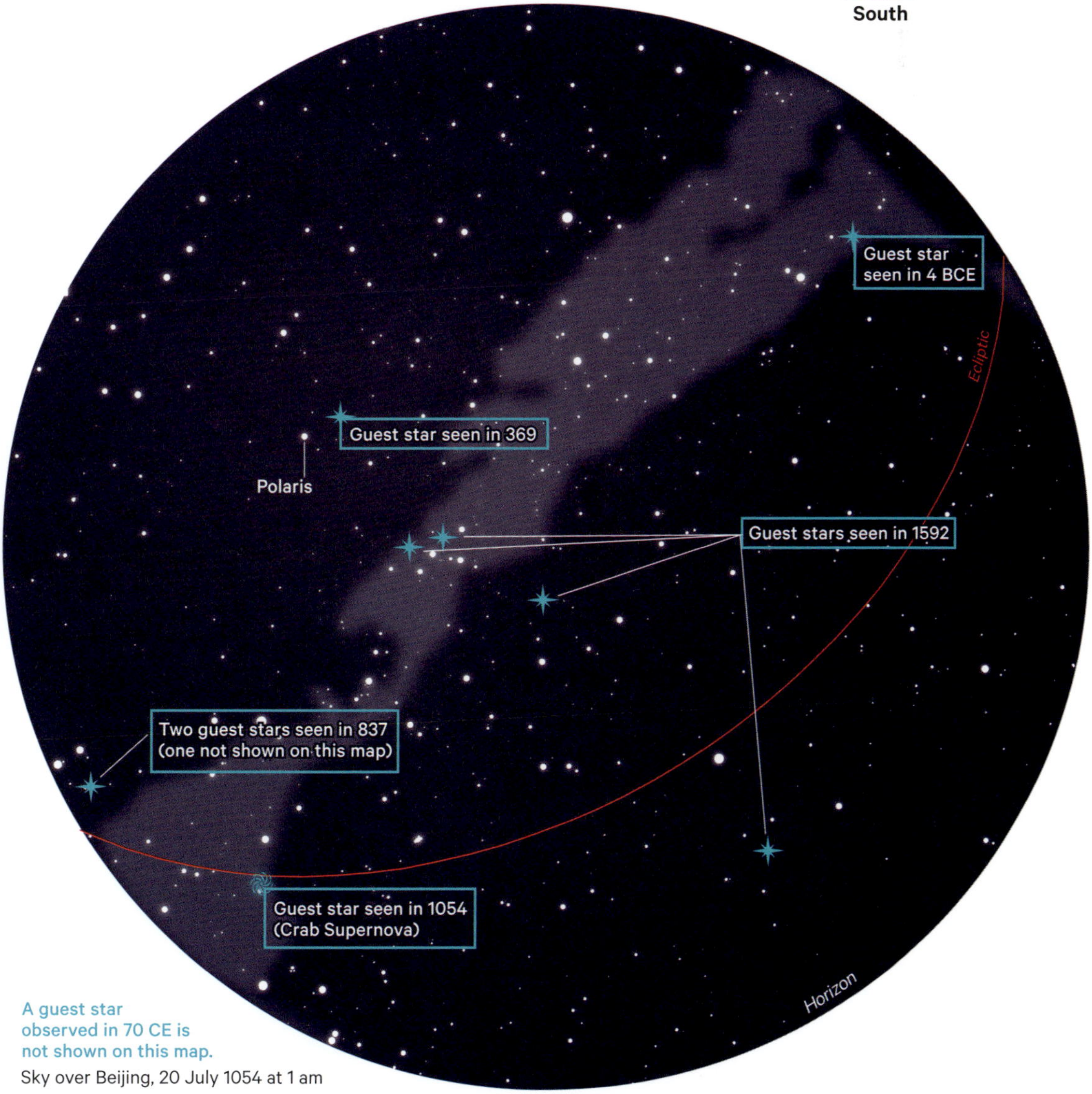

A guest star observed in 70 CE is not shown on this map.

Sky over Beijing, 20 July 1054 at 1 am

China and the West

China had very little contact with the West, aside from a few exchanges with the Islamic world, until the arrival of Jesuit astronomers in 1582. Westerners were expelled from China in 1742. Up until the Republic of China was proclaimed in 1912, the story was one of mutual suspicion, wars and separation that left China increasingly isolated from Western advances in science and technology.

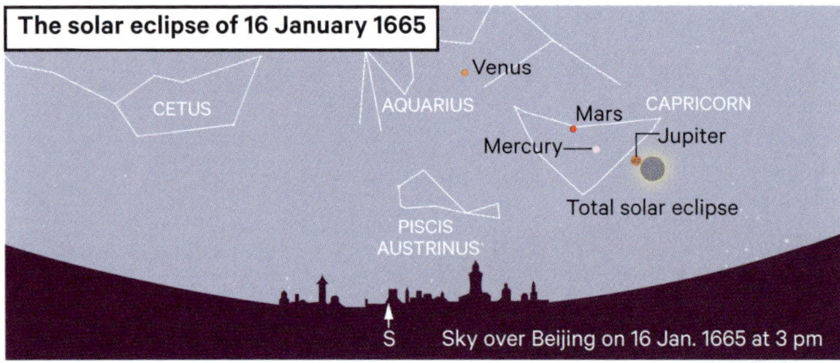

The Jesuits in China

Matteo Ricci, an Italian Jesuit who arrived in Macau in 1582, believed that preaching Christianity in China required the establishment of friendly relations. Ricci and his successors, including Adam Schall and Ferdinand Verbiest, used mathematics and astronomy as a way to establish a dialogue with the Ming and later the Qing emperors. Craftsmen built the first telescope, and Schall became a trusted figure in the new Manchu dynasty. With the Pope's approval, he was named president of the Board of Mathematics, which was founded under the Han dynasty, and Beijing observatory. Verbiest, who arrived in China in 1660, worked under Schall. Resentment of their influence led to attacks by Chinese astronomers, a trial (1664) and a test to predict the solar eclipse of 1665. The calculations of the two imprisoned Jesuits were more accurate by almost an hour, but this was not enough to win them their freedom. They were sentenced to death, but were saved by an earthquake, a meteorite and a fire at the palace. Instead, both were exiled to Canton, where Schall died. Verbiest was pardoned and returned to the imperial capital. He drew up a new calendar (1672) and was promoted to the highest rank of mandarin. The influence of the Jesuits slowly waned. Copernicus's model was little known in China, except by young astronomer Wang Zhenyi (1768–1797) and Western telescopes did not replace traditional instruments that relied on the naked eye for some time.

- Ⓐ Armillary sphere
- Ⓑ Celestial globe
- Ⓒ Planisphere
- Ⓓ Compass
- Ⓔ Quadrant
- Ⓕ Jacob's staff
- Ⓖ Astrolabe

Religious disputes

Throughout the 18th century, the Catholic Church was obliged to choose its stance on the relationship between Christianity and local cultures, and whether it would allow the practice of local rituals. According to Confucian traditions, which gave Chinese society its structure, every village had an ancestral shrine for ceremonies and acts of filial piety, which held records of the ancestors of the entire community, often going back centuries. These were a source of friction between Rome and its missionaries. The Church eventually outlawed them, which led to all Europeans being expelled from China (1742) and to the persecution of Christian converts.

The failed British diplomatic mission of 1793

Taking advantage of the troubles in France, King George III of England sent a delegation of over a hundred people to the Chinese emperor, led by ambassador George Macartney. Its mission was to establish an embassy in Beijing, to compete with the quasi-monopoly of the Portuguese in Canton. From Macau, Macartney travelled to Beijing, where he presented a gift of the most advanced technological instruments. After keeping them waiting for months, the emperor eventually received the group. Protocol dictated that they should bow, but Macartney refused and thus was forced to leave, taking his gifts with him. In a letter to the king, the Qianlong Emperor wrote that he refused to exchange Chinese products for 'barbarian' ones.

Macau becomes Portuguese trading post
1557

Matteo Ricci's first Jesuit mission to China
1582–1610

Jesuits suppressed by Rome
1773

Macartney Embassy to China
1793–1796

Jesuits return and build the Sheshan Observatory in Shanghai
1885

The astrolabe

The astrolabe is a relatively small instrument used both for calculations and for observation. Like a globe or a celestial sphere, it is a way of depicting the sky and analysing the movements of celestial bodies. It is simpler and less expensive to make than an armillary sphere, and has been in existence since the time of Plato and even Thales (6th century BCE). There are records of it in Alexandria that date to the 3rd century and it was introduced to the Islamic world. The astrolabe was constantly being developed and improved, and it was used widely, reaching the West around the year 1000.

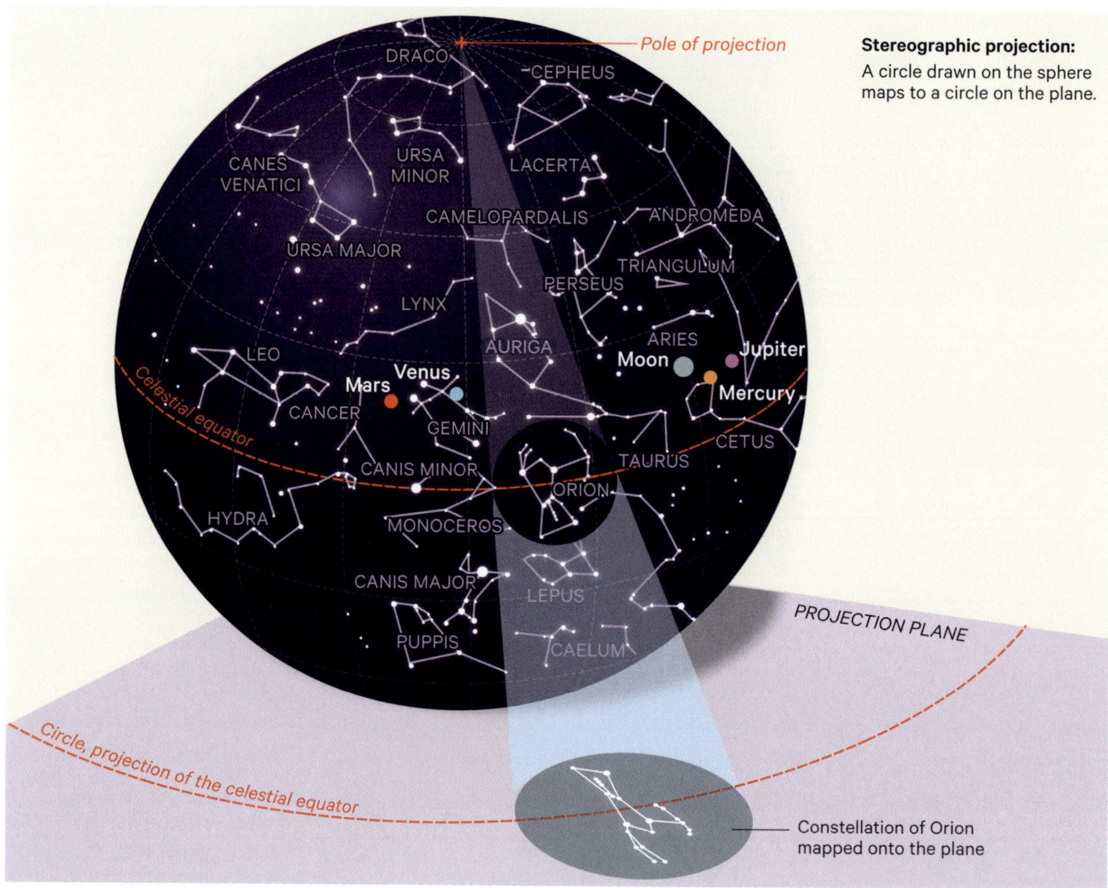

From three dimensions to two

When depicting the celestial or terrestrial sphere, there was one important question: how could the shapes and positions of stars and planets in three dimensions be depicted on a two-dimensional surface that was easy to work with, such as papyrus or a tablet? A fairly simple geometric trick (called stereographic projection) associates each point on the sphere with a point on the plane. This projection has two surprising properties: any circle traced on the sphere (the celestial equator or the ecliptic) becomes a circle on the plane; anything drawn on the sphere's surface close to one of its points is projected almost identically onto the plane.

Mul Apin
Babylonian astrology text
At least 686 BCE

Hipparchus and Ptolemy
Stereographic projection & first astrolabes
2nd century BCE–2nd century CE

Theon of Alexandria
Astrolabe treatise (lost)
4th century CE

John Philoponus
Astrolabe treatise
530 CE

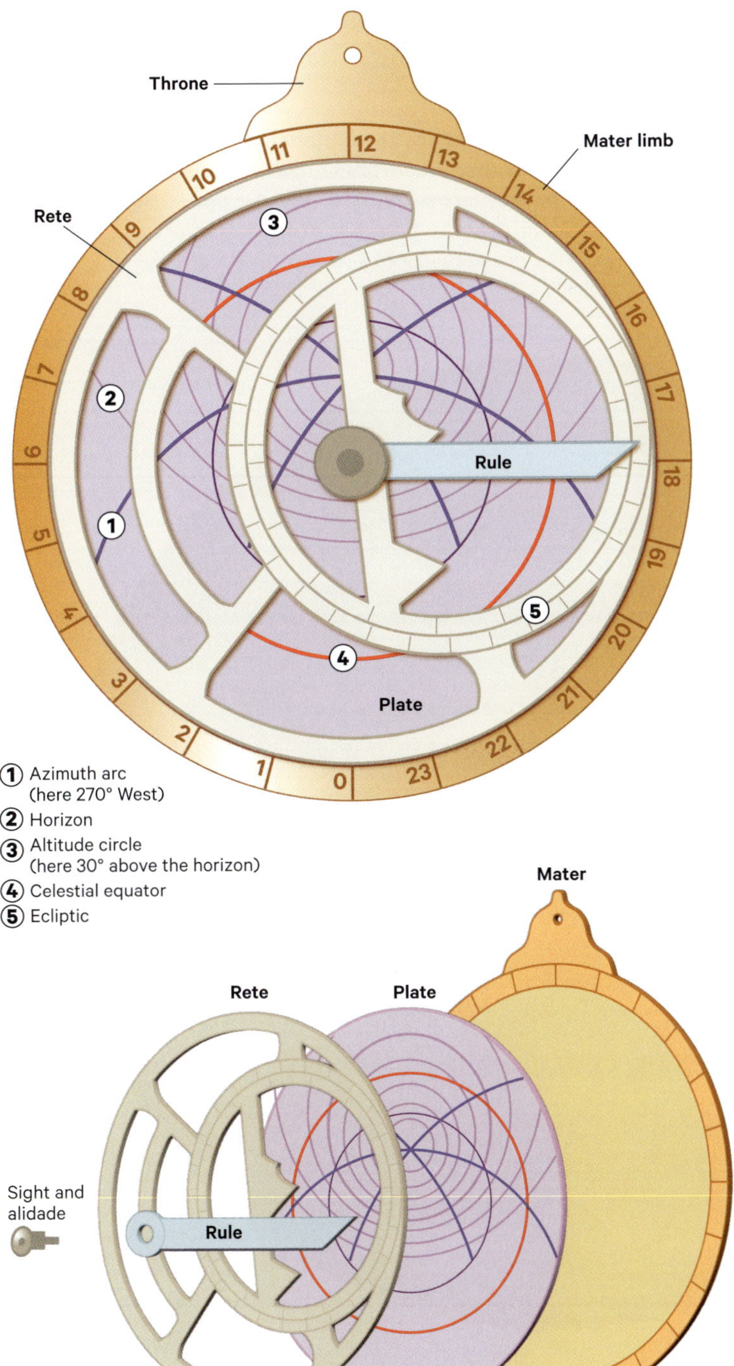

① Azimuth arc
(here 270° West)
② Horizon
③ Altitude circle
(here 30° above the horizon)
④ Celestial equator
⑤ Ecliptic

The structure and use of an astrolabe

An astrolabe is simple to use for observation (back face, not shown) and more complex when used for calculations (front face). Using the back face, by holding the disc vertical and pointing a small rotating ruler, we can measure the angle that gives the height of the Moon or a star above the horizon, to a level of accuracy a little greater than a degree. The shadow of the alidade is used to align the astrolabe with the sun. The back face is engraved with a calendar that converts the current date to the position of the Sun on the ecliptic. The front face is used to calculate the time of observation, which depends on the place where the observer is. The mater, which is fixed in place, is engraved with circles (called 'almucantars') that denote latitudes. Each one is the projection of a circle of the celestial sphere, from the horizon (0°) to the point that marks the zenith (90°). A second network of circles, matching the meridians of this sphere, indicate the direction (azimuth), between 0° (north) and 360° (north after one rotation). The moving rete represents the sky and its rotation. It shows the ecliptic, as well as the various positions of the Sun and brightest stars.

After moving the rule to the height of a star with the pointer on the mater, one rotation of the rete gives its azimuth and the time of observation. The astrolabe works like a compass or a watch. The rotation of the rete also makes it a kind of pocket planetarium. It makes it possible to calculate the time and direction of the rising and setting of the Sun and stars.

Al-Khwarizmi
Major innovations
825 CE

Al-Biruni & Al-Sufi
Improvements
10th & 11th centuries

Pope Sylvester II
Probable introduction to the Christian West
10th century

Manufactured
throughout Europe
15th & 16th centuries

Islam and astronomy

From 710, a new empire began to spread, from the borders of India to Spain, stretching as far as the southern Mediterranean. First Damascus, then Baghdad were the capitals where Islamic culture flourished. Rulers supported astronomers in their quest to understand the sky, as this was necessary for the practice of Islam, the shared faith of these nations. Astronomical writings from Babylon, Greece and India were translated, inspiring scholars for the next three centuries, and leading to the development of new mathematical principles.

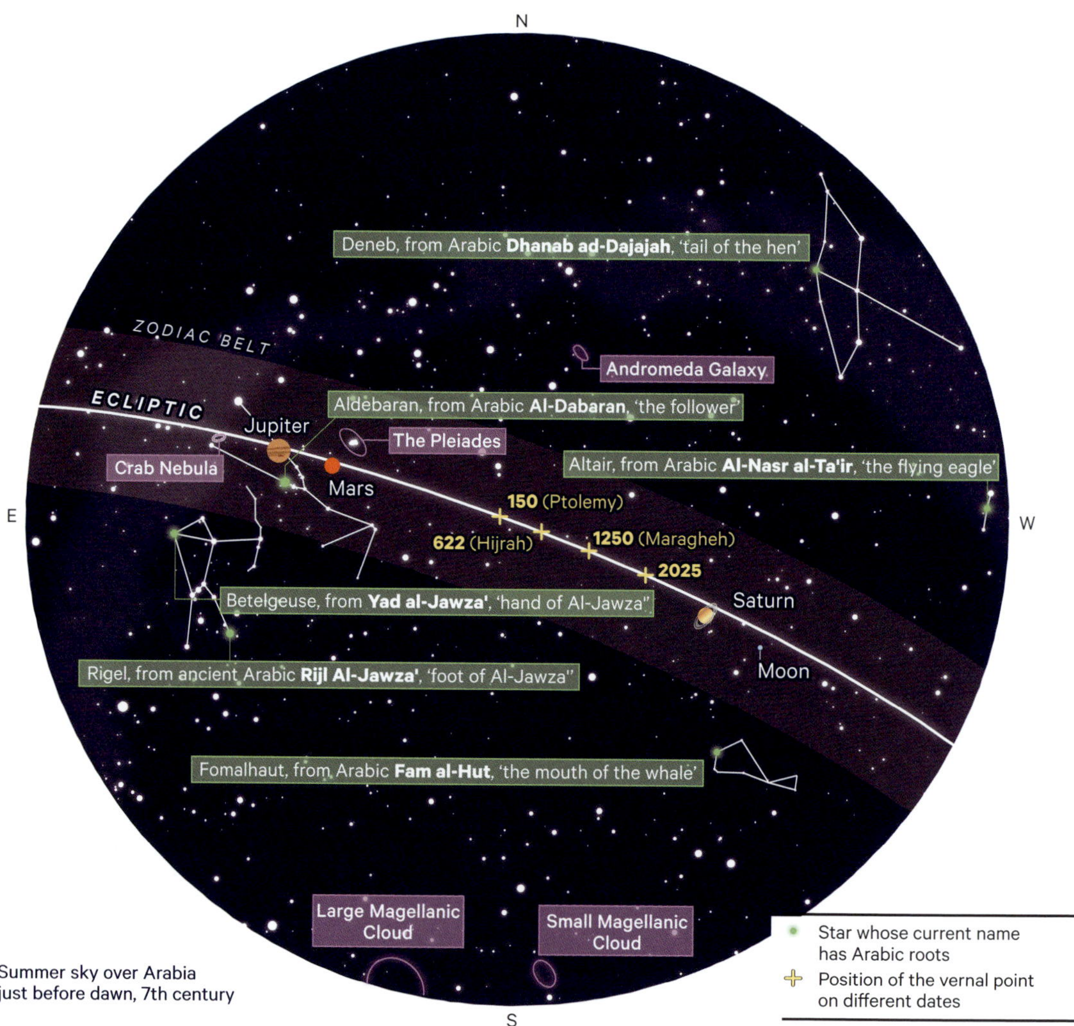

Summer sky over Arabia just before dawn, 7th century

Star names with Arabic origins

Drawing on Ptolemy's star catalogue, Al-Sufi published *The Book of Fixed Stars* (964) in Isfahan. He and his successors named 160 stars, most of which are still known by the same names today. He was the first to mention the swathe of light (nebula) in the constellation of Andromeda, as well as the Large Magellanic Cloud, which can easily be seen on the southern horizon from south Arabia. He wrote about the angle between the celestial equator and the ecliptic (obliquity). The supernova of 1006 was described by others, and in 1054, the Crab Nebula supernova was recorded by Christian physician Ibn Butlan.

Understanding the sky: a Muslim requirement

Although astrology was a controversial driving force behind the study of the skies, the practice of Islam played a more important role. Muslims must face the holy city of Mecca when praying, so they need to know which direction (*qibla*) it is in. This requires knowledge of cartography, longitude and latitude. The times of the five daily prayers (*salat*) are determined by the movement of the Sun from its rising to dusk, calculated by measuring the length of shadows. In the Islamic calendar, the first day of the month is the one in which the smallest crescent moon can be seen in the west, just after the new moon. It is sometimes difficult to determine this, which can lead to a discrepancy between local calendars. Ramadan (the month of fasting) is the ninth month, and its start date must be accurately determined. These everyday needs drove the work of astronomers, who developed mathematics and studied shapes drawn on a sphere. Al-Khwarizmi (whose name is the origin of the word 'algorithm') invented algebra (*al-jabr*, 'transposition') to solve problems as wide-ranging as inheritance, land surveying and the motion of celestial bodies. The sample problem (right) was originally expressed in words, as mathematical symbols first appeared in the West in the 16th century.

An algebra problem of inheritance

A woman dies, leaving a husband, one son and three daughters. She leaves 1/8 + 1/7 of everything she owns to a stranger. This inheritance is allocated first. The law then assigns a quarter of what's left to her husband, and the rest to her children, with a son getting twice as much as a daughter. How should the inheritance be divided up? The solution to this problem from Al-Khwarizmi's book is given on p. 225.

Discoveries from the Islamic world

The Islamic world, unified by the Arabic language, drew on the work of its predecessors from Andalusia to Samarkand. Although Aristotle, Ptolemy and geocentrism remained the accepted system, other models were studied, and observatories and the schools that they housed improved measurements and made significant contributions to science. From the 16th century, there were also developments in the Ottoman Empire, as well as in India after the fall of the Mongol Empire.

A long history of creativity

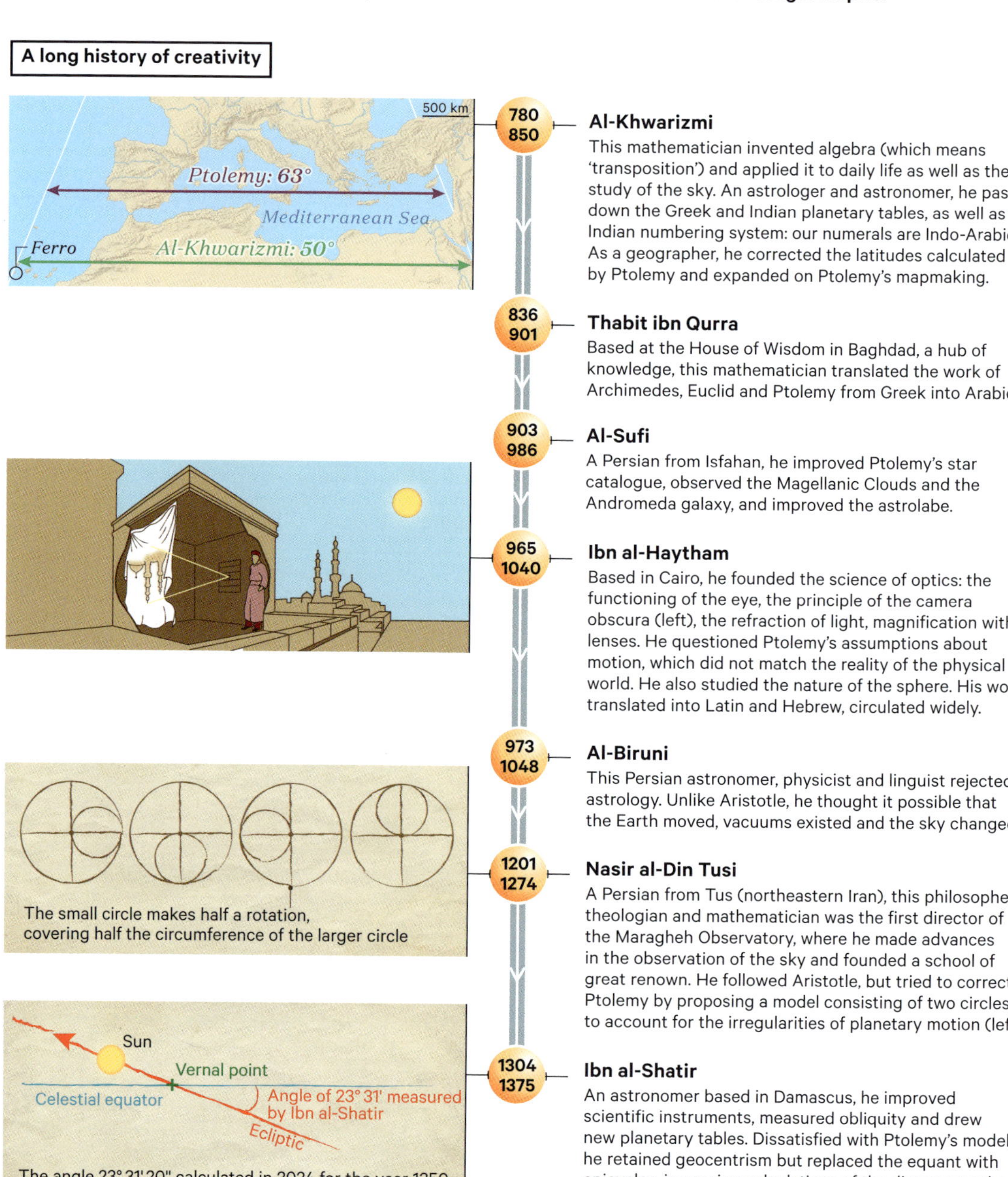

780 850 — Al-Khwarizmi

This mathematician invented algebra (which means 'transposition') and applied it to daily life as well as the study of the sky. An astrologer and astronomer, he passed down the Greek and Indian planetary tables, as well as the Indian numbering system: our numerals are Indo-Arabic. As a geographer, he corrected the latitudes calculated by Ptolemy and expanded on Ptolemy's mapmaking.

836 901 — Thabit ibn Qurra

Based at the House of Wisdom in Baghdad, a hub of knowledge, this mathematician translated the work of Archimedes, Euclid and Ptolemy from Greek into Arabic.

903 986 — Al-Sufi

A Persian from Isfahan, he improved Ptolemy's star catalogue, observed the Magellanic Clouds and the Andromeda galaxy, and improved the astrolabe.

965 1040 — Ibn al-Haytham

Based in Cairo, he founded the science of optics: the functioning of the eye, the principle of the camera obscura (left), the refraction of light, magnification with lenses. He questioned Ptolemy's assumptions about motion, which did not match the reality of the physical world. He also studied the nature of the sphere. His work, translated into Latin and Hebrew, circulated widely.

973 1048 — Al-Biruni

This Persian astronomer, physicist and linguist rejected astrology. Unlike Aristotle, he thought it possible that the Earth moved, vacuums existed and the sky changed.

1201 1274 — Nasir al-Din Tusi

A Persian from Tus (northeastern Iran), this philosopher, theologian and mathematician was the first director of the Maragheh Observatory, where he made advances in the observation of the sky and founded a school of great renown. He followed Aristotle, but tried to correct Ptolemy by proposing a model consisting of two circles, to account for the irregularities of planetary motion (left).

1304 1375 — Ibn al-Shatir

An astronomer based in Damascus, he improved scientific instruments, measured obliquity and drew new planetary tables. Dissatisfied with Ptolemy's model, he retained geocentrism but replaced the equant with epicycles, improving calculations of the distance to the Moon. It's not clear whether Copernicus was aware of this work when he was developing his heliocentric model.

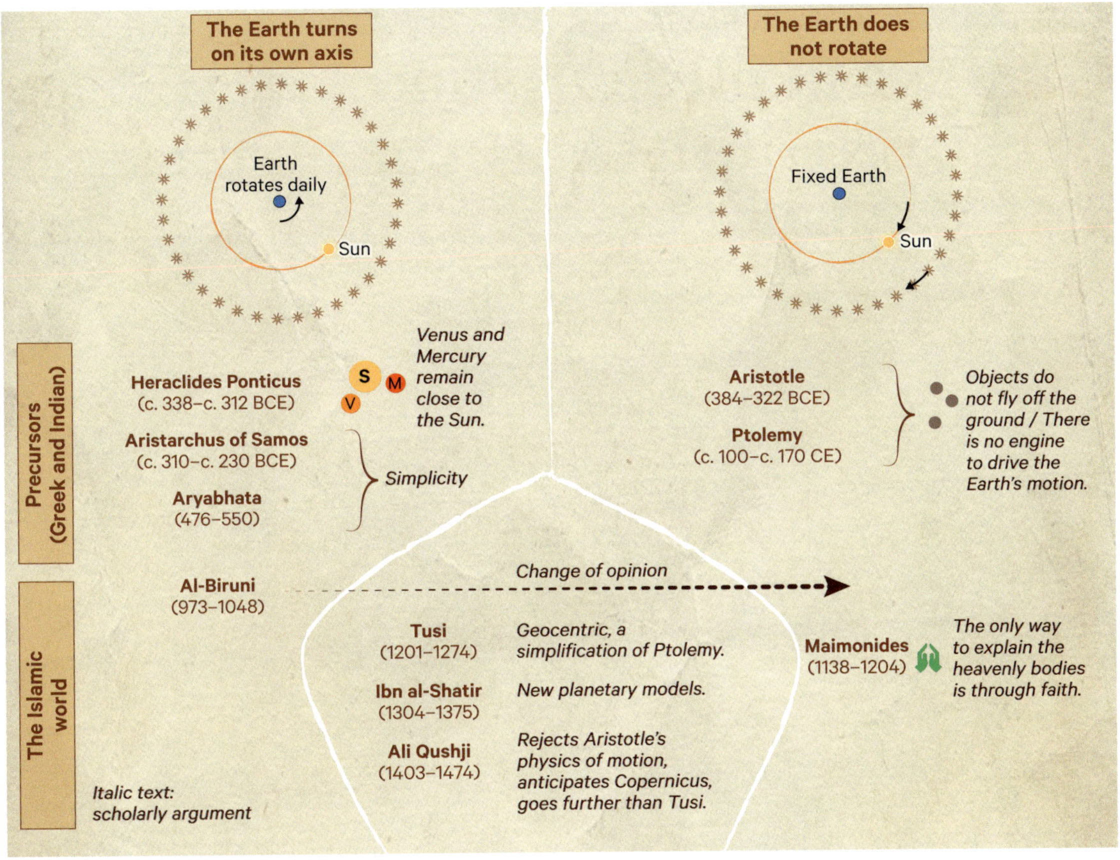

The Earth turns on its own axis
Earth rotates daily
Sun
The Earth does not rotate
Fixed Earth
Sun
Precursors (Greek and Indian)
Heraclides Ponticus (c. 338–c. 312 BCE)
S M V
Venus and Mercury remain close to the Sun.
Aristarchus of Samos (c. 310–c. 230 BCE)
Aryabhata (476–550)
Simplicity
Aristotle (384–322 BCE)
Ptolemy (c. 100–c. 170 CE)
Objects do not fly off the ground / There is no engine to drive the Earth's motion.
The Islamic world
Al-Biruni (973–1048)
Change of opinion
Tusi (1201–1274)
Geocentric, a simplification of Ptolemy.
Ibn al-Shatir (1304–1375)
New planetary models.
Ali Qushji (1403–1474)
Rejects Aristotle's physics of motion, anticipates Copernicus, goes further than Tusi.
Maimonides (1138–1204)
The only way to explain the heavenly bodies is through faith.
Italic text: scholarly argument

For or against a moving Earth

The ancient world, from Aristotle to Ptolemy, believed in a fixed Earth surrounded by moving stars and planets. The opposing concept of the universe, as proposed by Heraclides Ponticus and Aristarchus of Samos, was forgotten. However, in India, Aryabhata also embraced this concept and passed it on to the Islamic world. Aristotle's system was widely accepted by Islamic astronomers, although the hypothesis of a moving Earth was examined without prejudice. When the dynamism of the Islamic world began to decline from the 15th century, this debate remained unresolved and was taken up by the Christian West. The unexplained irregularities in the motions of the planets, which Ptolemy described with great precision in his tables, led to the development of other geometric models – still geocentric – such as those created by Al-Tusi or by Al-Shatir in Damascus.

Opening
Light
Wall or pylon
90°
Graduations (in degrees)
Quadrant
NORTH
25 metres
0°
SOUTH
The observatory ruins
Still extant
No longer extant

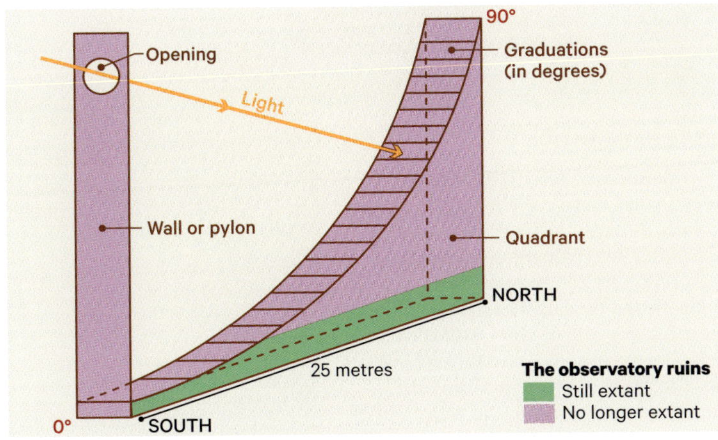

The Maragheh Observatory

This observatory in the northwest of modern Iran was founded in 1259 by Hulagu Khan, who wanted to use it for astrological predictions, and he chose Al-Tusi to oversee it. Its main instrument, a quarter-circle quadrant, improved the measurement of alignments, and therefore of the position of celestial bodies. It also boasted an armillary sphere and a celestial globe, which have survived. Playing host to a number of researchers before its decline in the late 13th century, this school was a hub of knowledge and served as a model for later observatories (Samarkand, Jaipur).

The sky in the Christian West: 6th–16th century

The knowledge of the ancient world, especially Plato, Aristotle and Ptolemy, was incorporated into the Christian conception of the world, which embraced a literal interpretation of the Bible. But new discoveries from the Arab and Greek worlds also found their way into the Christian West.

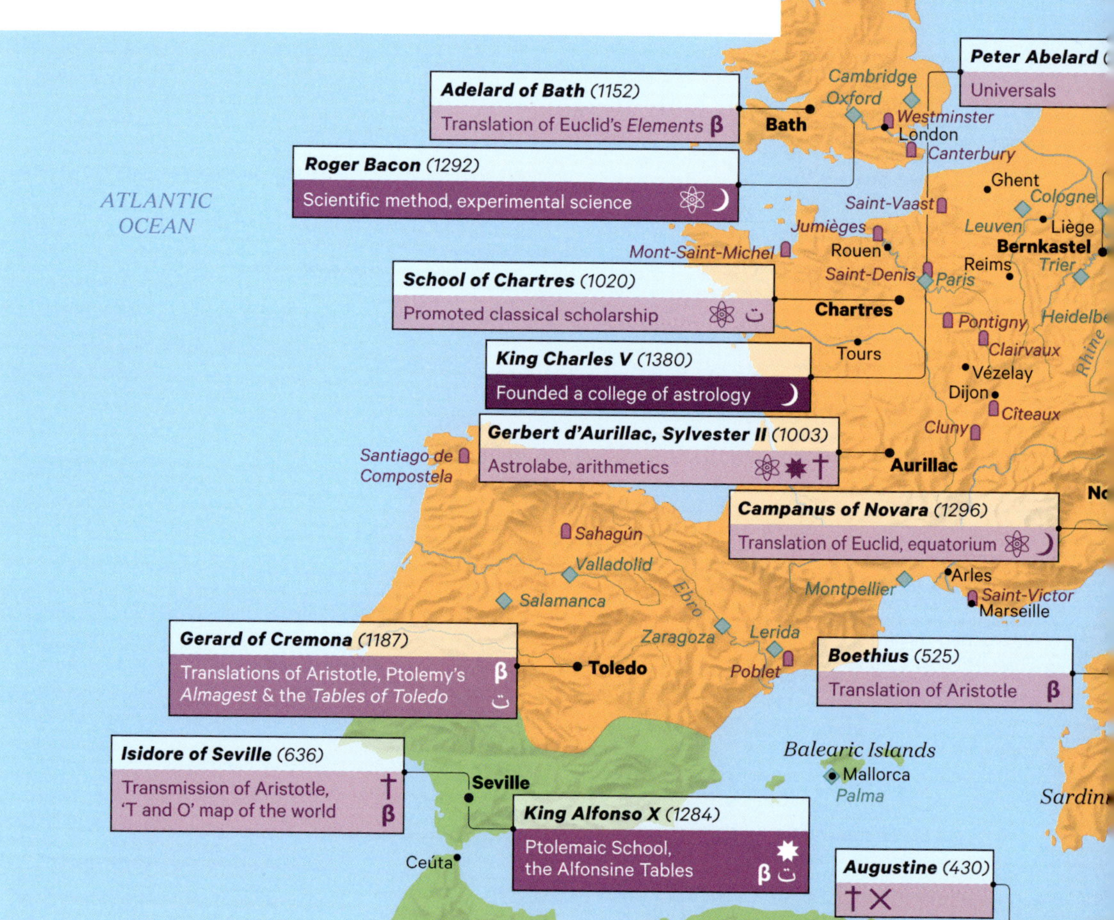

The development of European scientific thought

A large number of ancient texts, particularly the works of Aristotle, were preserved in monastery libraries. Since late antiquity, classical learning had been incorporated into scholarly thought, particularly through the work of Isidore of Seville (c. 560–c. 636). Isidore's *Etymologiae* was a very influential source of knowledge throughout the Middle Ages and contributed to the spread of Aristotelian thought. The astronomical discoveries of the Islamic world also reached Europe, mainly through places where it came into contact with the rest of the Mediterranean world (Andalusia, Sicily, Constantinople) and through Italian trade. This curiosity was also driven by rulers' interest in astrology. The 14th century saw the 'Ptolemaic revolution' as the West opened up to the wider world.

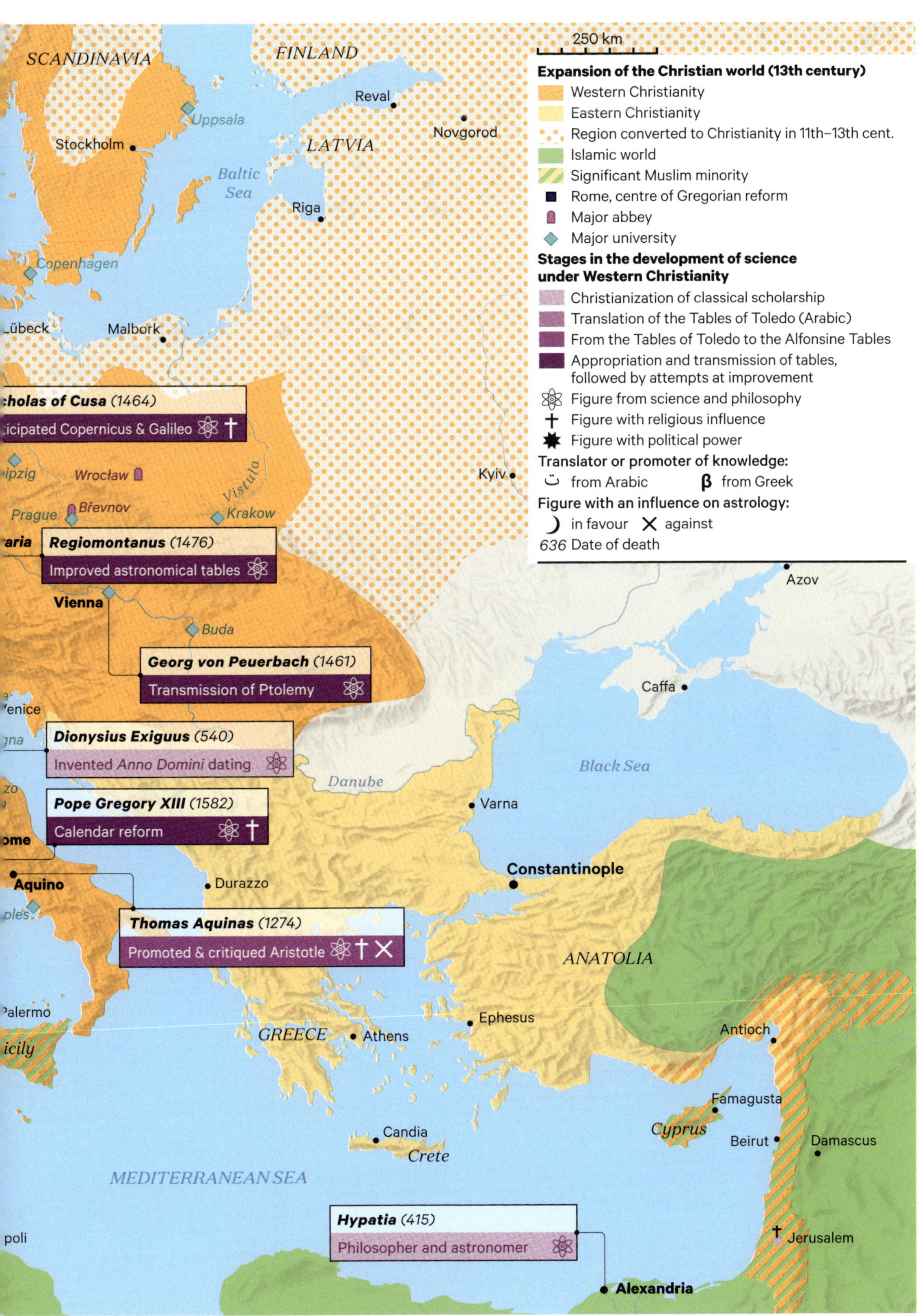
SCANDINAVIA
FINLAND
Reval
Uppsala
Stockholm
LATVIA
Novgorod
Baltic Sea
Riga
Copenhagen
Lübeck
Malbork
250 km

Expansion of the Christian world (13th century)
Western Christianity
Eastern Christianity
Region converted to Christianity in 11th–13th cent.
Islamic world
Significant Muslim minority
Rome, centre of Gregorian reform
Major abbey
Major university
Stages in the development of science
under Western Christianity
Christianization of classical scholarship
Translation of the Tables of Toledo (Arabic)
From the Tables of Toledo to the Alfonsine Tables
Appropriation and transmission of tables,
followed by attempts at improvement
Figure from science and philosophy
Figure with religious influence
Figure with political power
Translator or promoter of knowledge:
from Arabic   β from Greek
Figure with an influence on astrology:
in favour   ✕ against
636 Date of death

Nicholas of Cusa (1464)
Anticipated Copernicus & Galileo
Leipzig
Wrocław
Prague
Břevnov
Vistula
Krakow
Kyiv

Bavaria
Regiomontanus (1476)
Improved astronomical tables
Vienna
Buda

Georg von Peuerbach (1461)
Transmission of Ptolemy

Venice
Bologna
Dionysius Exiguus (540)
Invented Anno Domini dating

Rome
Pope Gregory XIII (1582)
Calendar reform

Aquino
Durazzo
Naples

Thomas Aquinas (1274)
Promoted & critiqued Aristotle

Azov
Caffa
Black Sea
Danube
Varna
Constantinople
ANATOLIA

Palermo
Sicily
GREECE
Athens
Ephesus
Antioch

Famagusta
Cyprus
Beirut
Damascus

Candia
Crete

MEDITERRANEAN SEA

Tripoli
Hypatia (415)
Philosopher and astronomer

Jerusalem

Alexandria

The location of Heaven

In the Abrahamic faiths, Heaven refers to the transcendent domain of God and the heavenly beings, angels. In Christianity, the chosen go to heaven or paradise, to live in the kingdom of God, after the Last Judgment; this heaven has long been associated with the sky. The original paradise was an earthly one: the Garden of Eden. In Hinduism, Buddhism and Jainism, souls are reincarnated until they achieve ultimate freedom (moksha or nirvana), but this is not linked with the sky.

The heavens and the sky

'In the beginning God created the heavens and the earth' (Genesis 1:1): the pairing of Earth and Heaven imposes order on creation. The heavens are not everything that is visible from Earth, which is where we get the use of the plural in the Hebraic and Christian traditions (Paul, 2 Corinthians). In the Qu'ran, the paradise of believers (Jannah), described as a magnificent garden, is not situated in the sky. Catholic iconography maintains a clear distinction between the sky as the Earth's atmosphere and the eternal Heavens, the place where God lives. In depictions of Christ's Ascension or the Assumption of the Virgin Mary, such as that by Titian (the Frari *Assumption*, 1518, far left), there are three distinct layers: the apostles on the Earth with the sky above them, then the Virgin being carried by angels, and finally the Empyrean (heaven) where God greets Mary.

Angels, heavenly beings

In the Paradise section of the *Divine Comedy* (1321), Dante, guided by Beatrice, who symbolizes theology, travels through the series of celestial spheres that made up the geocentric model of medieval astronomy. Beyond the fixed stars (8th sphere), he reaches the furthest sphere, the Primum Mobile. This is the home of the angels, who surround God in nine rings. Angels can be found in the holy texts of the three Abrahamic religions, as well as in the Zoroastrian Avesta, where their primary role is as intercessors between God and humans. Since the 5th century, Catholic angelology has maintained that there is a celestial hierarchy of nine choirs of angels, described by Thomas Aquinas in the 13th century. This echoes the ecclesiastical hierarchy in the human world. The 'Thrones' serve as God's seat and the foundation of the world; the 'Powers' fight demons; the 'Archangels' (including Michael, Gabriel and Raphael) are God's messengers.

First Order

Seraphim

Cherubim

Thrones

Second Order

Dominions

Virtues

Powers

Third Order

Principalities

Archangels

Angels

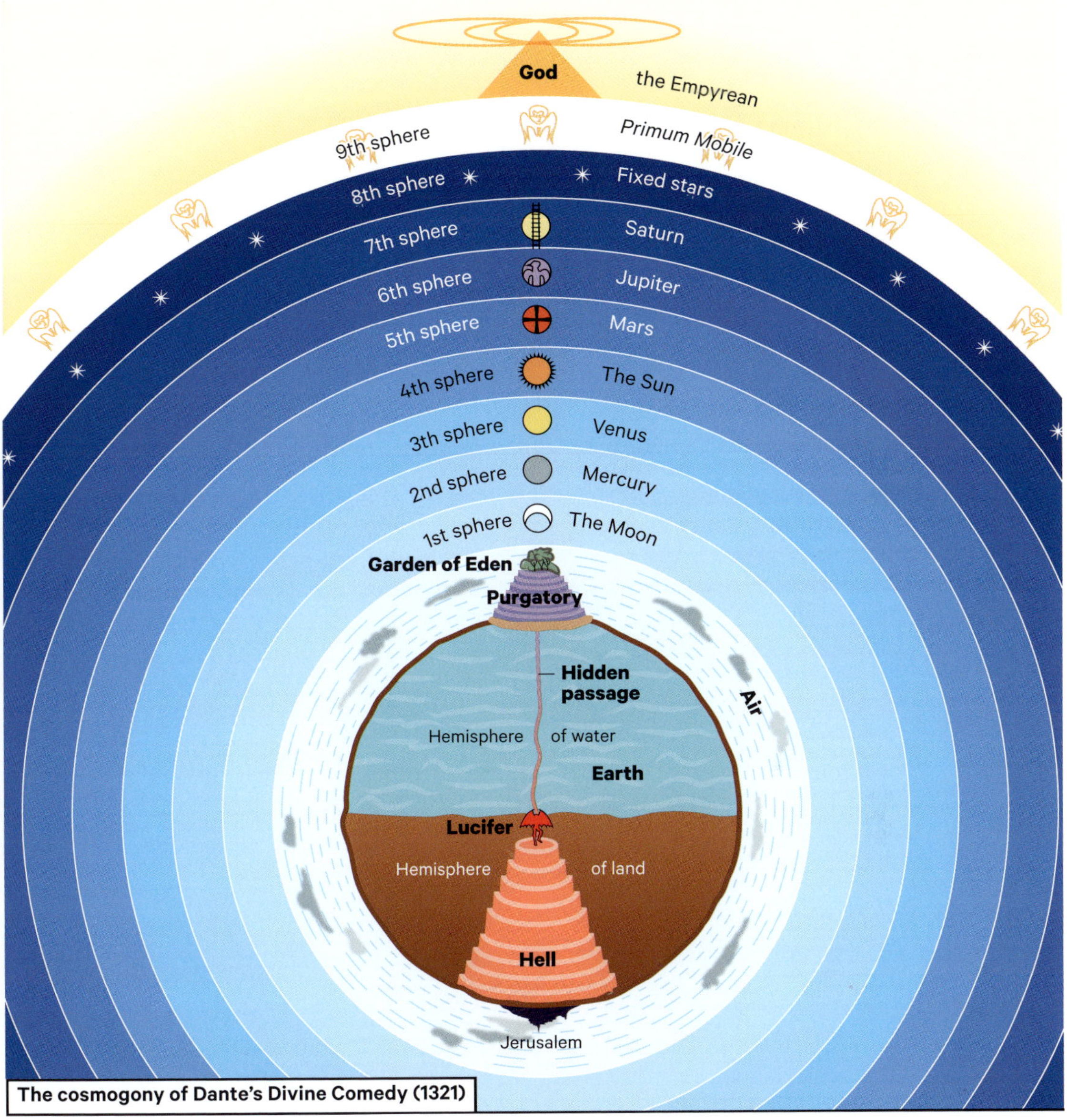

The cosmogony of Dante's Divine Comedy (1321)

At the centre of the universe

Earth: It stands at the centre of the universe and is made from the three primal elements of earth, fire and water.

Lucifer: when he was cast out of Paradise, he plunged down through the Earth, forming the circles of Hell.

Paradise

Air: surrounding the Earth.

1st sphere: the Moon, whose waxing and waning are associated with inconstancy.

2nd sphere: Mercury, which is associated with ambition because it's close to the sun; this sphere is filled with those who do good out of a desire for glory.

3rd sphere: Venus, the realm of love without temperance.

4th sphere: the Sun, source of the Earth's light and linked with the virtues of prudence, justice and temperance.

5th sphere: Mars, home of the warriors of faith.

Hell: made up of nine circles (the first being Limbo, where the virtuous pagans dwell, including Socrates).

Hidden passage: connects Hell and Purgatory.

The Garden of Eden

6th sphere: Jupiter, associated with just rulers because Jupiter was the ruler of the classical gods.

7th sphere: Saturn, the sphere of the contemplatives and of monasticism.

8th sphere: the Fixed Stars, the realm of faith, hope and love, and the church triumphant.

8th sphere: The *Primum Mobile* is the last sphere of the physical universe. It is moved directly by God and causes the motion of all the other spheres. It is where angels dwell.

The Empyrean: where God dwells, beyond the physical realm.

The skies of the southern hemisphere

None of the civilizations of the northern hemisphere could see all of the stars of the southern celestial hemisphere. Only the stars close to the celestial equator could be observed near the horizon in Greece, Egypt, and the Islamic world. Described and studied for centuries by Native American, African and Aboriginal civilizations among others, the southern sky was first observed by Westerners during voyages of exploration. They then mapped it and chose their own names for its constellations.

The skies over Polynesia

For the ancient peoples of Polynesia, the sky was a mirror of the ocean. The constellations were named after tools used for fishing or navigation, or after fish or birds. But the celestial bodies and their constellations were also personified as figures in historical tales, or by gods and heroes such as Maui.

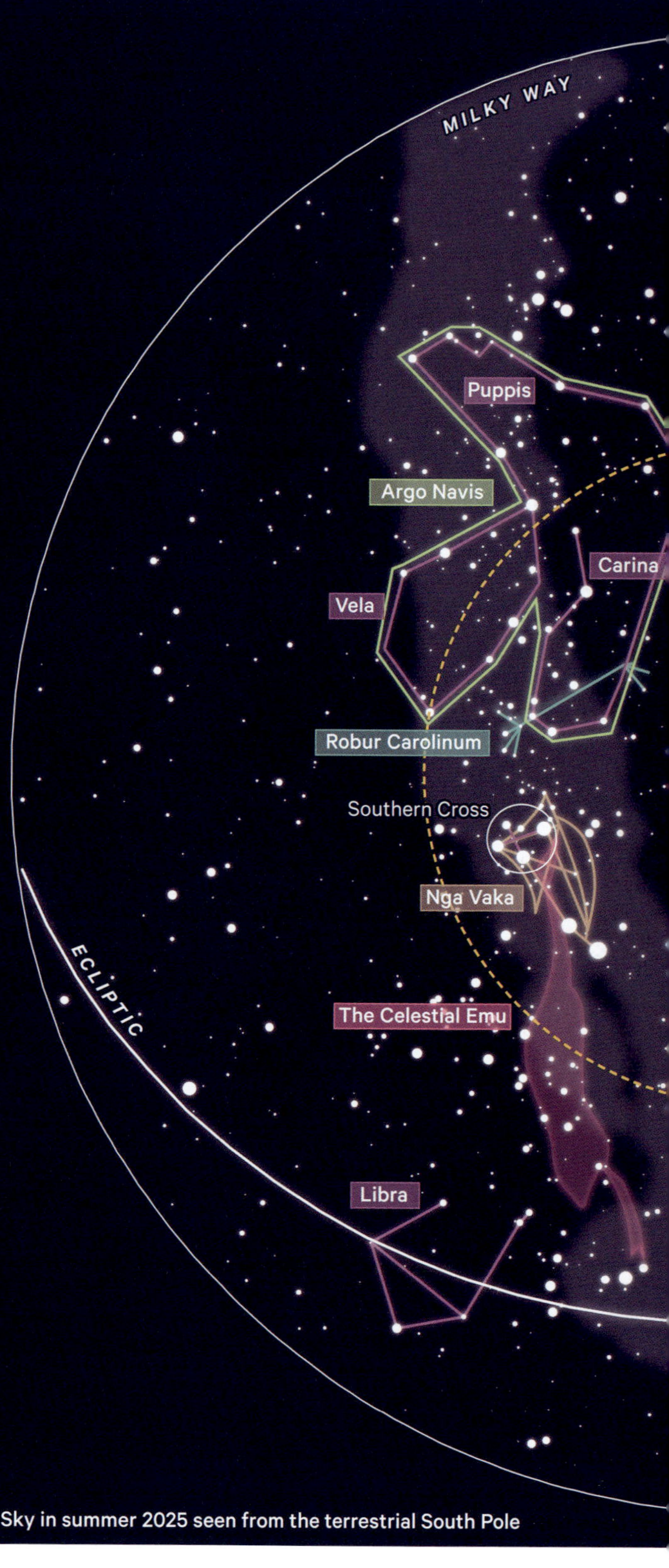

Current constellation
Constellation suggested by Europeans, but now no longer recognized
Former Greek constellation
Polynesian constellation
Aboriginal constellation
Not all constellations are shown.

Sky in summer 2025 seen from the terrestrial South Pole

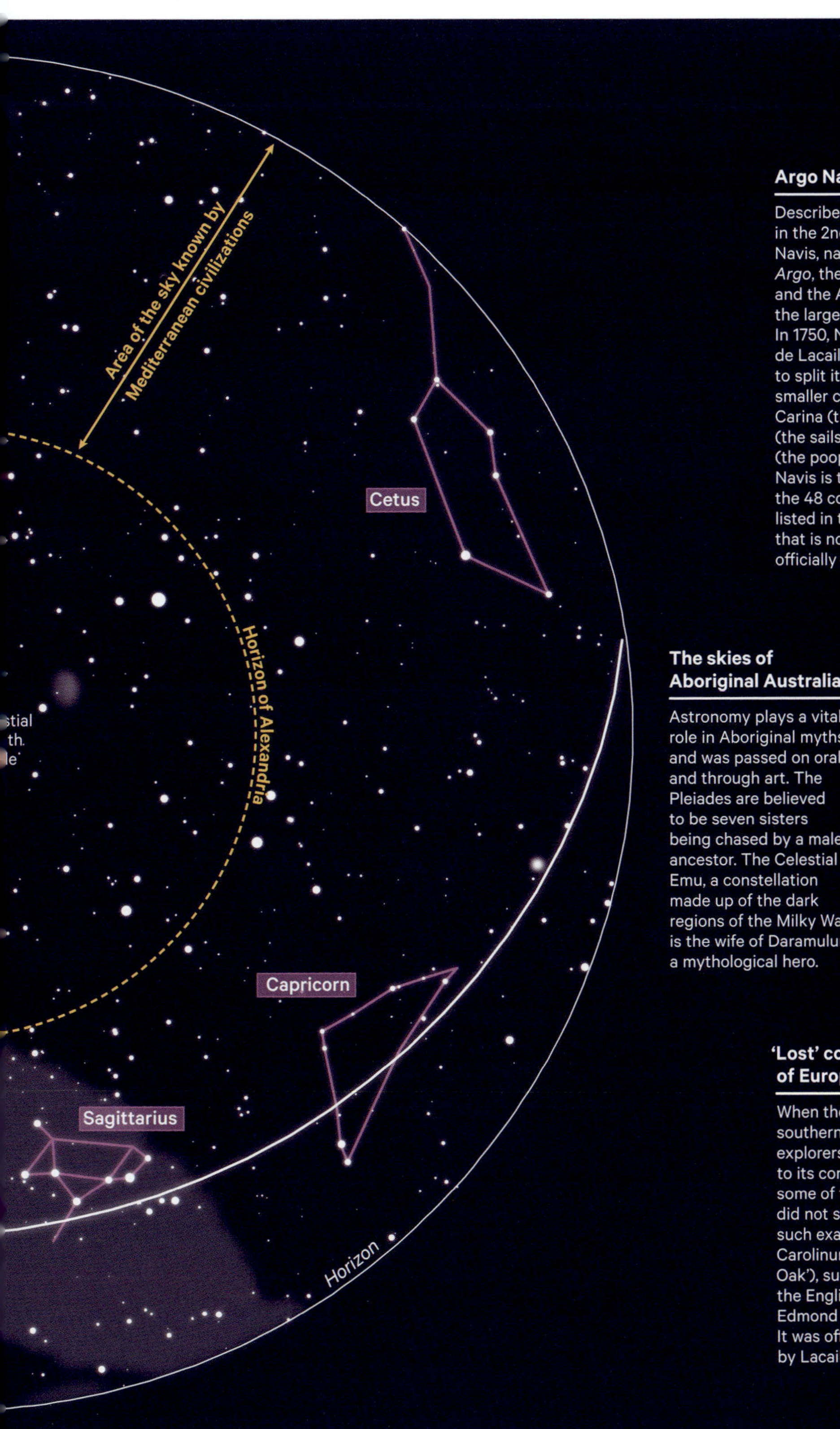

Argo Navis

Described by Ptolemy in the 2nd century, Argo Navis, named after the *Argo*, the ship of Jason and the Argonauts, was the largest constellation. In 1750, Nicolas-Louis de Lacaille decided to split it into three smaller constellations: Carina (the keel), Vela (the sails) and Puppis (the poop deck). Argo Navis is the only one of the 48 constellations listed in the *Almagest* that is no longer officially recognized.

The skies of Aboriginal Australia

Astronomy plays a vital role in Aboriginal myths and was passed on orally and through art. The Pleiades are believed to be seven sisters being chased by a male ancestor. The Celestial Emu, a constellation made up of the dark regions of the Milky Way, is the wife of Daramulum, a mythological hero.

'Lost' constellations of Europe

When they first saw the southern skies, European explorers gave names to its constellations, but some of these names did not survive. One such example is Robur Carolinum ('Charles' Oak'), suggested by the English astronomer Edmond Halley in 1679. It was officially rejected by Lacaille.

Nicolaus Copernicus: the Earth in motion

Nicolaus Copernicus (1473–1543) was born in Poland and studied in Krakow. He was part of a scientific tradition that drew on and developed ideas from ancient Greece, in the Islamic world and Christian Europe. He tried to find a physical description of the world that was 'true', in which the Earth moved and the organization of the planets was not arbitrary. This Renaissance man, drawing on ancient astronomy, proposed a heliocentric model that opened up a new understanding of the skies.

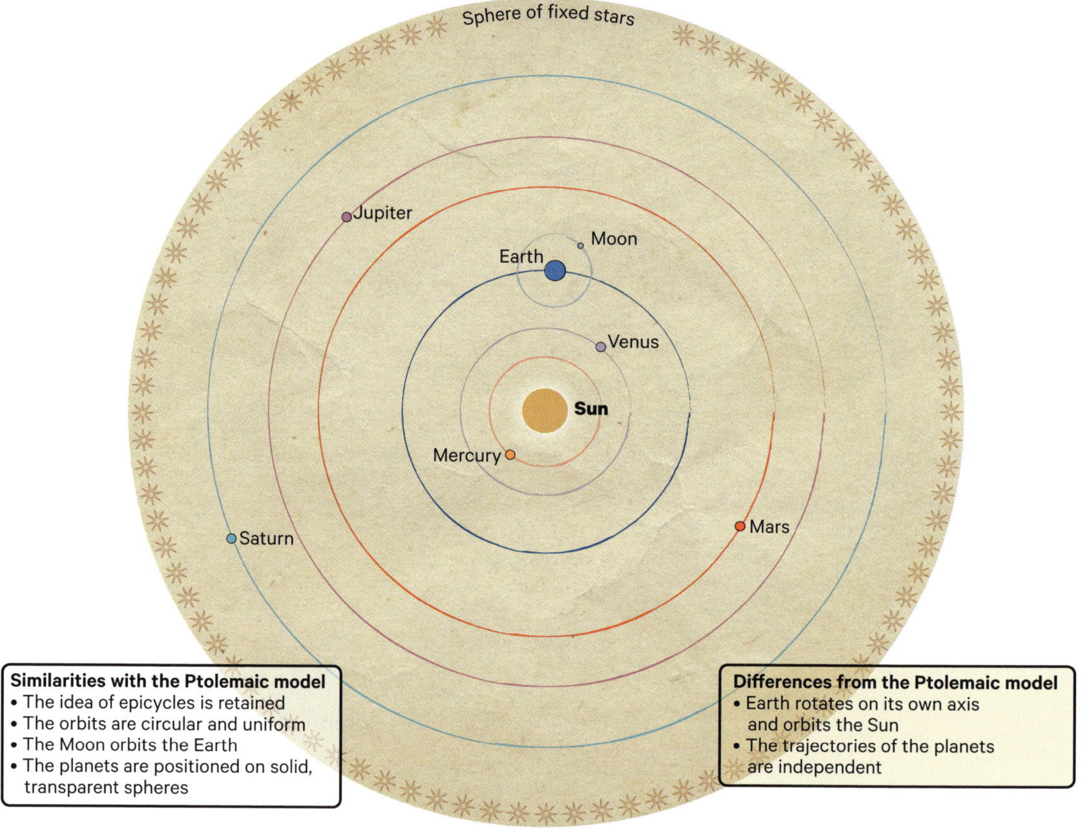

The Sun at the centre

An astronomer since his youth, Copernicus was involved in the reform of the Julian calendar. Perhaps inspired by earlier challenges to geocentrism, in the Islamic world as well as the West, he questioned the ability of Ptolemy's system to represent more accurate observations, which had been compiled over the centuries. Completed in around 1530, Copernicus's manuscript *On the Revolutions of the Heavenly Spheres* was not published until after his death. He retained some of Aristotle's principles, such as epicycles, but simplified the Ptolemaic model. The Earth orbits the Sun, as do all the planets. Its daily rotation means that the sphere of the stars does not turn. This sphere must be very far away, because the movement of the Earth around the Sun, which changes the angle of sight (parallax), does not change the constellations. The distances and orbital periods of the planets are organized simply, and this system explains their retrograde motion as being caused by the movement of the Earth.

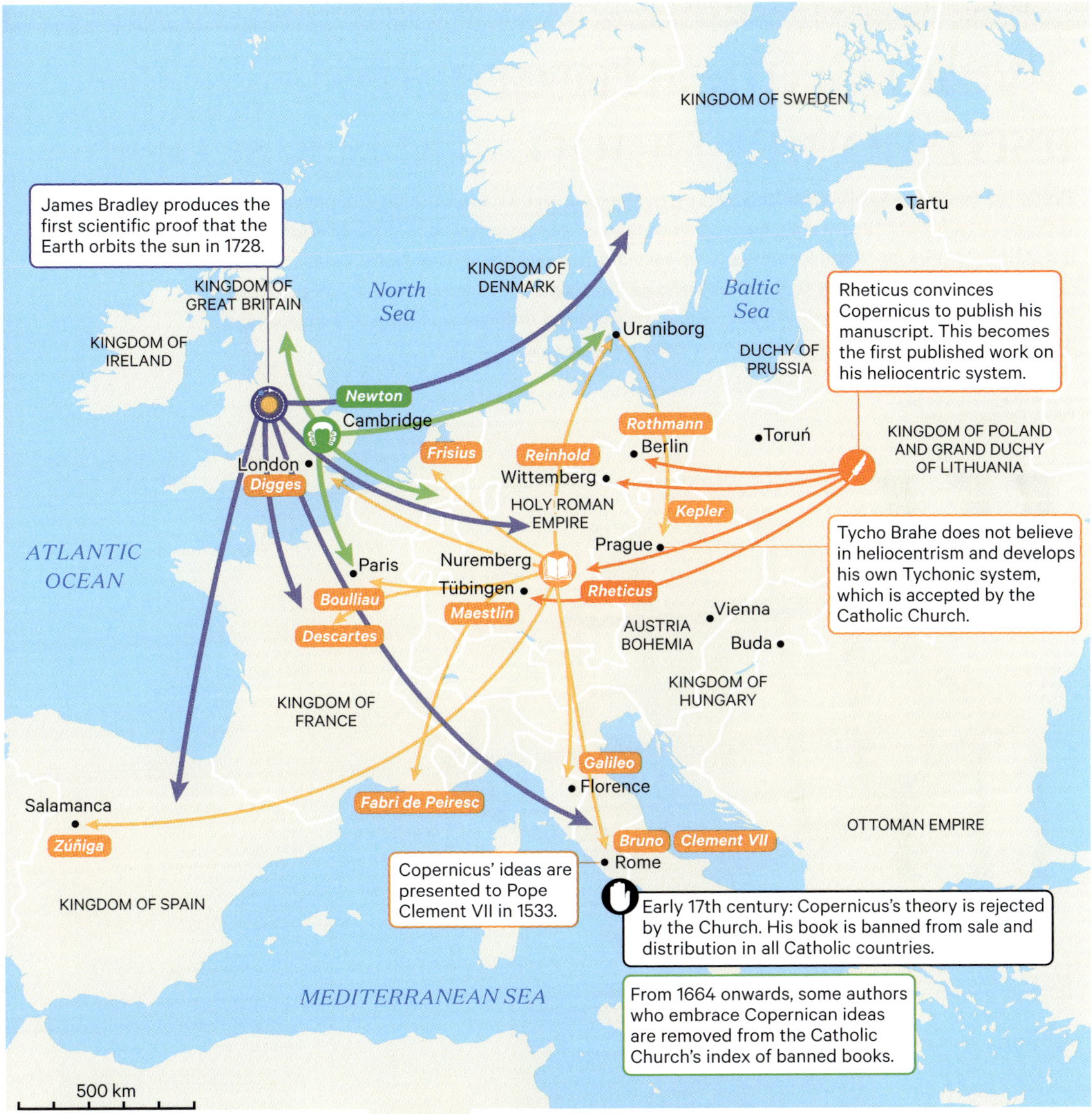

Major events in Copernican astronomy

 First manuscript version of *De revolutionibus orbium coelestium* by Nicolas Copernicus in 1530

 Printed edition in 1543

Spread of Copernican astronomy and rejections

→ Spread of theories in 16th century by great scholars of Europe

 The Church refutes Copernicus' theory

Isaac Newton spreads the ideas in the late 17th century

 Ideas spread through Europe in the mid-18th century.

Galileo Influential figure

Copernicus inspires Europe

The idea for his manuscript, written in Latin, was presented to Pope Clement VII in 1533. His work gradually spread across all of Europe, although his tables were not that extensive. After 45 years, the manuscript reached the greatest astronomer of the time, Tycho Brahe, who recognized the simplicity of the model, but rejected the idea that the Earth orbited around the Sun – a hypothesis that physicists of the time were not prepared to accept. He foresaw possible religious conflicts, and came up with his own system. At the end of the century, his student, Kepler, who had abandoned the idea of uniform circular motion due to observations of Mars, adopted Copernicus's heliocentric system. In the 17th century, Galileo, Descartes (cautiously) and particularly Newton overturned the ancient conception of the sky, while the Roman Church clung on to a literal interpretation of the Bible, which led to Galileo's condemnation. For many years, the Jesuits would continue to teach the Tychonic system.

Tycho Brahe: the greatest astronomer before the telescope

Tycho Brahe (1546–1601), from Denmark, was the greatest astronomer and observer of the 16th century, before the invention of the telescope. The instruments that he developed made it possible to take more precise measurements. He observed two rare astronomical events, challenging Aristotle's view that the sky was unchanging. However, he did not accept the hypothesis of a moving Earth. An astrologer, he believed that there was a link between the death of Suleiman the Magnificent and a lunar eclipse seen in 1566.

Tycho's quadrant

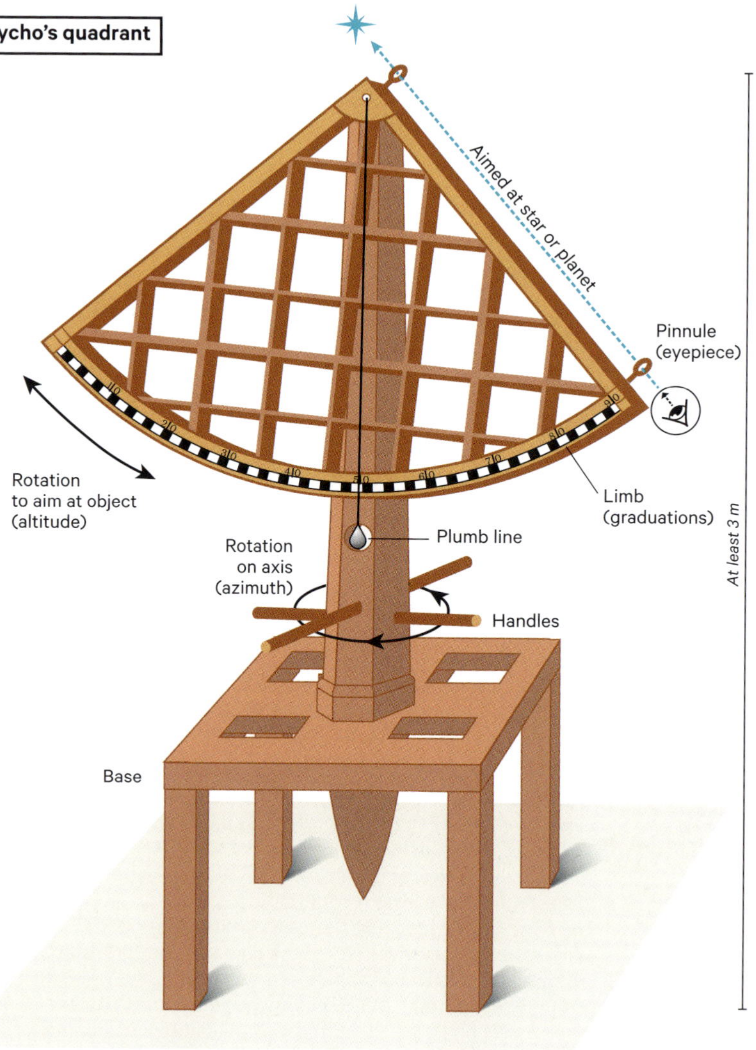

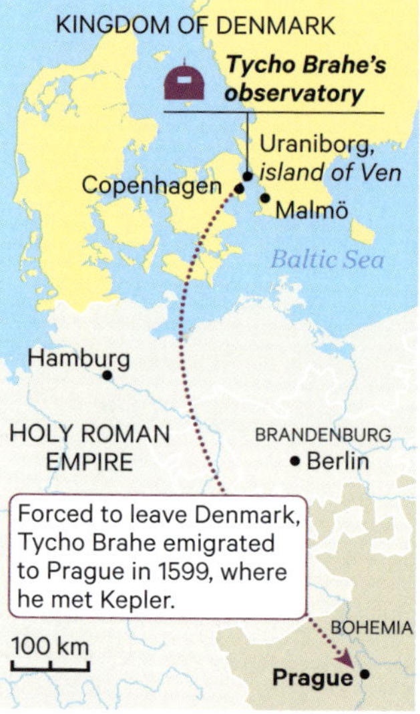

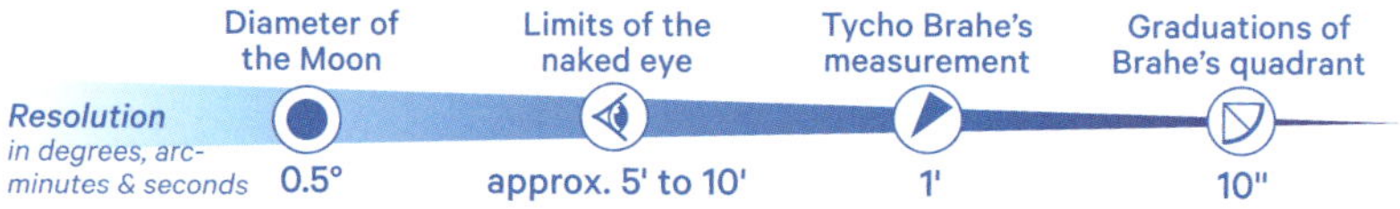

Precise observations

As a young man, Tycho realized that solar eclipses and planetary conjunctions were poorly described by the existing tables, and identified a need for more precise observations. In Uraniborg, the observatory that he founded on the island of Ven, he built a quadrant (left) that allowed him to take measurements with the naked eye that were accurate to one arc-minute, more than ten times as precise as those of his predecessors. His catalogue recording the positions of 1,004 stars became a definitive work. Exiled to Prague in 1599, he taught Kepler. He did not accept the idea of a moving Earth and came up with his own system, which would later be disproved.

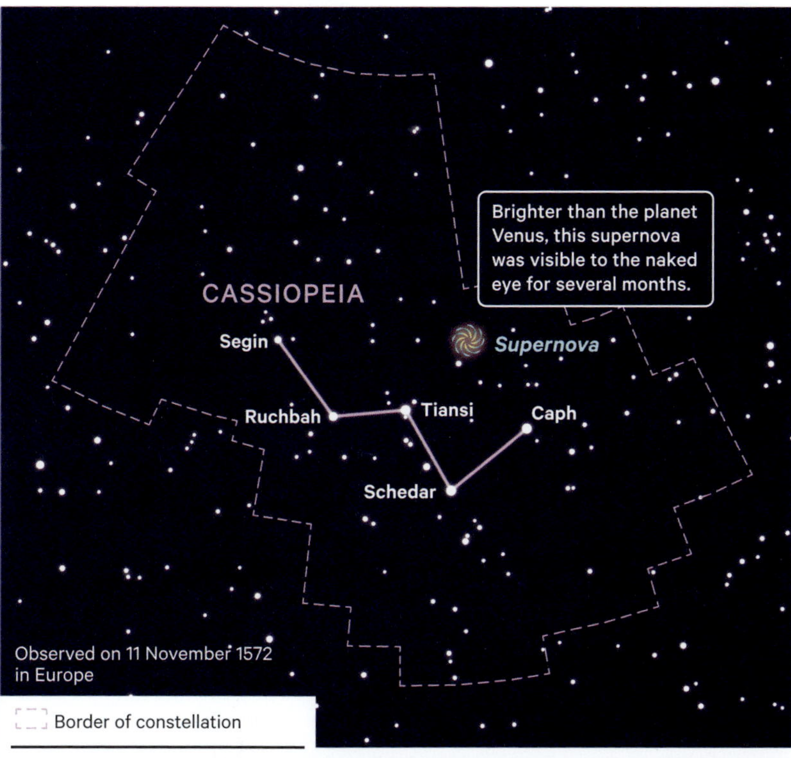

Tycho's supernova

On 11 November 1572, at Herrevad Abbey in southern Sweden, Tycho observed a new star, and continued to observe it over the course of a few weeks. The star was initially as bright as Venus, then it started to grow dimmer, eventually becoming invisible two years later. It was seen all across Europe and was always in the same position in the sky, even when seen from different places. This showed that it did not have a parallax, and must be situated beyond the Moon's orbit. This disproved Aristotle's theory that the sky was unchanging. Modern measurements have identified it as a supernova (SN 1572) and measured it at 6,400 light-years away. The remnant of the supernova can still be seen using a telescope.

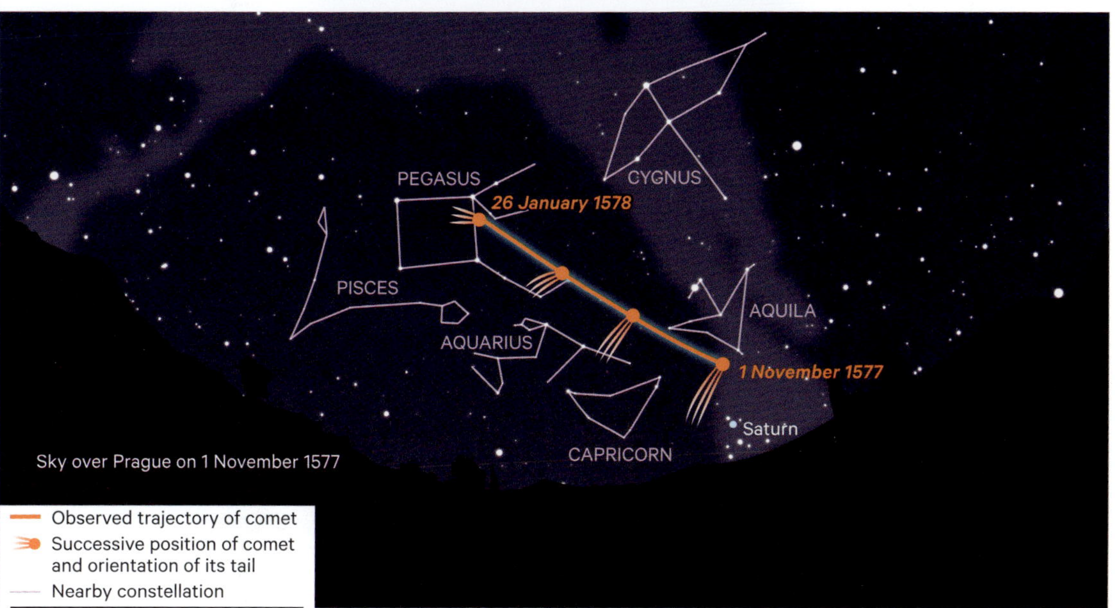

The Great Comet of 1577

In November 1577, a bright comet appeared in the skies above Europe and beyond. Using a quadrant, Tycho measured its position night after night. Comparing his own measurements to those taken in Prague, both of the Moon (which shows a parallax) and of the comet (which did not), he deduced that the comet must be at least three times as far away as the Moon, supporting his belief that the skies were not immutable. The comet sparked speculation about potentially imminent catastrophes, which Tycho contested, publishing his own astrological prediction of the downfall and death of Ivan the Terrible in Russia (1584).

Johannes Kepler: carrying the torch

German mathematician Kepler (1571–1630) accepted Copernicus's model, but sought to understand it better. In Prague, he met Tycho Brahe, who entrusted him with his very precise observations of Mars. Eight years of hard work led him to reject the idea of uniform motion and circles in favour of an ellipse, and to develop two of the three laws of the motion of Mars, which he also applied to the other planets. At the turn of the century, Kepler improved Copernicus's model through observation and paved the way for Isaac Newton.

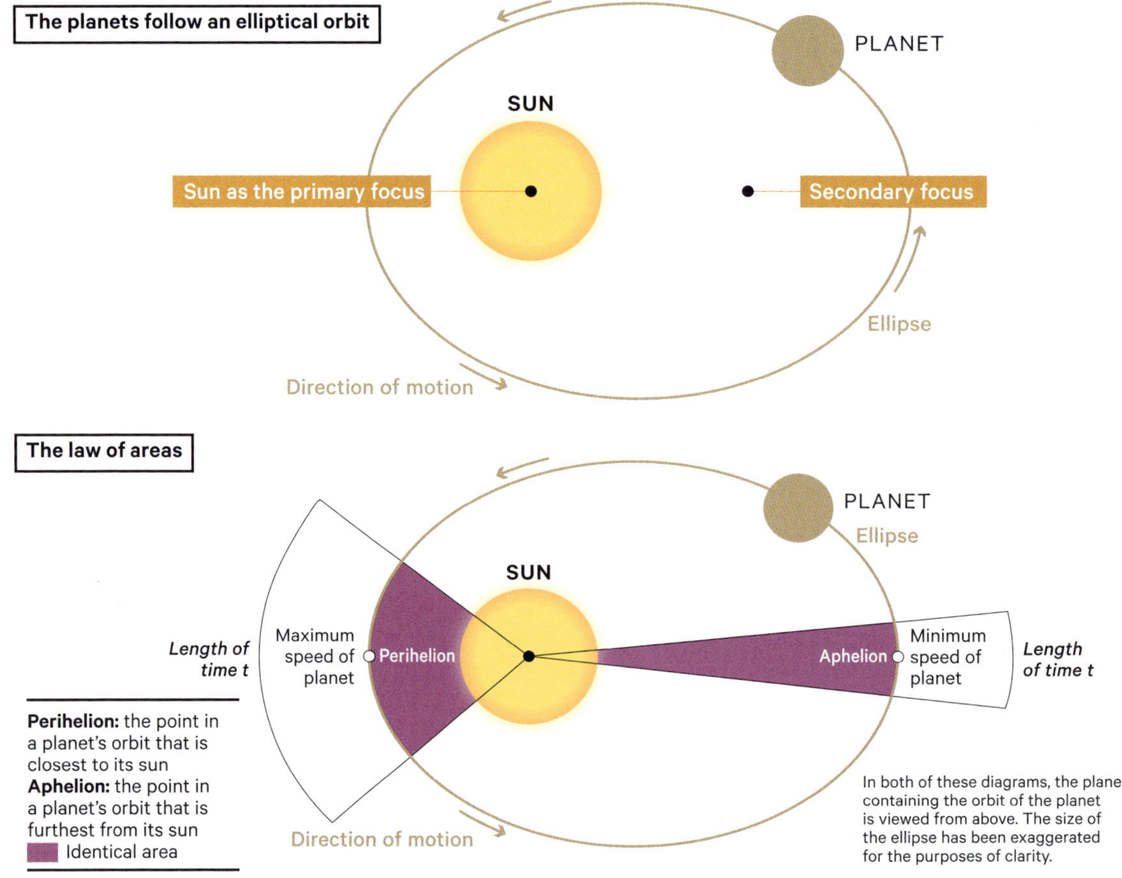

Three laws of planetary motion

As a mathematician and Copernican, first Kepler tried, without success, to find a connection between the planetary orbits and the five regular solids, known since the time of Plato. A Lutheran, he sought refuge in Prague in 1600 and became a student of Tycho Brahe, who shared with him his series of observations of the planet Mars. Kepler tried in vain to reconcile these with Copernicus's circles and epicycles, but eventually abandoned this system in favour of his two laws of motion (1609): the planets trace an elliptical orbit, with the Sun as one of the two foci (see diagram above); the change in the velocity of their orbit is determined by the law of areas (see diagram below). After ten years of work, he established the third law, which sets out the relationship between the planet's orbital period and the size of its ellipse.

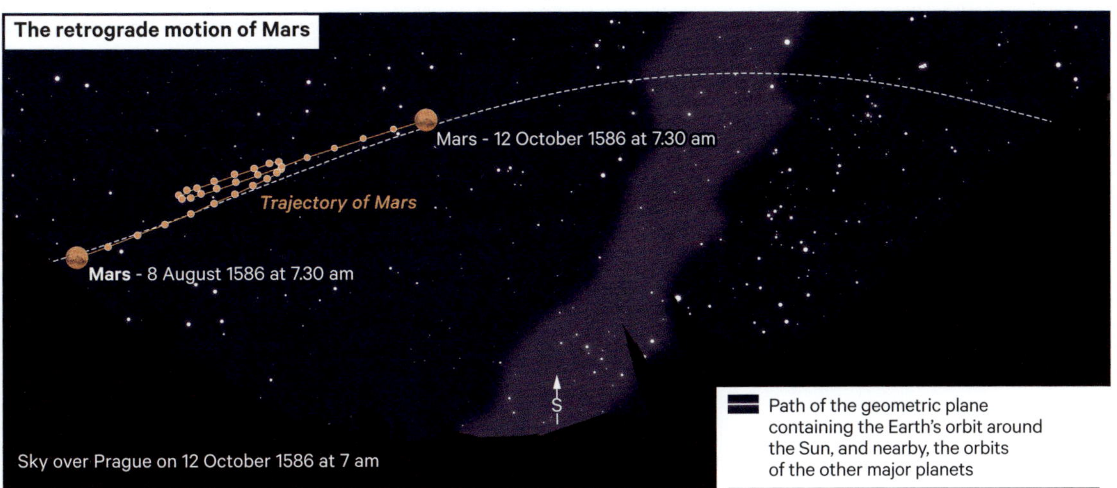

Path of the geometric plane containing the Earth's orbit around the Sun, and nearby, the orbits of the other major planets

The retrograde motion of Mars

Roughly every two years, as it moves along the ecliptic, Mars seems to 'move backwards' across the sky for a few months, then sets off again (see above). Using Copernicus's model, it is possible to trace the path it follows among the constellations of the zodiac, as seen from Earth (centre diagram), as well as its changing distance from the Earth, which causes significant differences in brightness. During an opposition, Mars is on the opposite side of the Earth to the Sun, therefore it is at its closest to the Earth and is brighter.

The centre diagram showing the loops of Mars was developed by Kepler, using Tycho's observations between 1580 and 1596, and published in his 1609 work, where he set out his two first laws. It shows both the regularity of Mars's motion and the variable length of time between two loops. Kepler had to introduce the ellipse to account for this motion.

The bottom diagram shows why, when seen from Earth, Mars seems to move backwards, as its orbital velocity is almost half that of the Earth.

Once he had explained these loops of Mars with his first two laws, Kepler turned his mind to another task. Ptolemy had believed that there was a relationship between the distance of a planet and its orbital period. It was Kepler who, in his third law, explained the relationship between the elliptical path followed by a planet and the time taken to complete it.

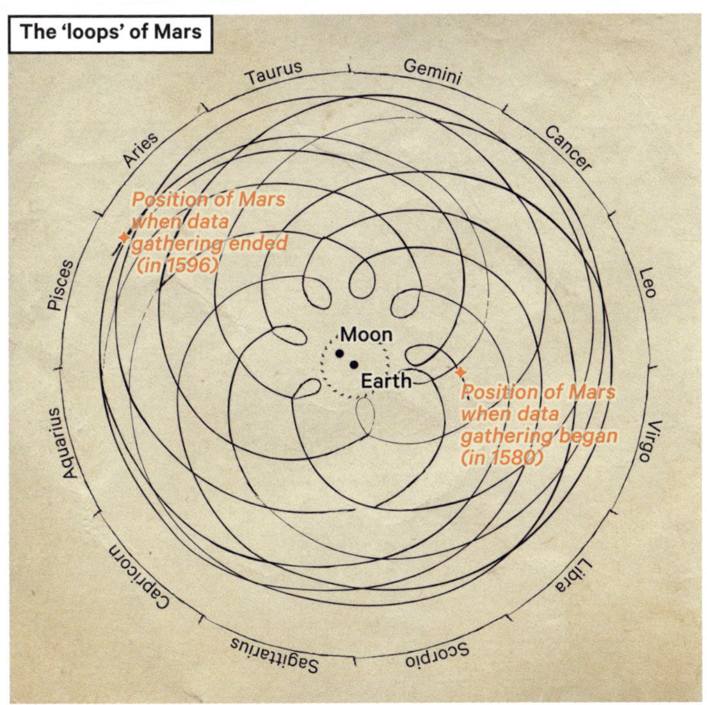

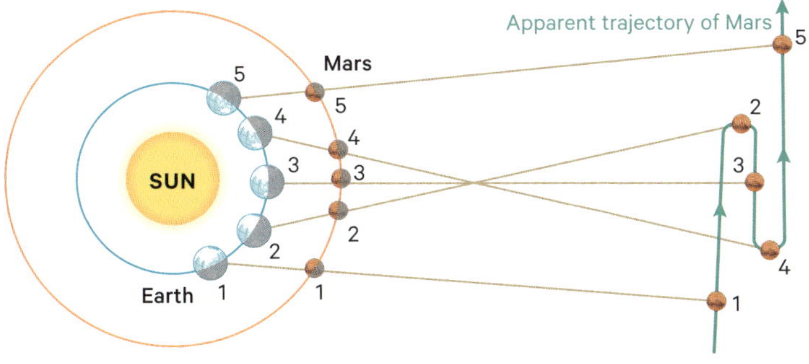

3

A broader sky
(17th–19th century)

Between the 17th century and the end of the 19th, a major scientific shift took place in Europe. The sky was suddenly bursting with new stars, thanks to three advances in observation: the telescope, photography and spectroscopy. These improved on the human eye, making it possible to capture more light and see objects that were less bright; the images had an incredible level of detail; with photography, we could see objects that were very far away. At the same time, the discovery of universal gravitation made it possible to understand the motion of the planets and the stars. There were developments in mechanics, physics and mathematics alongside astronomy. This knowledge completely transformed every human's conception of the world and their place in it. European empires spread overseas, where local knowledge and scientific traditions were often suppressed.

A timeline of science (1610–1920)

1610

Galileo's telescope allowed him to see beyond the naked eye. He observed the mountains on the Moon, the phases of Venus and the moons of Jupiter.

Johannes Kepler
1571–1630

A Copernican and student of Tycho Brahe, Kepler used Brahe's measurements of Mars to determine that the planets followed elliptical orbits. He set out three empirical laws governing planetary motion. The idea of uniform circular motion was abandoned.

1629

Kepler predicted the transits of Mercury (in 1631) and Venus (in 1639 and 1761). Observations and calculations of these events were used to determine the size of their orbits.

1676

By observing a moon of Jupiter at the Paris Observatory, Ole Rømer concluded that the speed of light was not infinite, and produced the first estimate of it.

Edmond Halley
1656–1742

After observing a comet in 1682 and analysing its previous appearances, Halley predicted it would return in 1758. The comet did indeed appear again in 1759. This astronomer drew up the first map of the southern sky.

1675

Greenwich Observato was found

1727

Observations of stellar aberration (the apparent shift in a star's position depending on the position of the observer and the speed of light) provided direct proof that the Earth orbits the Sun.

1735–1736

Expeditions to Peru (Pierre Bouguer, Charles de La Condamine) and Lapland (Pierre de Maupertuis) to measure two meridians, near the equator and near the North Pole.

1737

Confirmation that the Earth is a flattened sphere, in accordance with Newton's laws of gravity.

1756

Mathematicia and physicist Émilie du Cha translated Ne Philosophiae *Naturalis Prir Mathematica* into French.

1840–1850

First astronomical photographs (the Sun and Moon).

1838

First measurement of distance using the parallax method on the star 61 Cygni (Friedrich Wilhelm Bessel).

Friedrich Wilhelm Bessel
1784–1846

A mathematician and expert in geodesy, in 1838 he produced the first precise measurements of the distance of a star, and discovered a white dwarf star, an invisible companion of the star Sirius.

1846

Prediction of the existence of the planet Neptune through calculations (Urbain Le Verrier).

Angelo Secchi
1818–1878

A Jesuit, he was one of the pioneers of spectroscopy, which he applied to the Sun and stars. He drew up a classification of the stars, based on the spectrum of their light, and determined the nature of the solar corona, observed during an eclipse.

1859

Gustav Kirchho established the of thermal radia which can be us to work out the temperature of stars.

Galileo Galilei
1564–1642
He pointed a telescope at the sky, amassed a wealth of discoveries and revolutionized the basic principles of astronomy. His study of falling bodies led to the development of a new mechanics. His method of seeking proof through experiments laid the foundations of modern science.

1660
The Royal Society was founded in London.

Christiaan Huygens
1629–1695
He built his own telescopes and discovered Titan, a moon of Saturn. He was the first person to conceive of light as a wave and he understood centrifugal force. His invention of pendulum clock made it possible to measure time with greater precision.

Isaac Newton
1642–1727
A mathematician, physicist and theologian, Newton established mechanics, the laws of motion, and the laws of gravitation in the universe. Although his understanding of light was flawed, he separated white light into its component colours. He invented the reflecting telescope.

1667
The Paris Observatory was founded.

1666
The Académie des Sciences was founded in Paris.

759
The return of Halley's Comet was observed.

William Herschel
1738–1822
The discoverer of a new planet, Uranus, Herschel and his sister Caroline built telescopes with larger diameters. Together they produced the first estimate of the structure and dimensions of the Milky Way. Herschel also discovered infrared light.

1774–1781
Charles Messier compiled a catalogue of around a hundred nebulae, objects that appear different from stars because they are spread across a large area.

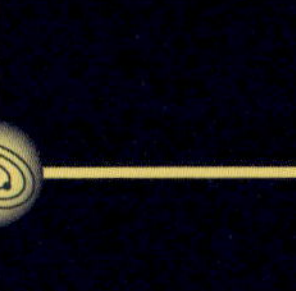

821
First spectrum of the Sun (Joseph von Fraunhofer).

1815
The change in the orientation of the Earth's axis relative to the stars over 26,000 years (precession) and its additional slight oscillation (nutation) were measured by Friedrich Wilhelm Bessel.

1795
Pierre-Simon Laplace established the stability of the Solar System over time.

1781
Discovery of the planet Uranus and first estimate of the size of our galaxy (William Herschel).

1867
First classification of stars by colour.

1884
International adoption of Greenwich Mean Time (GMT) as the standard by which local times are defined.

1887
Global project to draw up an Astrographic Catalogue.

1912
Creation of Universal Time and founding of the International Time Bureau in Paris.

1919
The Hooker Telescope (with a diameter of 2.5 metres) was installed on Mount Wilson, near Los Angeles.

Europe from the Renaissance to 1914

Europe was not politically unified in this time period, but its countries had close cultural and economic ties. It used globalization to its advantage. In this context, an independent science of the sky developed, which went beyond classical scholarship and and made use of new technology.

A new science of the sky

European humanism, which drew on ancient texts, notably the work of Claudius Ptolemy, transformed our understanding of the sky, which up until then had been shared throughout the west of the Old World. Long-distance voyages made Europeans aware of a larger sky and demanded a greater understanding of celestial markers for the purposes of navigation. Advances in optics, timekeeping and calculation techniques, as well as increasing freedom from religious control, led to the development of science. The community of intellectuals across Europe, which had already been strong in the Middle Ages, grew stronger, although there was some competition between nations. The most powerful nation-states in the West transformed ancient universities, founded new institutions (academies, learned societies) and funded increasingly expensive equipment (observatories) and scientific expeditions. In the late 19th century, this scientific community spread across the Atlantic before becoming globalized in the 20th century.

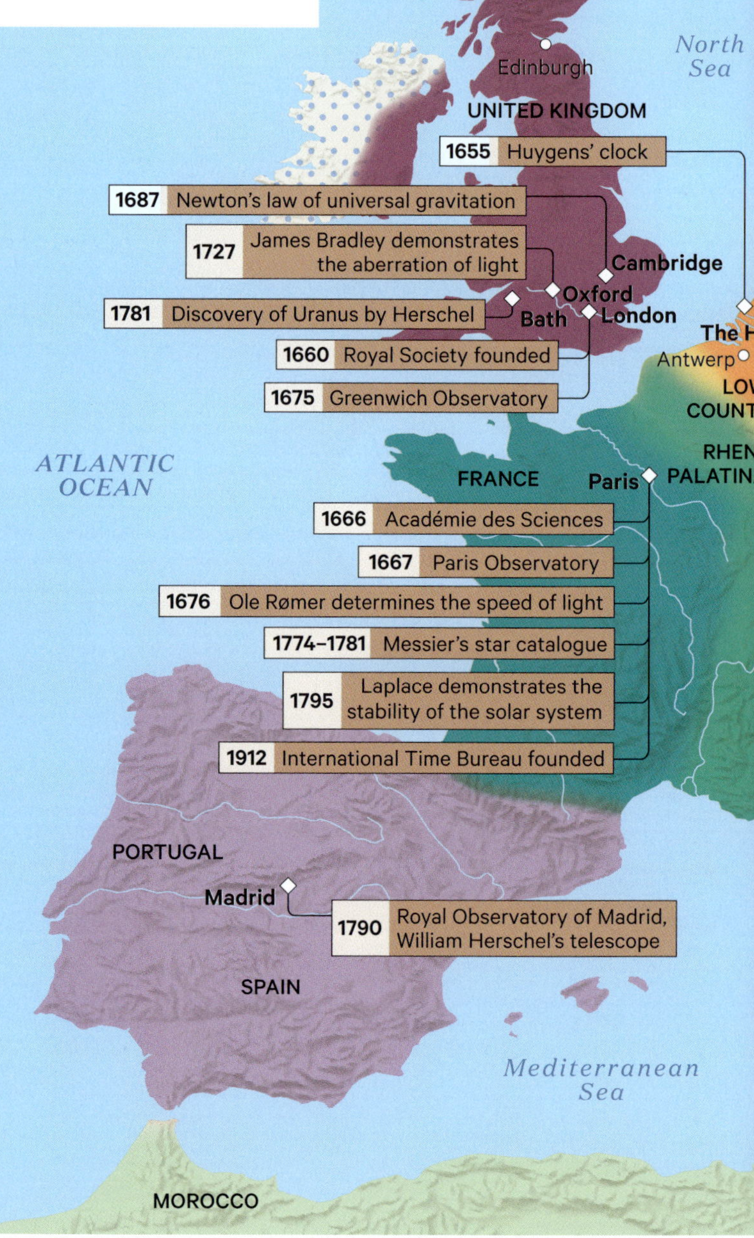

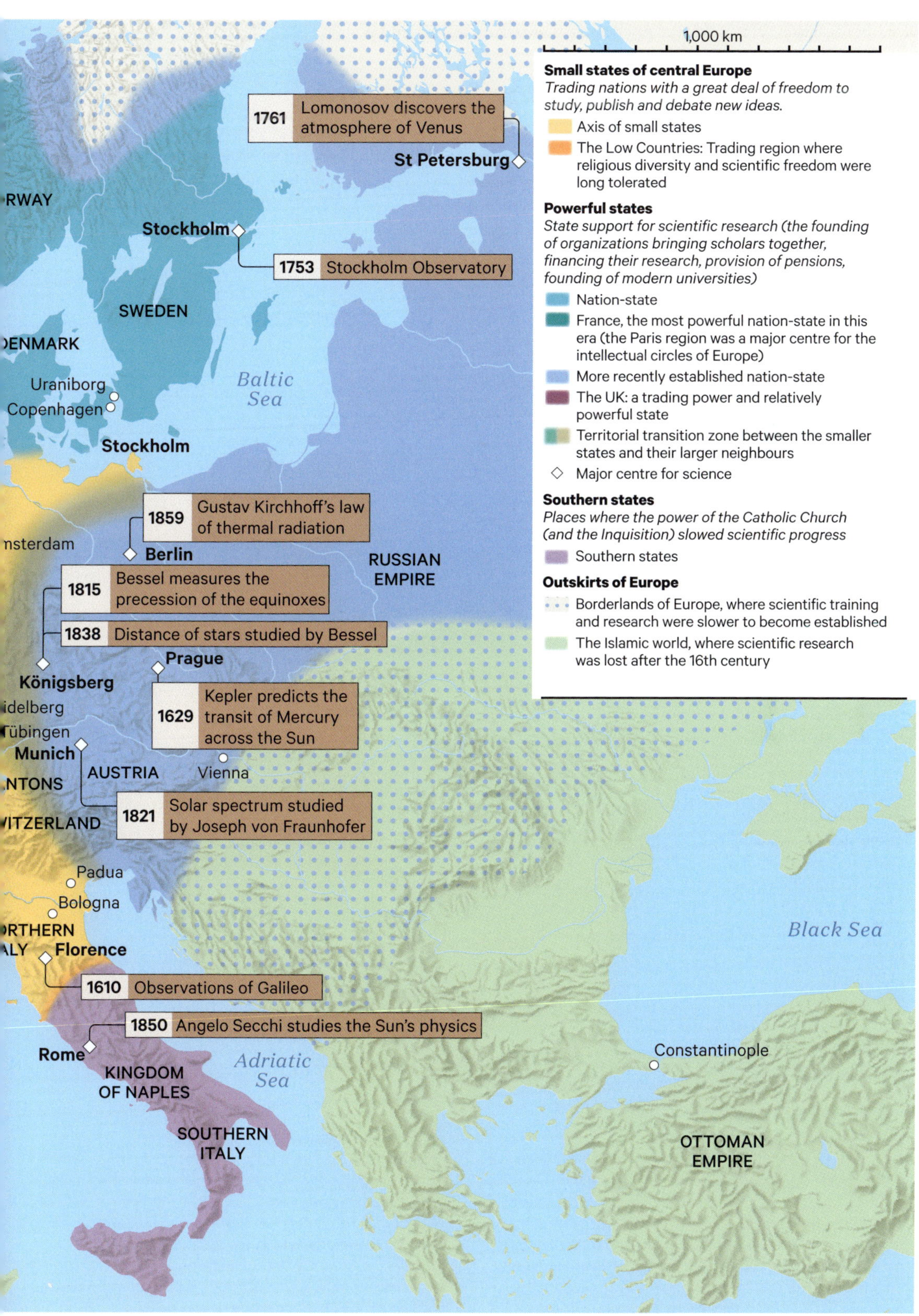
1,000 km

1761 Lomonosov discovers the atmosphere of Venus
St Petersburg

Stockholm
1753 Stockholm Observatory

SWEDEN

NORWAY
DENMARK
Uraniborg
Copenhagen
Stockholm

Baltic Sea

1859 Gustav Kirchhoff's law of thermal radiation
Berlin
RUSSIAN EMPIRE
1815 Bessel measures the precession of the equinoxes
1838 Distance of stars studied by Bessel
Prague
Königsberg
Heidelberg
Tübingen
Munich
1629 Kepler predicts the transit of Mercury across the Sun
AUSTRIA   Vienna
CANTONS
SWITZERLAND
1821 Solar spectrum studied by Joseph von Fraunhofer
Amsterdam

Padua
Bologna
NORTHERN ITALY
Florence
1610 Observations of Galileo
1850 Angelo Secchi studies the Sun's physics
Rome
KINGDOM OF NAPLES
Adriatic Sea
SOUTHERN ITALY
Constantinople
Black Sea
OTTOMAN EMPIRE

Small states of central Europe
Trading nations with a great deal of freedom to study, publish and debate new ideas.
Axis of small states
The Low Countries: Trading region where religious diversity and scientific freedom were long tolerated

Powerful states
State support for scientific research (the founding of organizations bringing scholars together, financing their research, provision of pensions, founding of modern universities)
Nation-state
France, the most powerful nation-state in this era (the Paris region was a major centre for the intellectual circles of Europe)
More recently established nation-state
The UK: a trading power and relatively powerful state
Territorial transition zone between the smaller states and their larger neighbours
Major centre for science

Southern states
Places where the power of the Catholic Church (and the Inquisition) slowed scientific progress
Southern states

Outskirts of Europe
Borderlands of Europe, where scientific training and research were slower to become established
The Islamic world, where scientific research was lost after the 16th century

Telescopes

In 1610, Galileo turned a telescope towards the sky, bringing an end to thousands of years of looking at the sky with the naked eye. First refracting telescopes, then reflecting telescopes were developed. Astronomers' understanding of the sky was revolutionized, as they could now observe celestial bodies that were much less bright and perceive far more detail. After 1850, the eye behind the telescope was gradually replaced by photography with long exposure times.

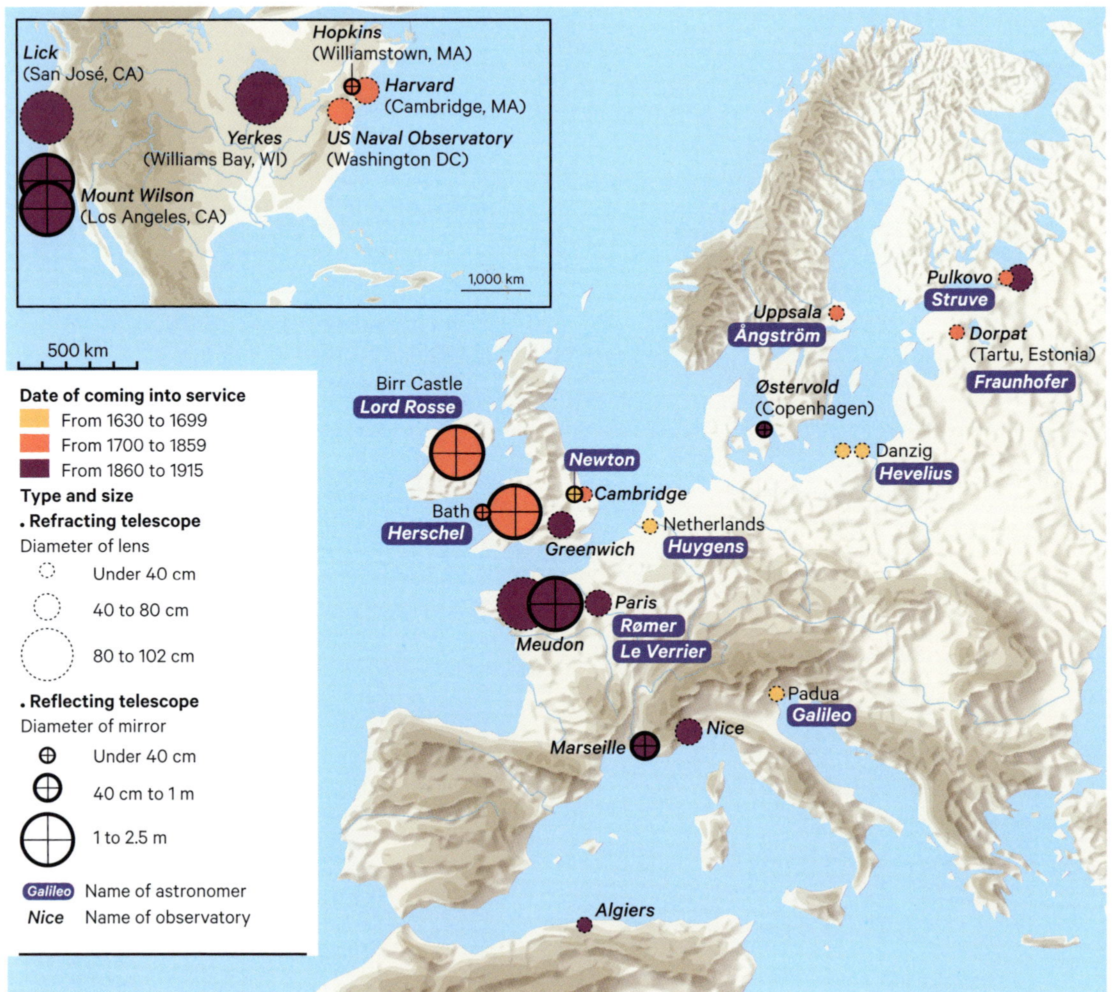

Telescopes after Galileo

The first person to look at the sky through a telescope in 1610, Galileo paved the way for a more detailed study of the heavens. At first, advances in observation were made by individuals, sometimes amateur astronomers, then by state-supported institutions. As well as refracting telescopes, reflecting telescopes were developed. In the early 18th century, the Dutchman Christiaan Huygens's invention of the pendulum clock made it possible to measure time more precisely. In 1860, Léon Foucault made a major advance at Paris Observatory. He replaced the bronze mirror in telescopes, which was not very reflective and easily became tarnished, with a mirror made of silvered glass. After the last refracting telescopes, reflecting telescopes with ever-larger mirrors were built, culminating in the Hooker Telescope – installed at Mount Wilson, California, in 1915 – which had a mirror with a diameter of 2.5 metres. It remained the largest optical telescope in the world until 1948.

Galilean telescope

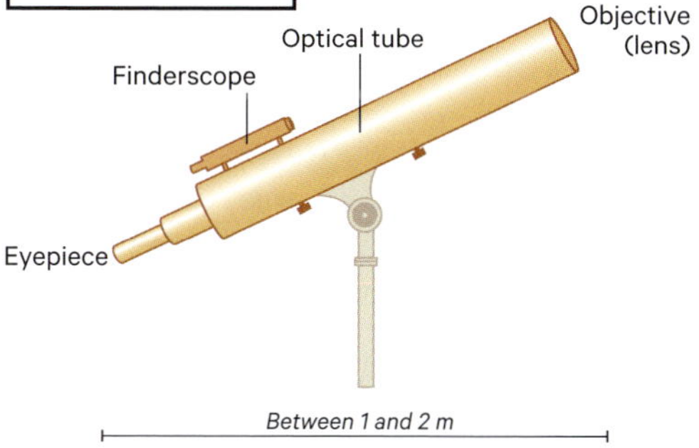

Finderscope
Optical tube
Objective (lens)
Eyepiece

Between 1 and 2 m

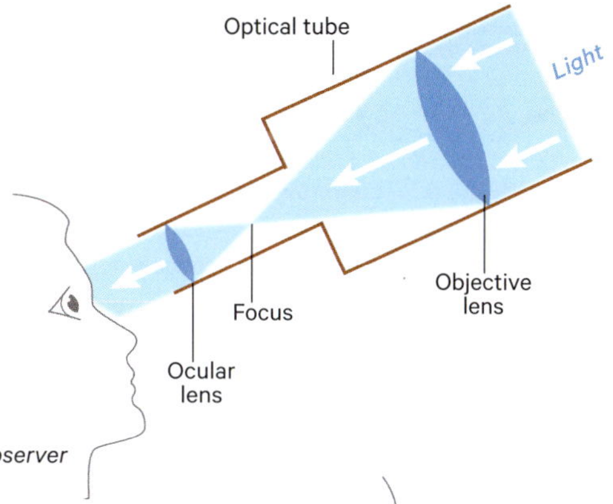

Optical tube
Light
Objective lens
Focus
Ocular lens
Observer

Newtonian telescope

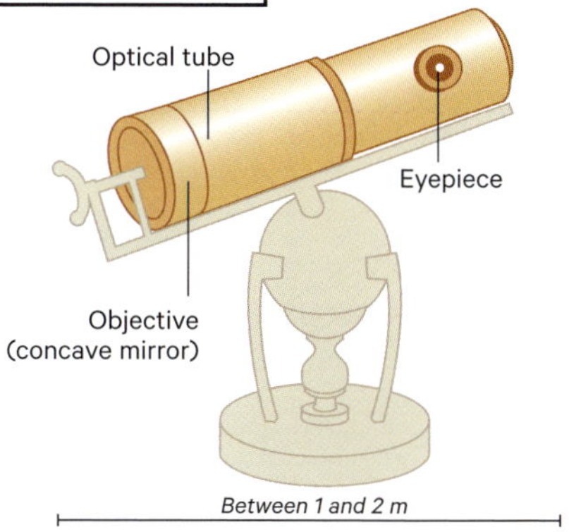

Optical tube
Eyepiece
Objective (concave mirror)

Between 1 and 2 m

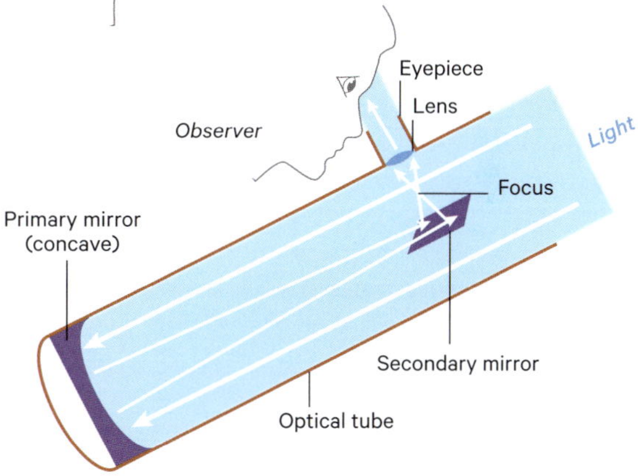

Observer
Eyepiece
Lens
Light
Focus
Primary mirror (concave)
Secondary mirror
Optical tube

Transit telescope

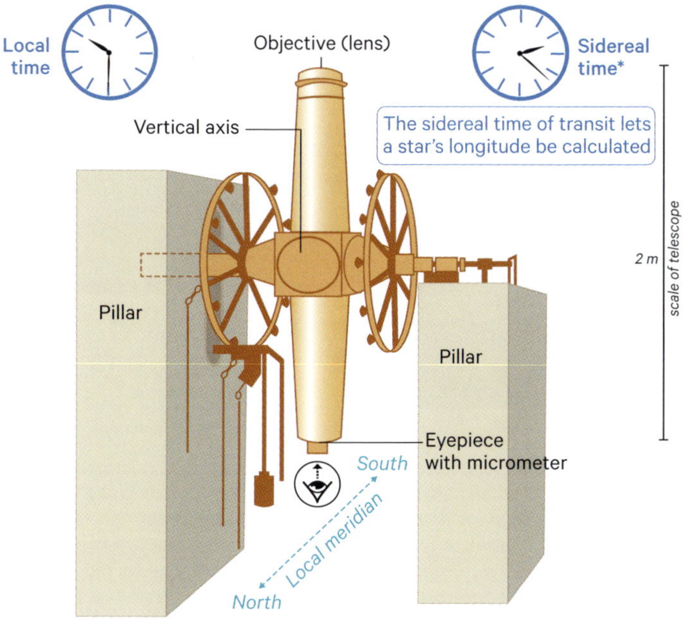

Local time
Objective (lens)
Sidereal time*

The sidereal time of transit lets a star's longitude be calculated

Vertical axis
Pillar
Pillar
Eyepiece with micrometer
South
Local meridian
North

2 m
scale of telescope

The evolution of telescopes

In 1609, Galileo developed and improved a Dutch invention: a spyglass which increased the brightness and refined the details of the stars or planets being observed. He used it to observe the Moon and the Milky Way and discovered the phases of Venus and the presence of moons around Jupiter. Inspired by these discoveries, other astronomers, such as Johannes Kepler, built their own telescopes. These grew to huge sizes, sometimes up to 30 metres long, so were therefore difficult to reposition to follow the movement of the stars. The quality of the images suffered, with the large lenses warping under their own weight and dispersing the colours. Reflecting telescopes with mirrors therefore became the norm. Meanwhile, transit telescopes allowed the position of the stars to be measured with greater precision as they passed over a local meridian.

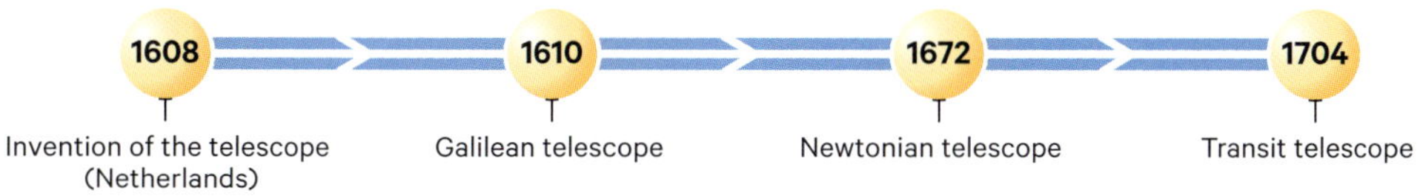

1608 — Invention of the telescope (Netherlands)

1610 — Galilean telescope

1672 — Newtonian telescope

1704 — Transit telescope

The sky seen by Galileo

This was the first time in history that a human could look at the sky with an instrument that collected at least ten times more light than the naked eye. Galileo soon discovered that the Moon, the Milky Way, the Sun and Saturn were different from how we imagined. Jupiter was surrounded by moons, and Galileo was increasingly convinced by Copernicus's heliocentric system. The Aristotelian model, which was the accepted authority, no longer held up. In a trial that would shape history, Galileo fought for the movement of the Earth around the Sun to be accepted, although this was not proven beyond doubt until much later.

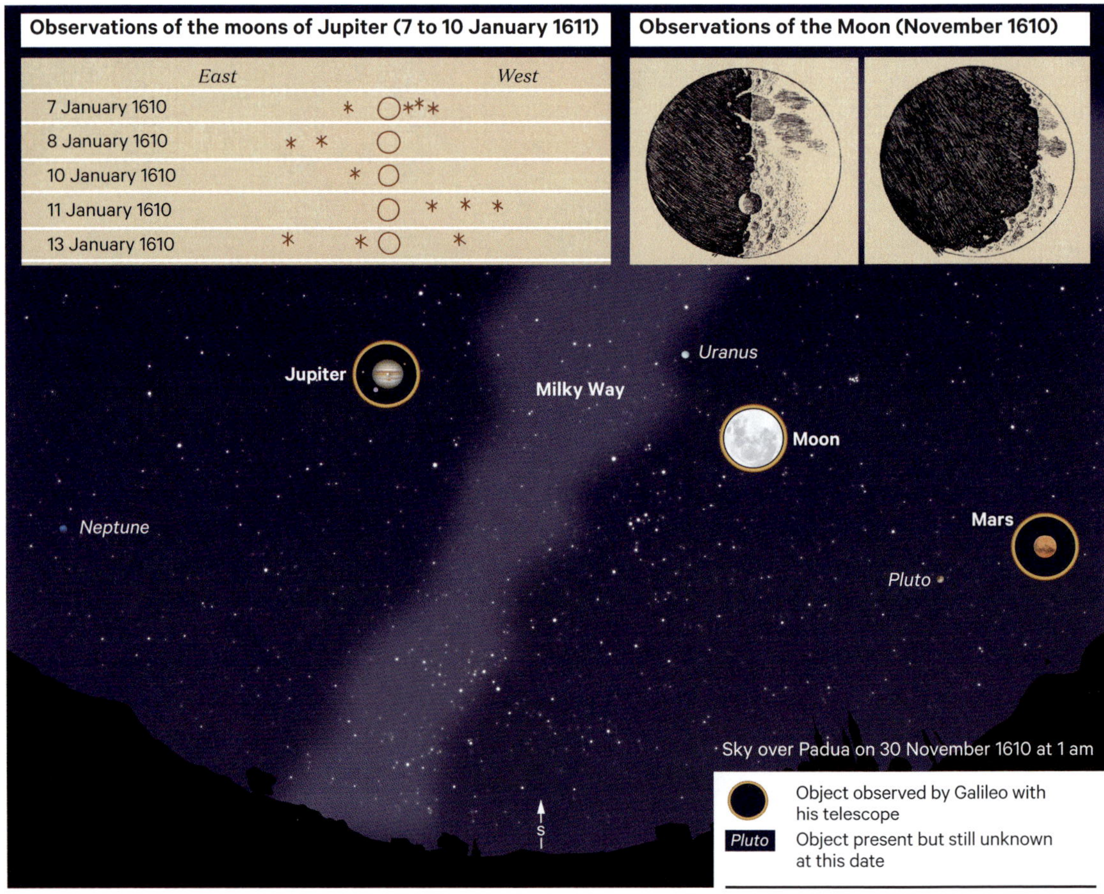

The skies over Padua, winter 1609–1610

Born in Pisa in 1564, Galileo initially studied mathematics and mechanics, then taught at the University of Padua until 1610. In 1604, still looking at the sky with the naked eye, he observed a 'new star' (Kepler's supernova), which challenged his Aristotelian view of the world. In 1609, he learned that a Dutchman, Hans Lippershey, had invented a spyglass that made it possible to see distant objects more clearly. He built one himself, improved it and presented it to the Doge of Venice. In 1610, he observed the Moon and discovered that it had mountains and was therefore imperfect.

He also discovered swathes of stars forming the Milky Way, and the moons that orbited Jupiter. These moons proved that not everything orbited around the Earth, and that Ptolemy's crystalline spheres did not exist: in other words, Copernicus was right. In March, he published his discoveries in *Sidereus Nuncius* ('Starry Messenger'), which was an immediate success and brought him fame. His telescopes improved on the resolution of the naked eye by a factor of at least ten. In July, he left Padua and went to Pisa and Florence, where the Medici family were in power.

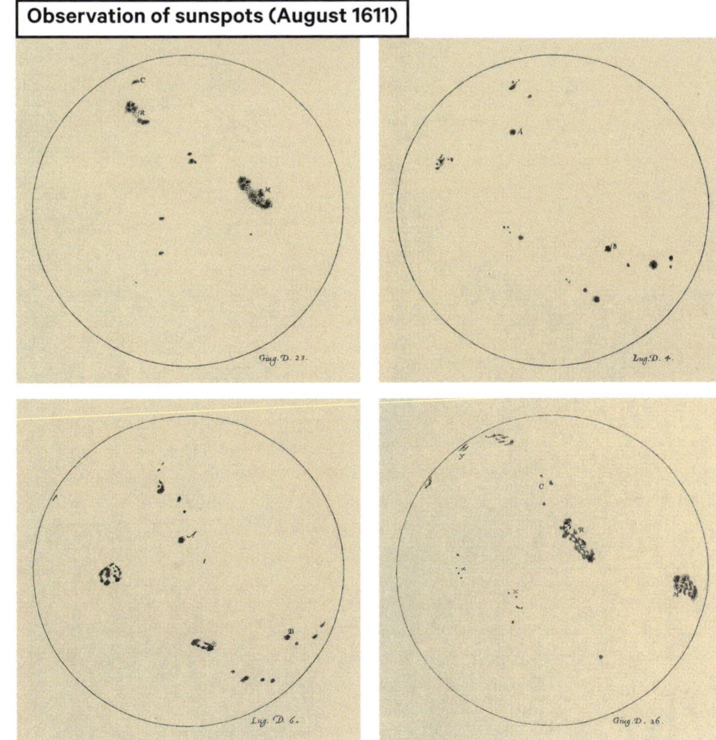

Observation of the phases of Venus (July 1611)

Observation of the rings of Saturn (September 1611)

Observation of sunspots (August 1611)

All sketches are by Galileo.

The skies over Florence, summer 1611

In Florence, Galileo observed the sky through a telescope and noticed spots on the Sun, which had previously been noted but ignored. He concluded from the movement of these spots that the Sun must rotate on its axis. Through observing Venus, he discovered its phases – 'first quarter', 'full Venus' – clearly proving that it orbited the Sun, obeying the heliocentric model.

When observing Saturn, he was surprised by its elongated shape, but his telescope could not make out the rings, so he theorized that Saturn was a 'three-bodied system'. In 1613, when challenged on the contradictions between his observations and the Bible, he said that science and faith were separate things. In 1616, Pope Paul V banned the heliocentric model and the idea of a moving Earth. In 1632, initially with the approval of Pope Urban VIII, Galileo published *Dialogue Concerning the Two Chief World Systems*, in which he supported Copernicus's thesis. Convicted in 1633 and held under house arrest, he died in 1642.

The science of motion

The 17th century saw the birth of a science concerned with the causes of motion, based on observations of the planets and experiments measuring falling bodies. Long-held, fixed ideas were swept aside and the idea of force was born. Magnets showed that a force could act at a distance, while circular motion demonstrated the centrifugal effect. Finally, Newton set out the precise laws of motion, as well as universal gravitation acting in a vacuum between celestial bodies, as on Earth.

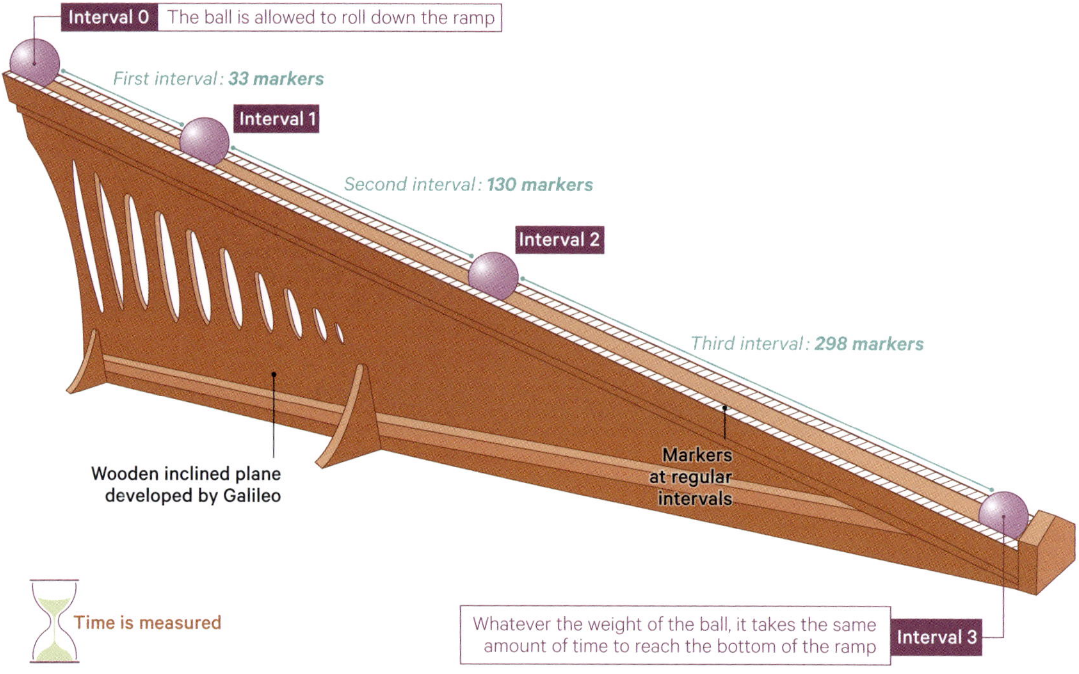

▲ Discovering and proving the laws of gravity

Aristotle had asserted that an object in freefall is attracted by the Earth, its 'natural place', and falls at a constant speed proportional to its weight, but testing this assertion through experiments proved complicated. Aristotle's claims were challenged, and there were a number of studies of gravity and motion, as early as the 6th century in Byzantium, then throughout the Islamic world and Europe. In 1585 in Delft, the Dutchman Simon Stevin dropped two lead balls of different masses and they hit the ground at the same time. In 1589, Galileo, seeking to disprove Aristotle, slowed down the fall, releasing balls that rolled along an inclined plane, under the effect of constant gravity. He measured how long they took to travel a certain distance and drew two main conclusions: they did not fall at a constant speed, but accelerated; the distance travelled increased in proportion to the square of the time spent falling.

▶ Newtonian mechanics

In 1659, Christiaan Huygens was studying circular motion (the stone in a slingshot). He realized that it was the force exerted by the string that maintained the motion. If the string broke, the stone did not fly in the direction of the circle's radius, but on a tangent. He discovered what is incorrectly called 'centrifugal force'. In 1687, Newton published the three laws of motion. The first came from Galileo, who said '[uniform] motion is like nothing'. Building on the work of Huygens, the second law set out the dynamics of motion, introducing the mass of an object – i.e. the quantity of matter it contains, as distinct from its weight. To apply this law to planetary motion, Newton drew on a key mathematical tool: differential calculus, which made it possible to carry out calculations about changes in distance and speed, even when these were very small. Finally, the third law deals with forces acting either through contact or at a distance, such as gravity.

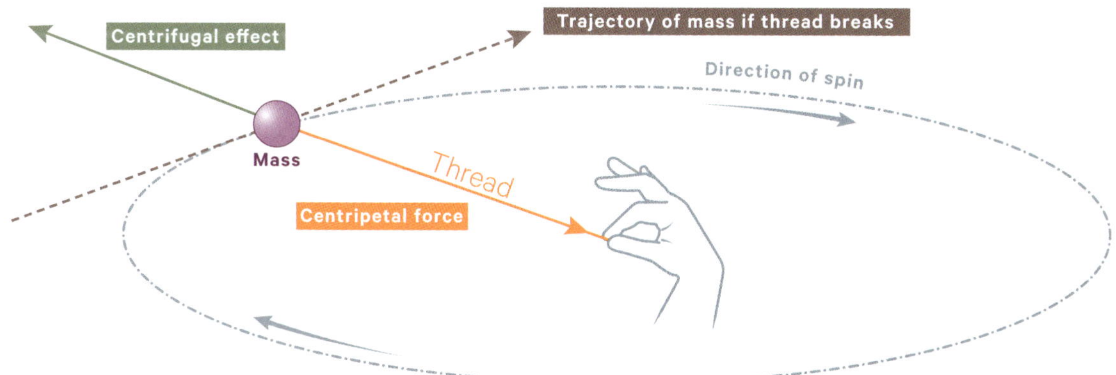

Newton's three laws

	Applied to physics	Applied to astronomy

First law
Inertia

'A body remains at rest, or in motion at a constant speed in a straight line, unless it is acted upon by a force.'

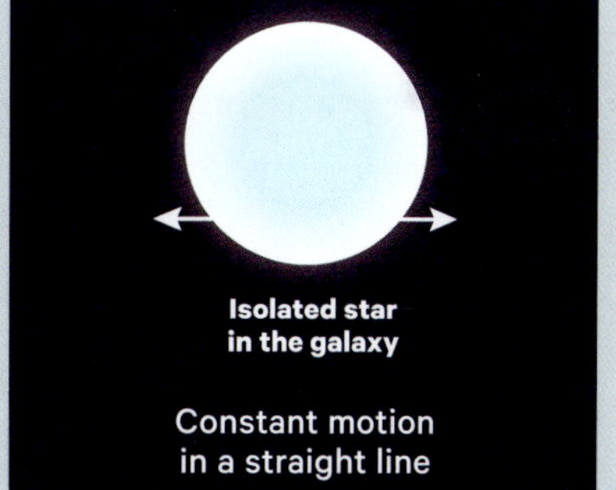

Second law
Momentum

'The acceleration of a body is proportional to the force exerted on it, but inversely proportional to its mass.'

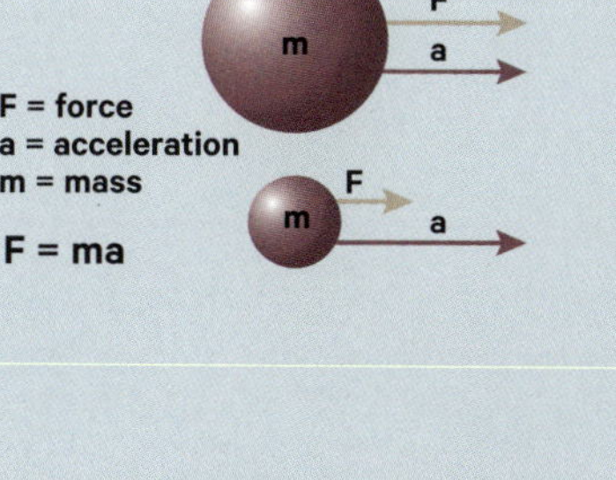

Elliptical orbit of a planet around its star

Third law
Action/reaction

'For every action, there is an equal and opposite reaction. The actions of two bodies on each other are always equal and in opposite directions.'

The Earth and the Moon are attracted to each other by gravity

Universal gravity and the shape of the Earth

In order to explain the precession of the equinoxes (the gradual shift of the spring equinox), Newton theorized that the Earth was flattened at the poles. He calculated this flattening, but Huygens and the anti-Newtonians, who rejected universal gravitation, disputed this. The measurements taken by two expeditions supported Newton's hypothesis. From the mid-18th century onwards, Newtonian gravity was no longer disputed. During the French Revolution, the universal metre was established, defined in relation to the dimensions of the Earth. It was widely adopted as a standard unit of measurement and led to more accurate maps.

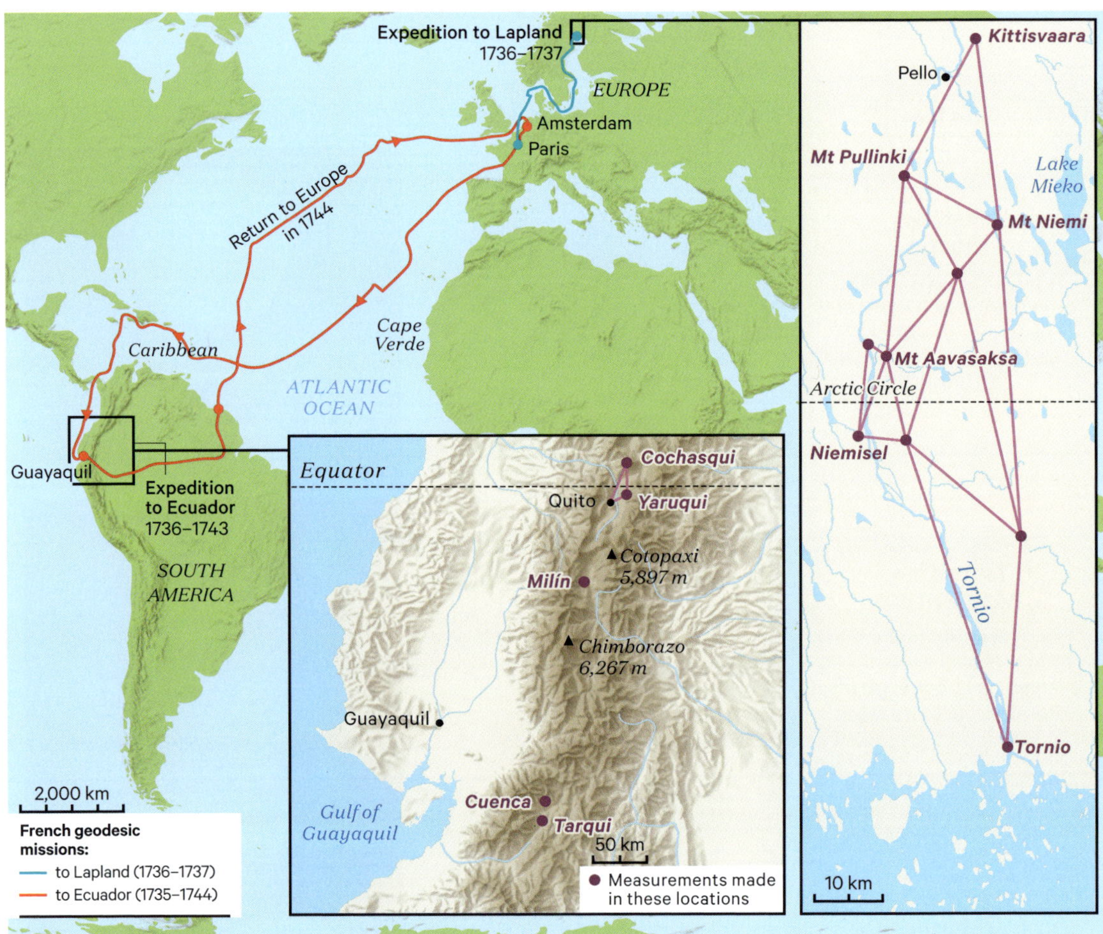

Measuring distant meridians

On the surface of a sphere, the length of any arc of one degree, as seen from the centre, is always the same. This is not the case if the sphere is flattened or stretched. On the orders of the King of France, the Académie des Sciences in Paris launched two expeditions: one to Peru led by Pierre Bouguer, Joseph de Jussieu and Charles de La Condamine, and another to Lapland led by Pierre de Maupertuis and Alexis

Clairaut. In Lapland, the arc of one degree measured 111.153 km, greater than the same arc measured in France, using the new, more accurate units. In 1744, after a number of mishaps, the expedition to Peru returned, having measured the arc at the equator as 110.613 km. There was no more room for doubt: the Earth is flattened at the poles by around 1/200. Newton's theory was accepted over Huygens's.

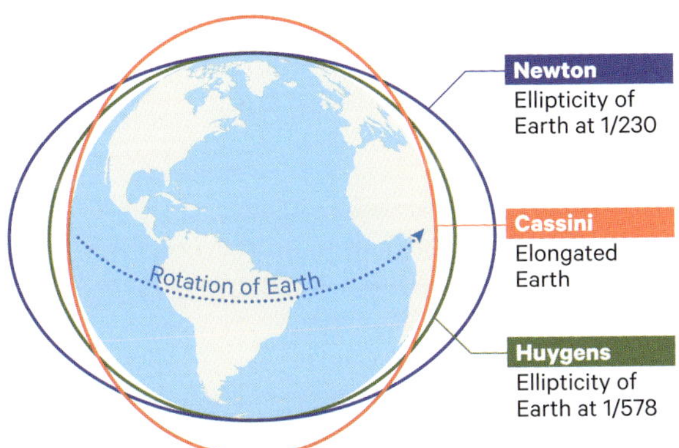

Controversy over the shape of the Earth

In 1686, Hooke believed that the rotating Earth must shaped like a flattened ellipsoid, because of the centrifugal effect. Applying the universal gravitation of 1687 to an Earth that was previously believed to be fluid, Newton calculated the degree of flattening. Along with many of his contemporaries, Huygens disputed this idea of a mysterious force acting at a distance that crossed empty space, so he calculated a value below half of Newton's figure. The only way to determine who was right was to take measurements of the Earth's flattened shape.

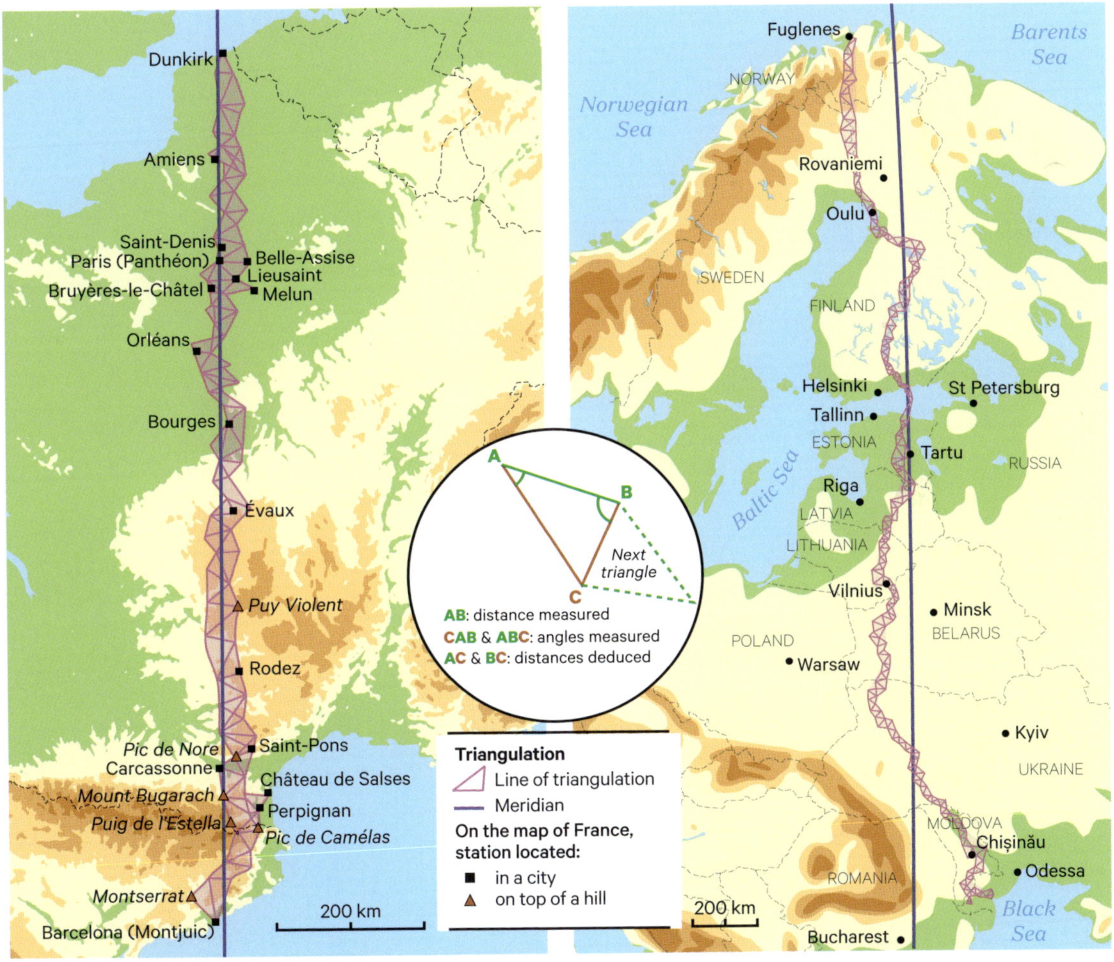

Measuring a meridian in Europe

The debate about the shape of the Earth had to be settled: was Newton right or wrong? The solution was to measure the length of a meridian arc (north–south) on the ground at different latitudes. If the Earth was a sphere, the length would be the same; if it was flattened, the length would be shorter in the north. In 1669, the priest Jean Picard invented geodesic triangulation and measured the length of an arc of the Paris meridian as 111 km. Others followed and extended the arc, confirming a lengthening. With the question decided in favour of a flattened Earth by the French expeditions, the debate moved on to the history of the Earth, and whether it was fluid or solid. More measurements were required, for example that of the Struve Arc, from Norway to the Black Sea, measured by Friedrich von Struve between 1816 and 1855.

Major sea voyages

Up until the 15th century, civilizations were only familiar with their own particular section of the sky. The voyages of the 15th to 17th centuries meant that, for the first time, scientists could study the sky in its entirety. This knowledge was particularly important for navigating beneath the skies of the southern hemisphere. In the 18th century, scientific expeditions were carried out to better understand the shape of our planet and its various land masses, completing the mapping of the Earth and sky.

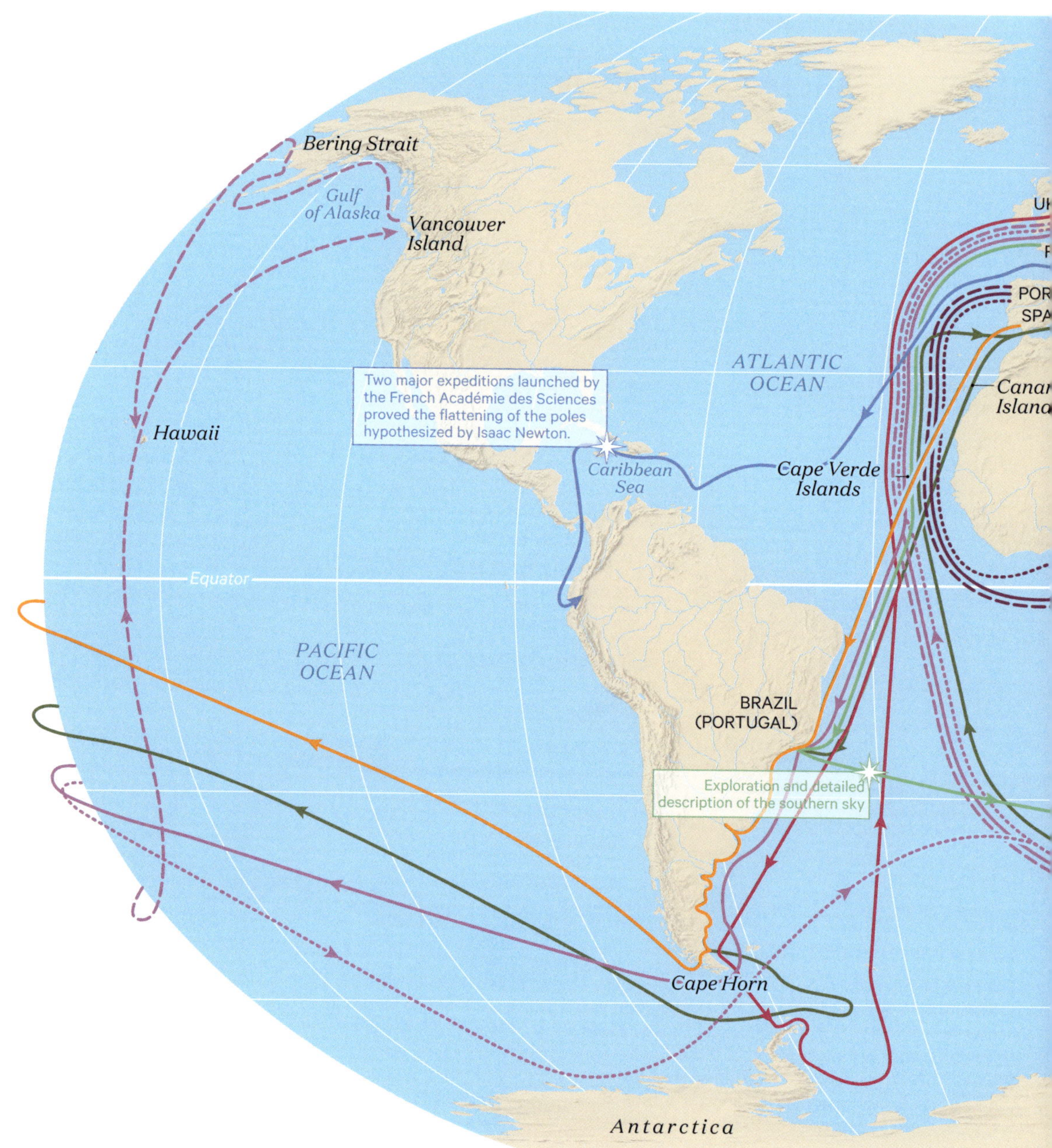

Voyages of Zheng He (1405–1433)
Portuguese explorers travel the coast of Africa (15th century)
Voyage of Lopes Gonçalves
Voyages of Diogo Cão
Voyages of Bartolomeu Dias
Voyages of Magellan (1519–1522)
Magellan–Elcano circumnavigation

Geodesic expeditions
Voyage of Pierre Louis Moreau de Maupertuis (1736–1737)
Voyage de Charles Marie de La Condamine (1736–1743)
Voyage of Nicolas-Louis de Lacaille (1750–1754)
Voyage de Guillaume Le Gentil (1760–1771)

Voyage of James Weddell (1821–1824)
Voyage of Jules Dumont d'Urville (1826–1829)
Voyages of James Cook
first (1768–1771)
second (1772–1775)
third (1776–1779)

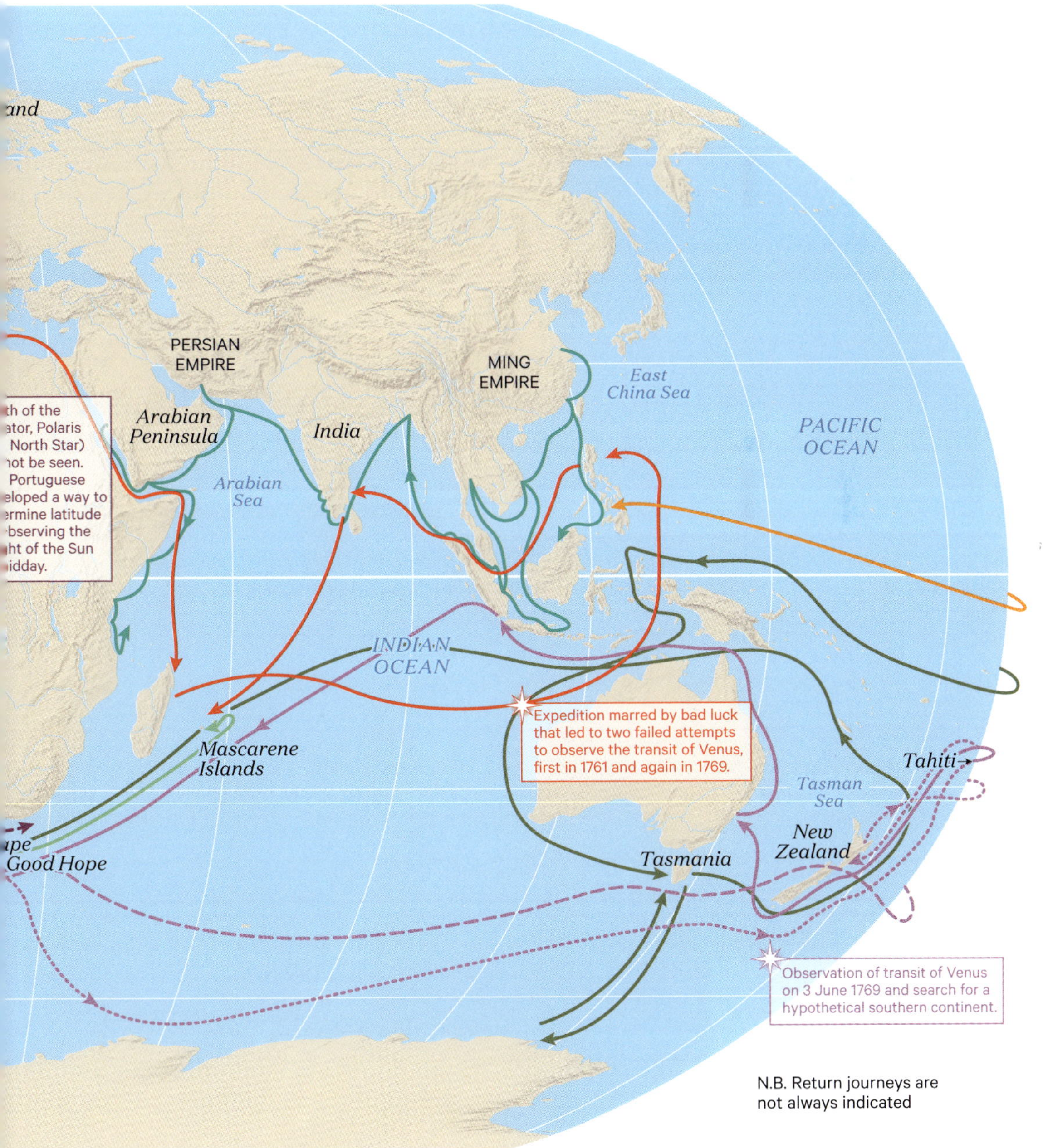

The Newtonian mechanics of the Solar System

In 1705, Edmond Halley theorized that the comet that had been seen in 1682 was the same as those recorded in 1531 and 1607. Drawing on Newton's universal gravitation (as set out in 1687), he predicted it would return in 1758. From that point on, Newton's law was used to calculate all celestial movements and led to the discovery of Neptune in 1842. The transit of Venus across the Sun in 1769 sparked a rivalry between France and Britain, with both nations vying to calculate the precise distance between the Earth and the Sun.

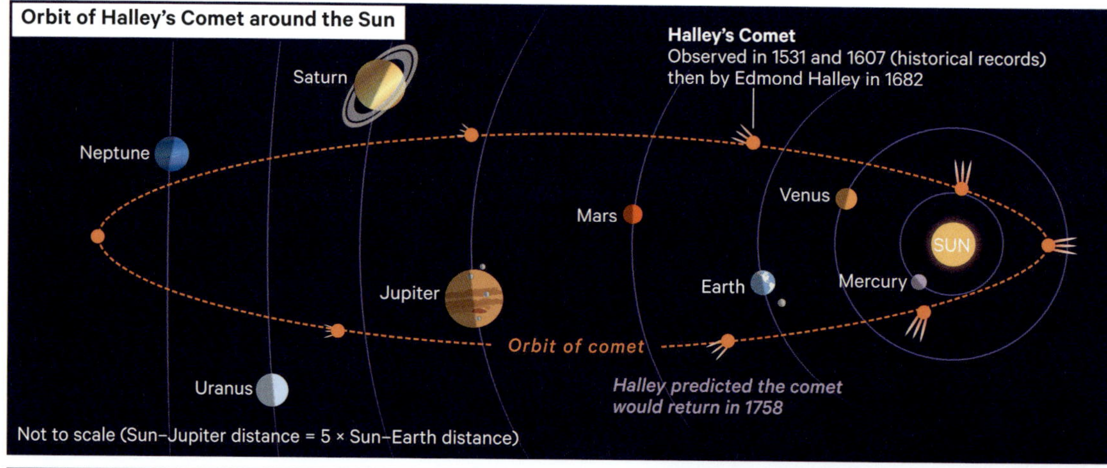

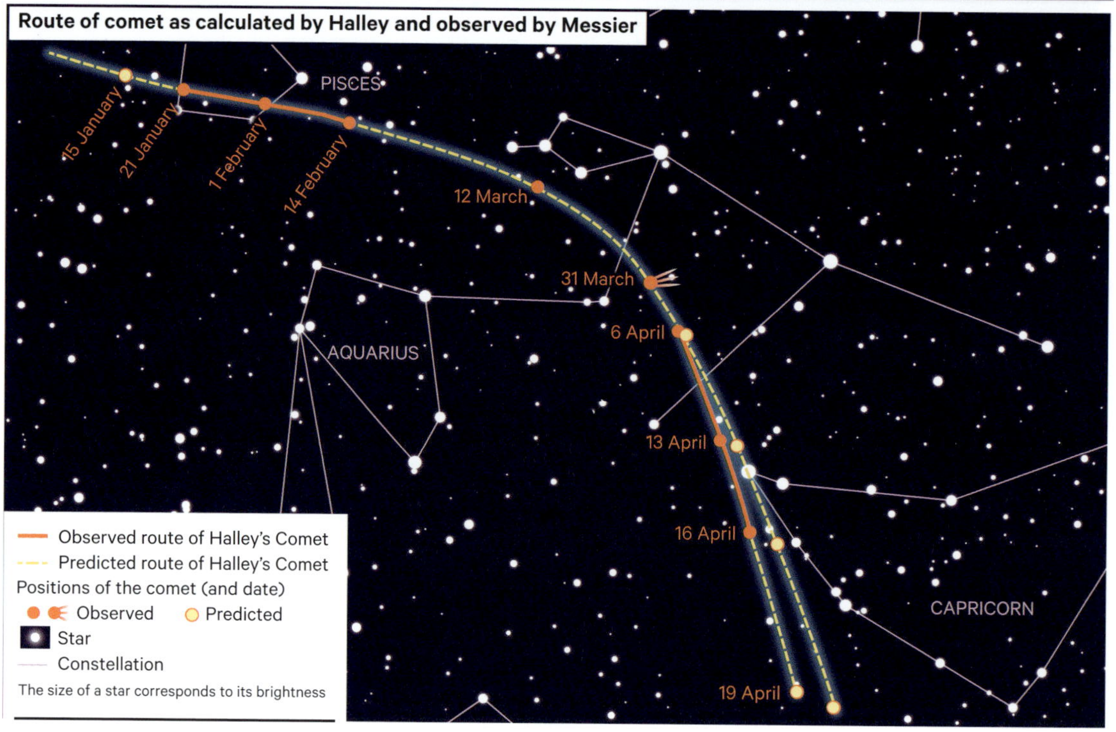

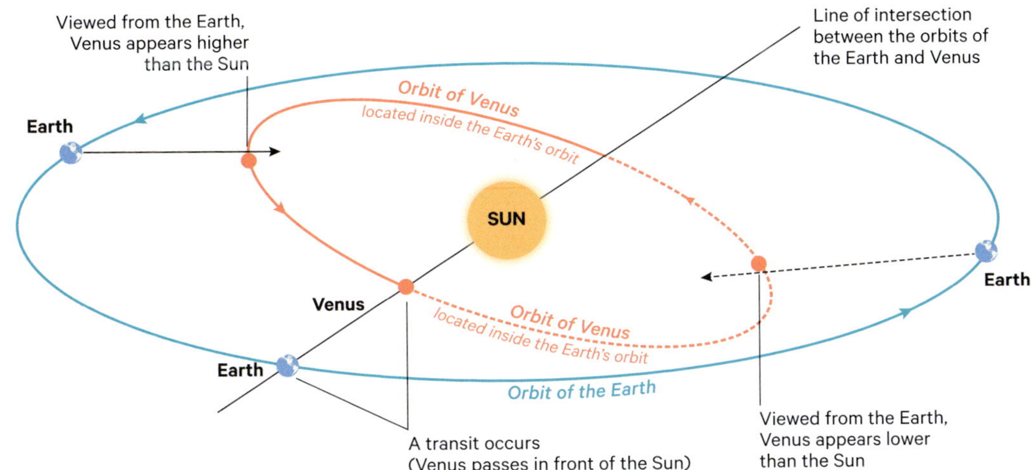

▲ The distance from the Earth to the Sun

In the late 16th century, calculations of the distance between the Earth and the Sun relied on Ptolemy's incorrect estimate. Scottish scientist James Gregory suggested using Venus's transit of the Sun, seen from two places on Earth that were far apart, to calculate it. Two transits, eight years apart, recur every 110 years, as the orbital planes of Venus and the Earth do not coincide. In 1677, Halley observed a transit of Mercury. He wanted to improve the accuracy of his calculations by observing a transit of Venus, which he predicted would take place in June 1761, but he died too soon to see it. The transit of 1769 was the impetus for James Cook's expedition to Tahiti and Guillaume Le Gentil's expedition to India. The distance to the Sun was measured more precisely: Jérôme de Lalande calculated the distance to be 153 million km, close to the true value.

▼ From 1655 to 1846: two new planets

In around 1655–1656, in the Netherlands, Christiaan Huygens discovered Titan, a moon of Saturn, using his improved telescope and realized that Saturn had a ring around it. In 1781, in England, William Herschel, using a telescope with a diameter of 17 cm, discovered the planet Uranus, which is visible to the naked eye, but which was previously thought to be a star. It was the first new planet to be discovered since ancient times. Its orbit was correctly calculated using Newton's laws. Could there be other planets, even further from the Sun? In 1846, in Paris, Urbain Le Verrier (1811–1877) observed small deviations in the calculated orbit of Uranus. Wanting to find the gravitational cause, he predicted the area of the sky where the object responsible would be found. Not long afterwards, Neptune was discovered by astronomer Johann Galle.

The discovery of Neptune

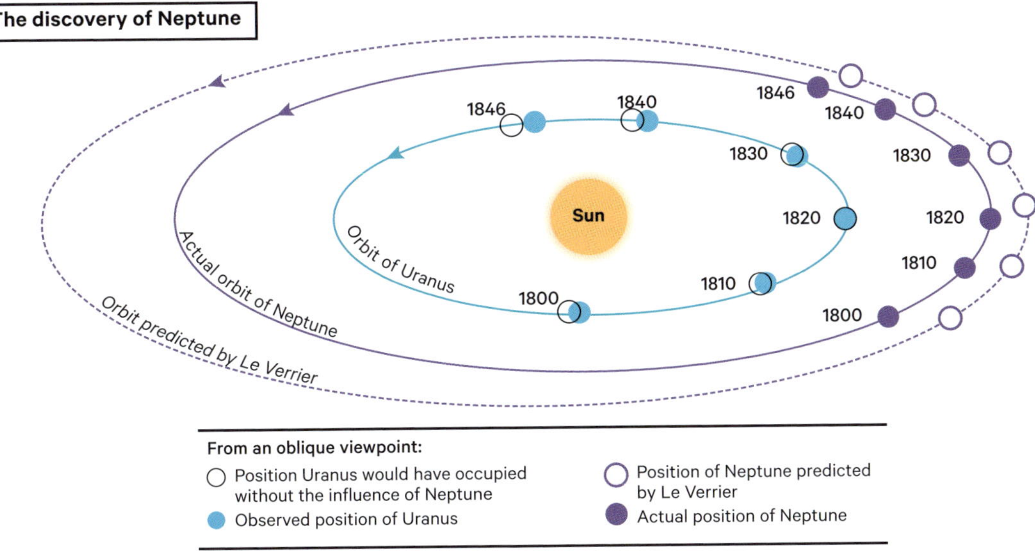

Measuring gravity

Newton stated that the gravitational force acting between two bodies is proportional to their mass, but did not know the value of the multiplying factor or constant in this equation. While this remained unknown, observing the movements of celestial bodies was not enough to calculate their masses. Using the force of attraction of a mountain, two British scientists found the value of this constant and, in 1775, calculated the mass of the Earth. Then Henry Cavendish measured the tiny gravitational force between two objects and calculated a more precise value for this constant. This made it possible to calculate the mass of celestial bodies from their movements.

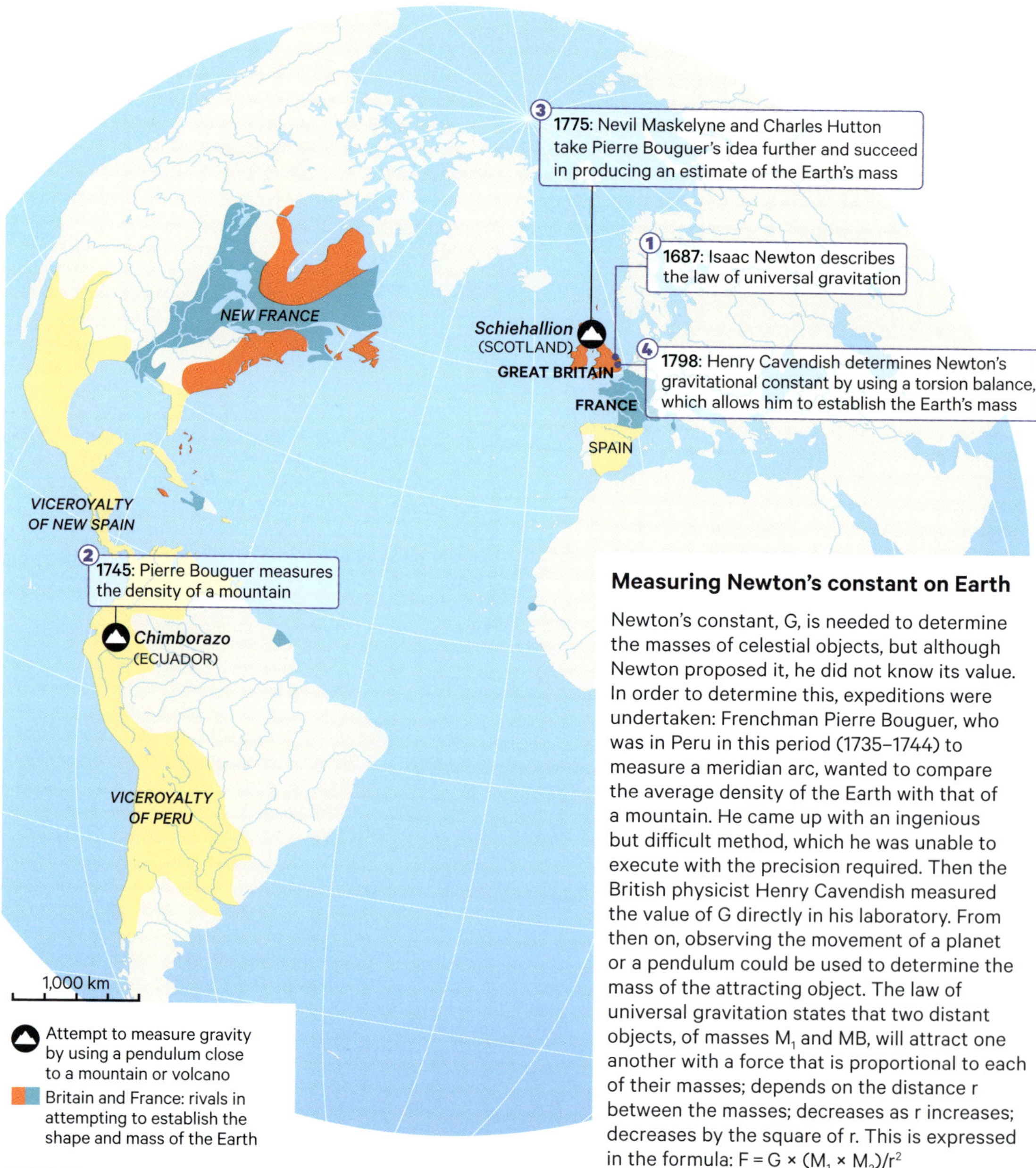

Measuring Newton's constant on Earth

Newton's constant, G, is needed to determine the masses of celestial objects, but although Newton proposed it, he did not know its value. In order to determine this, expeditions were undertaken: Frenchman Pierre Bouguer, who was in Peru in this period (1735–1744) to measure a meridian arc, wanted to compare the average density of the Earth with that of a mountain. He came up with an ingenious but difficult method, which he was unable to execute with the precision required. Then the British physicist Henry Cavendish measured the value of G directly in his laboratory. From then on, observing the movement of a planet or a pendulum could be used to determine the mass of the attracting object. The law of universal gravitation states that two distant objects, of masses M_1 and MB, will attract one another with a force that is proportional to each of their masses; depends on the distance r between the masses; decreases as r increases; decreases by the square of r. This is expressed in the formula: $F = G \times (M_1 \times M_2)/r^2$

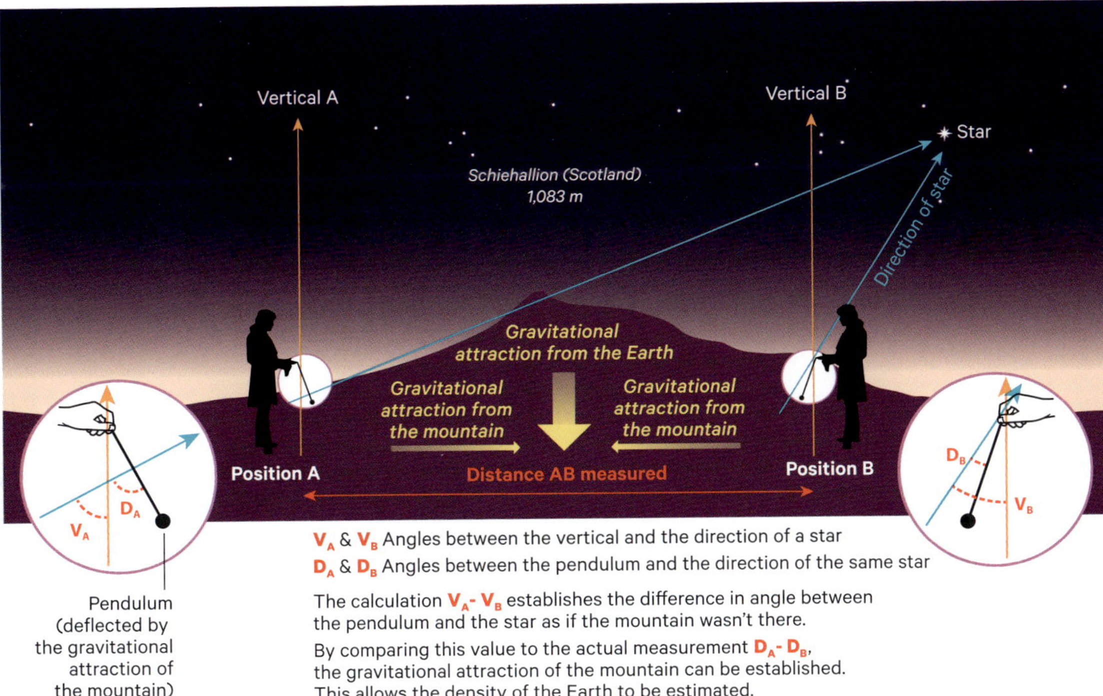

First measurement of the Earth's mass

Nevil Maskelyne (1732–1811) travelled to the island of Saint Helena in 1761 to observe the transit of Venus. Named British Astronomer Royal in 1765, in 1774 he used Bouguer's method to determine the Earth's mass. He chose Mount Schiehallion in Scotland, which has a regular conical shape, because this made it easy to estimate its volume and thus its mass. A pendulum was placed in two places A and B on the meridian, on either side of the mountain. A bright star was observed at the same time from A and B. The angles formed between its direction and that of the pendulum were then measured. The difference in latitude between A and B was determined by pacing out the arc AB. If there was no attraction from the mountain, the angle measured would be exactly equal to this difference. The gravitational attraction of the mountain pulls the pendulum slightly out of true. Measuring those angles can therefore be used to establish the relationship between the mountain's gravitational attraction and that of the Earth. The estimate of the mountain's mass was used to determine the mass of the Earth, as calculated by the mathematician Charles Hutton.

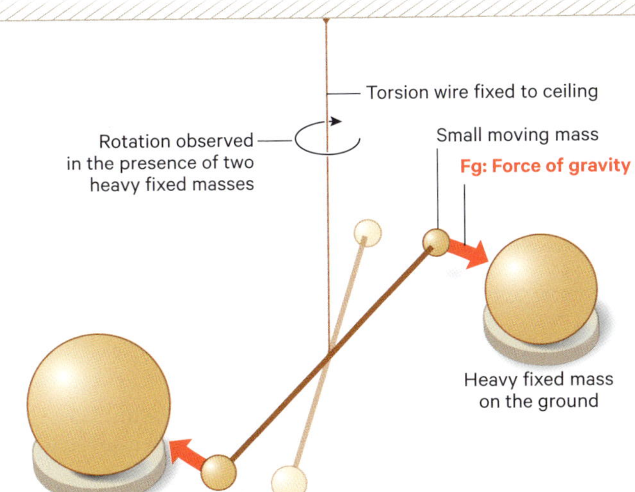

Measuring gravity in a laboratory

British chemist Henry Cavendish (1731–1810) discovered the gas hydrogen in 1776. A talented physicist, he set his sights on directly measuring Newton's constant G in a laboratory. This was difficult, as the gravitational forces between two objects are very weak: a falling stone only looks dramatic because of the large mass of the Earth, which attracts it. Using a torsion pendulum to observe the force between two objects of known mass at a known distance from one another, he measured the effect. This enabled him to calculate G, as well as the Earth's density, to a remarkable degree of accuracy (1%).

Universal gravity and the motion of the stars

Does Newtonian gravity govern the movement of the stars beyond the Solar System? The answer can be found by observing the position and velocity of binary stars, which mutually attract one another. This requires very precise measurements of time. James Bradley used the aberration of light to prove that the Earth orbited around the Sun. Subsequently, his observations of a number of double stars confirmed that Newtonian gravity applied to the known universe, which finally made it possible to calculate the mass of stars.

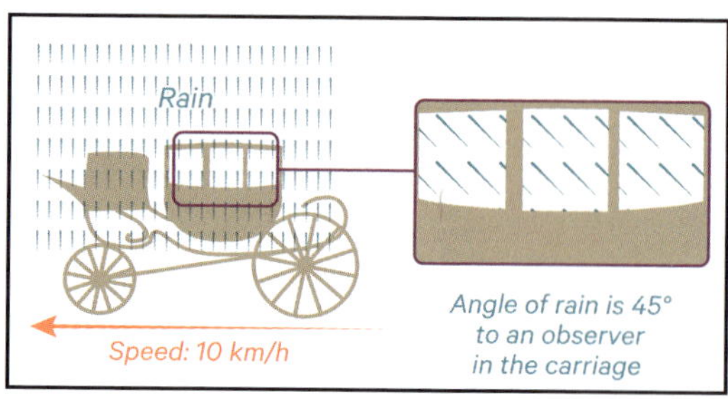

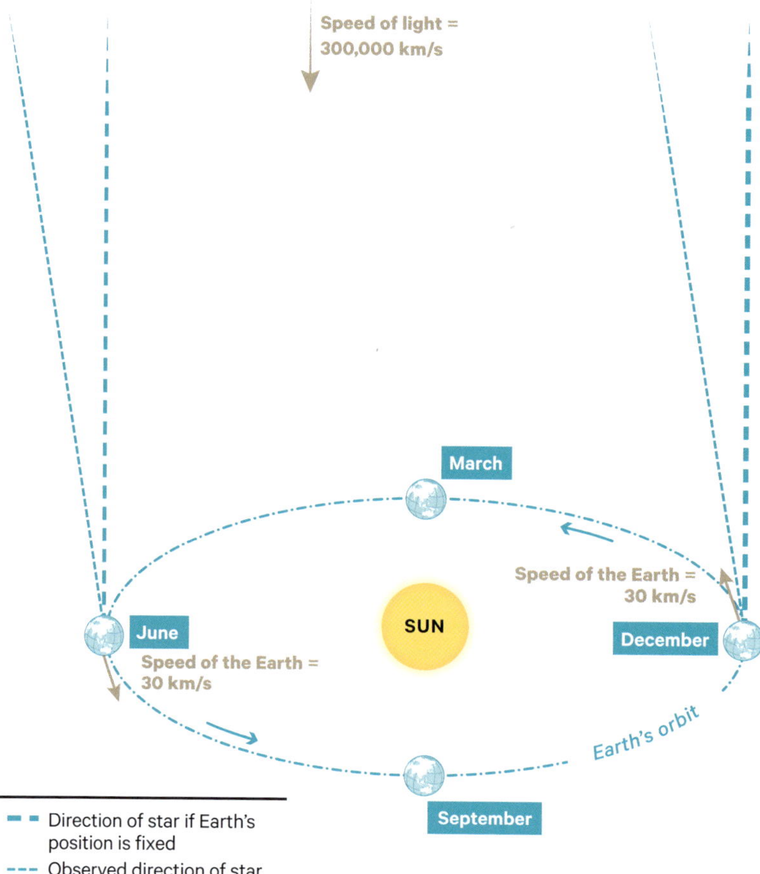

The motion of the Earth and the aberration of light (1725–1727)

Galileo believed that the Earth orbited around the Sun, but he was not able to prove this. As instruments became more advanced, the English scientist James Bradley provided this proof, using the aberration of light. Rain falling vertically on the window of a moving carriage leaves diagonal tracks; if it is falling at the same speed as the carriage is moving, for example 10 km/hr, the angle is 45°, and it increases in proportion with the velocity. The light of a star situated near the pole of the ecliptic falls perpendicular to the plane of the Earth's orbit, with the Earth travelling at 30 km/s, one ten-thousandth of the speed of light. The direction in which the star is seen is therefore slightly different, off by an angle equal to 20 arc-seconds ("), in a direction that depends on the position of the Earth. Therefore, over the course of a year, this star traces a small circle, with a radius of 20".

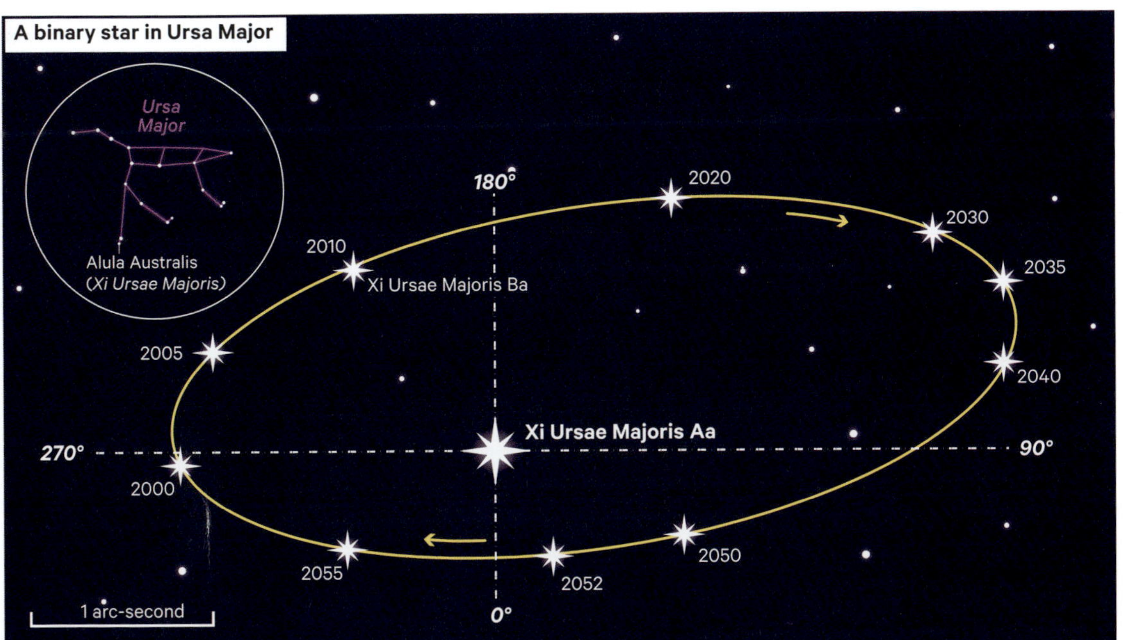

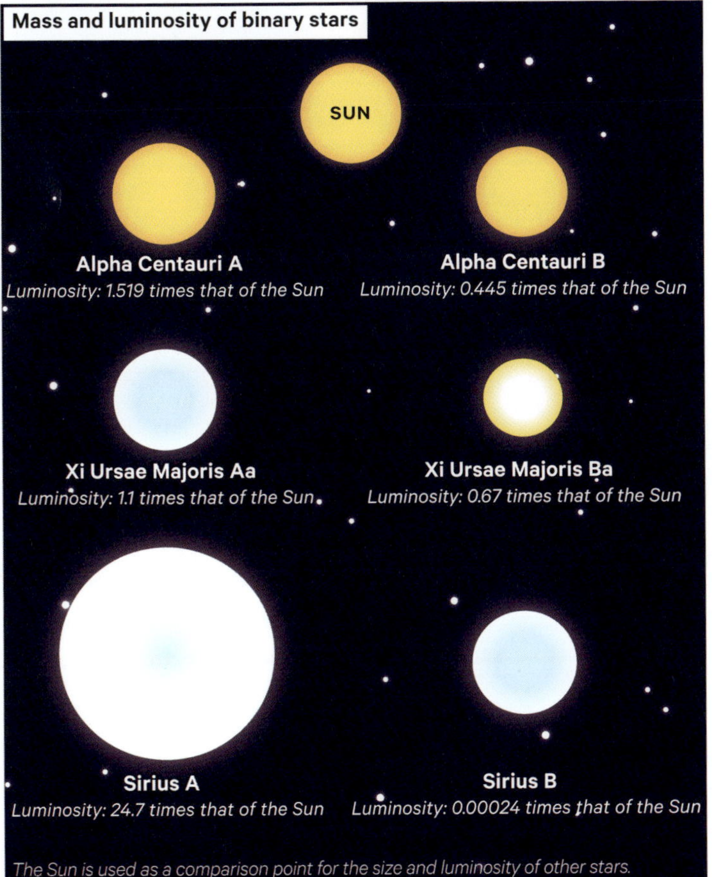

The Sun is used as a comparison point for the size and luminosity of other stars.
The colour of stars matches their apparent colour.

The diversity of binary stars

Xi Ursae Majoris is a binary star, a classification suggested in 1802 by William Herschel, who drew up a catalogue of 700 objects that, when observed through his telescope, seemed to be double rather than single objects. Xi, which is a naked-eye double star, became the first of these objects to have its orbit calculated in 1828. In Tartu, Estonia, Friedrich Georg Wilhelm von Struve published an even larger catalogue of binary stars in 1827. The two stars of Xi Ursae Majoris (Aa and Ba), can only be distinguished with the aid of a telescope, and follow an orbit determined by the law of universal gravitation. The diagram above shows the elliptical trajectory of the star Ba, which it takes sixty years to complete, with the star Aa unmoving at one of the foci of the ellipse. By observing the trajectories that Aa and Ba traced in the sky over many years, it became possible to determine the mass of each star, a new and very important measurement which revealed just how much stellar masses can vary.

The diagram on the left shows several stars with three of their characteristics: mass, luminosity and size compared to the Sun. Astronomers were not able to measure the luminosity and size of stars accurately until the late 19th century.

Measuring the distance to a star

How can the motion of the stars be represented on a map? With the help of more precise clocks and transit telescopes, scientists gradually developed markers to use as reference points, in the form of distant stars. Measurements of the precession of the Earth's axis became more accurate. The Sun moves in relation to the stars, so the changing position of nearby stars over a year demonstrates how far away they are. In the late 19th century, spectroscopy revealed the links between the brightness, mass, colour and distance of stars: from then on, maps could show how far away the stars were.

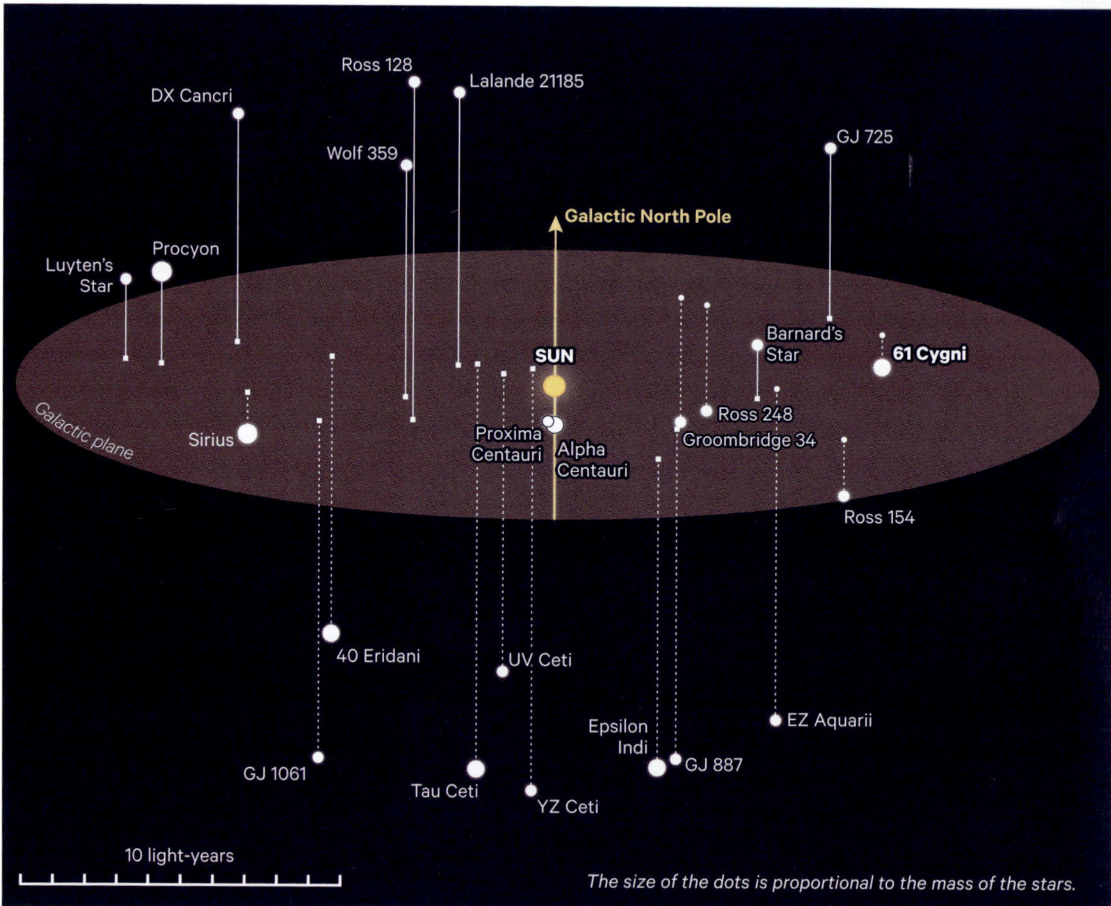

The closest stars to the Sun

This diagram shows the bright stars that are situated just a few light-years away from the Sun. Their distances were measured after 1838 by using their parallax (see opposite). This allowed scientists to draw an approximate map of the galaxy. These distances could then be included in star catalogues. The way their position was noted changed over time: coordinates measured from the ecliptic, as used by the Greeks, were replaced by coordinates measured from the celestial equator. The precession of the equinoxes (due to the change in the orientation of the Earth's axis), which gradually alters these coordinates, was measured precisely by Friedrich Bessel in 1818. Star catalogues are dated because the stars move within the galaxy, with nearby stars moving further. In 1718, using Hipparchus's catalogue from two thousand years earlier, Edmond Halley studied Sirius and Arcturus and learned that the stars move relative to one another. A century later, the Sun's movement relative to nearby stars was measured.

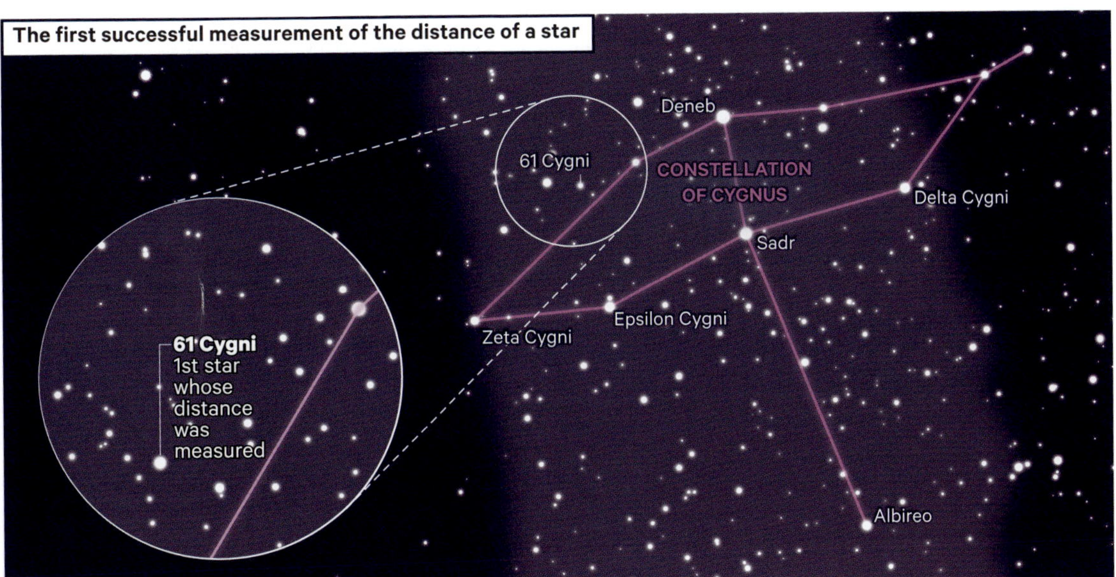

Using parallax to measure distances

If you close first your right eye, then your left, you will see that nearby objects seem to move, relative to their background. The closer the object is to you, the larger this apparent displacement. Now imagine an observer on Earth, looking at the same nearby star at six-month intervals. The angle, called the parallax, created by the star's apparent movement is small and can be measured in arc-seconds ("). Because it seems to move quickly relative to neighbouring stars (5" per year), the star 61 Cygni has been of interest to scientists since 1806. In 1838, German scientist Friedrich Bessel measured its parallax as 0.33" – at the limit of what a contemporary telescope could perceive. This established its distance as 10.5 light-years away. Before Bessel, Friedrich von Struve, working in Tartu, Estonia, measured the parallax of the bright star Vega as 0.125". This measurement was disputed at the time, but it was correct. Until the advent of more precise instruments in the 20th century, distances could only be measured using the parallax method if they were less than 20 to 30 light-years away, because otherwise the angle of displacement of the star became too small for the instruments to detect and was lost in the atmospheric distortion of the images.

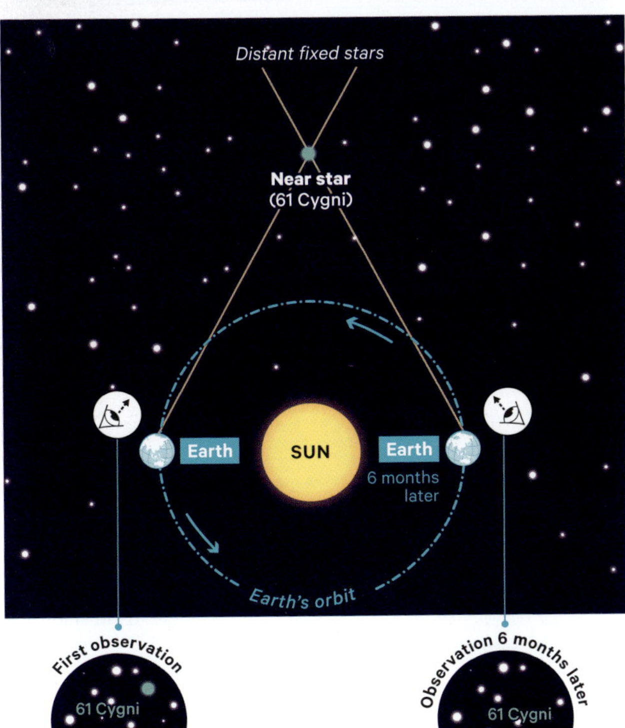

The shape of the Milky Way

In 1755, Immanuel Kant theorized that the stars in the Milky Way were connected by gravity and that this cluster of stars was not unique in the universe. William Herschel drew up the first map of our galaxy and calculated its size from the brightness of stars, which was believed to reflect their distance. He placed the Sun at the centre of this flat disc shape, an error that was not corrected until 1920, when Harlow Shapley's model positioned the Sun far from the centre of the galaxy. Spiral-shaped objects, such as the Andromeda galaxy, which can be seen with the naked eye, fuelled this debate.

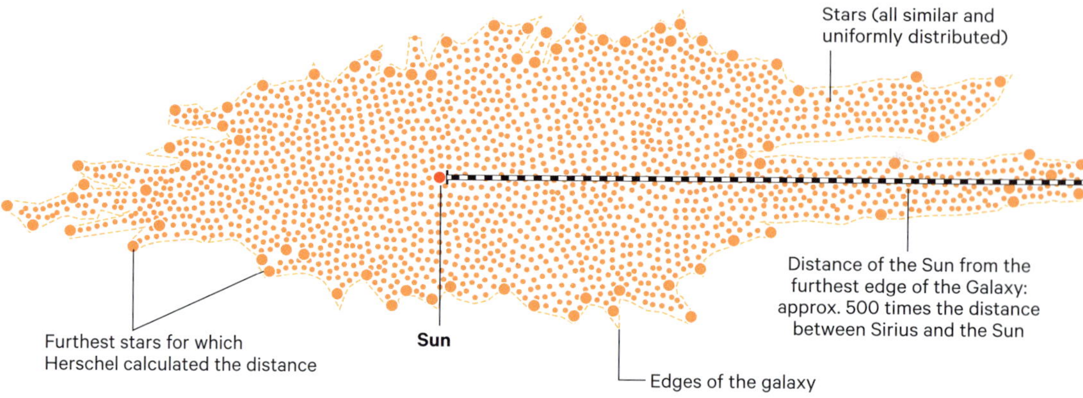

▲ The galaxy according to Herschel (1785)

With the help of his sister Caroline, William Herschel observed the Milky Way and concluded that it was disc-shaped and made up of stars that were evenly distributed. He estimated its size and believed that the Sun was at its centre – a mistake that could have been avoided by noting that the Milky Way contains more stars in summer than in winter. In 1785, he published a historic document that featured the first depiction of the galaxy as it would appear from the outside.

▼ The galaxy according to Shapley (1918)

By observing variable stars, US astronomer Henrietta Swan Leavitt came up with a method for calculating their distance. By applying this method, Harlow Shapley was able to work out the shape of the Milky Way. He concluded from his observations that the Sun was 50,000 light-years from the galactic centre. Later, this number was adjusted down to 30,000 light-years, but Shapley was the first person to show that the Sun was not at the centre of the Milky Way and to come up with a realistic estimate of the galaxy's size.

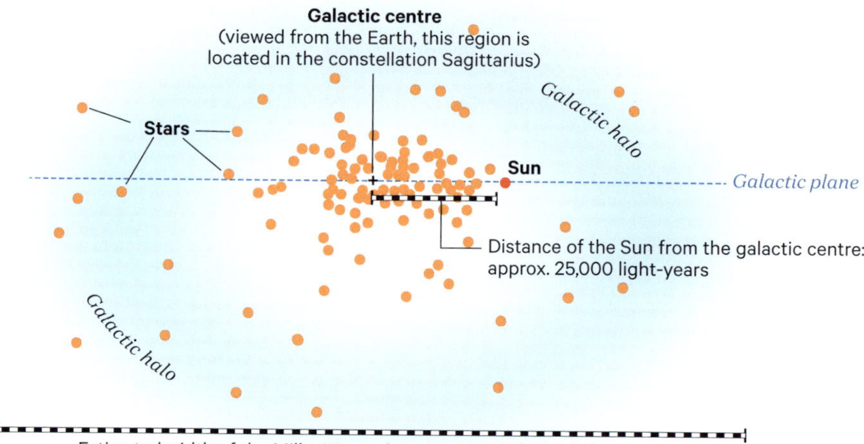

Estimated width of the Milky Way galaxy: approx. 300,000 light-years

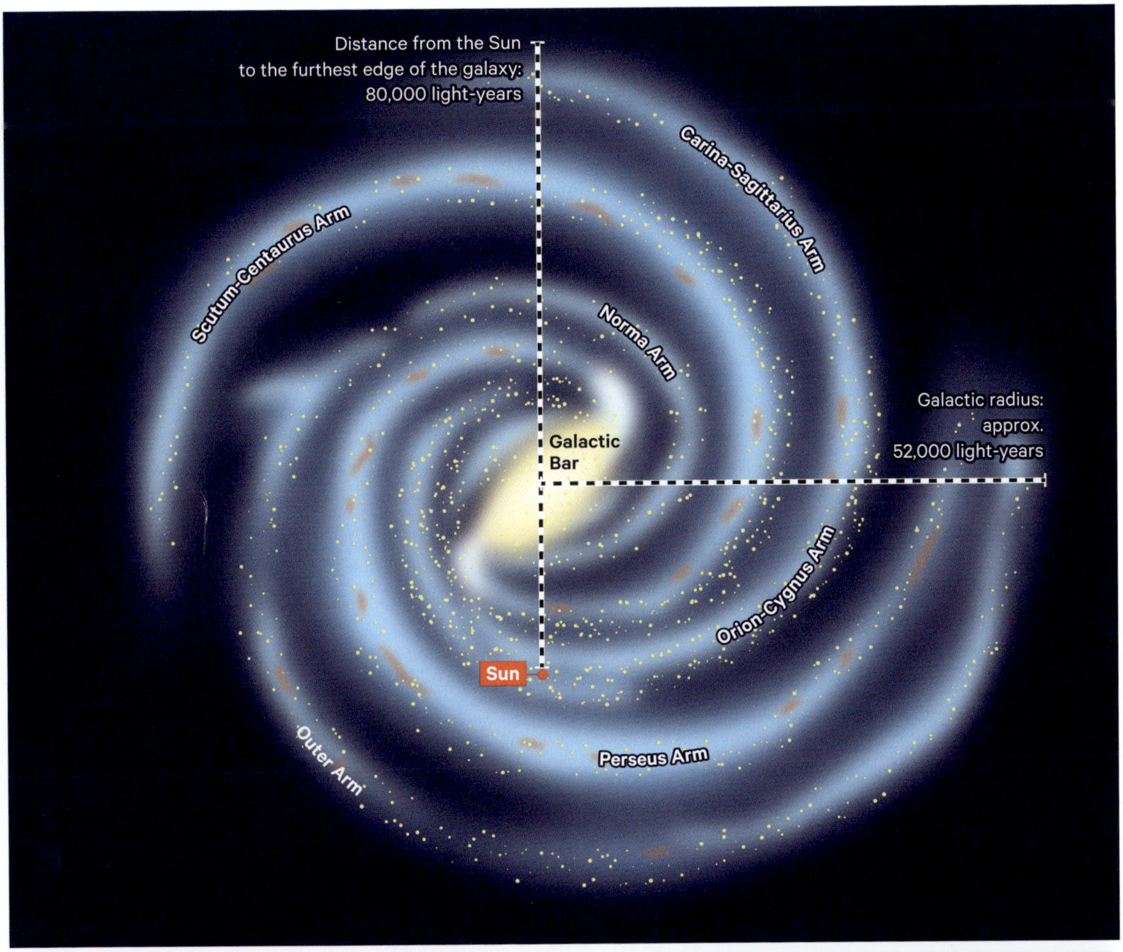

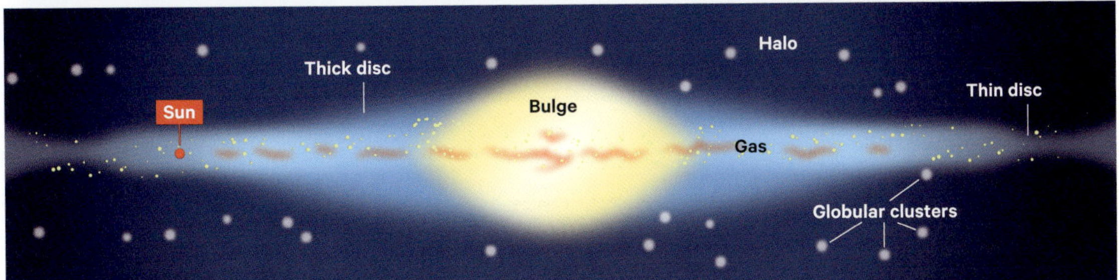

A current view of the galaxy

Advances in observation and digital simulations have made it possible to build up a more accurate picture of our galaxy. Seen from above (top diagram), it is a disc with a diameter of at least 100,000 light-years, in a rotating spiral shape. Like two thirds of spiral galaxies, the Milky Way has a bar-shaped structure at its centre: a denser area of stars and gas. The Sun is 26,600 light-years away from this central bar and takes roughly 250 million years to orbit around it. Seen from the side (bottom diagram), our galaxy looks like two discs, one thin and the other thicker. The thicker disc is mainly made up of older stars, which appeared during the early stages of the formation of the galaxy. There is also a halo, which extends beyond the discs. It contains around a hundred globular clusters, each of which contains hundreds of thousands of stars. The most recent estimate of the total mass of the Milky Way, obtained using the Gaia and Hubble observatories, is between 200 and 500 billion solar masses, more than the mass of the stars and interstellar gas. This difference is attributed to the presence of dark matter.

Nebulae

The advent of telescopes revealed bright objects that do not appear in the form of single points like stars. These are called nebulae. In 1752, Nicolas de Lacaille drew up an inventory of nebulae in the southern hemisphere. Later, around a hundred of them were catalogued by Charles Messier in 1771. They included objects, sometimes visible to the naked eye, such as the Magellanic Clouds. In time, scientists would come to understand just how diverse these nebulae were. In 1864, the invention of spectroscopy showed that some of them formed an interstellar medium made up of gas and dust.

Globular cluster
M15 (NGC 7078)

Galaxy
M31 Andromeda Galaxy (NGC 224)

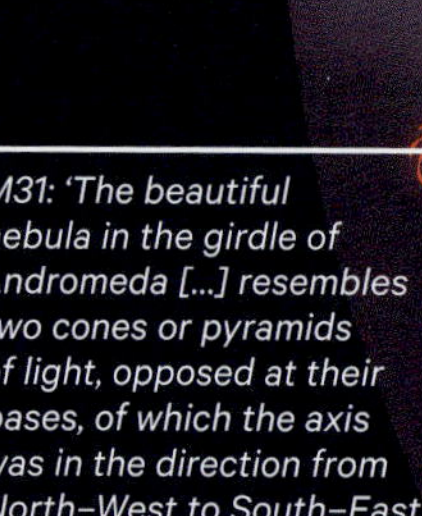

M31: 'The beautiful nebula in the girdle of Andromeda [...] resembles two cones or pyramids of light, opposed at their bases, of which the axis was in the direction from North–West to South–East.'
Messier, 3 August 1764

Supernova remnant
M1 Crab Nebula (NGC 1952)

M1: 'Nebula above the southern horn of Taurus, it doesn't contain any star; it is a whitish light, elongated in the shape of a flame of a candle, discovered while observing the comet of 1758.'
Messier, 3 August 1764

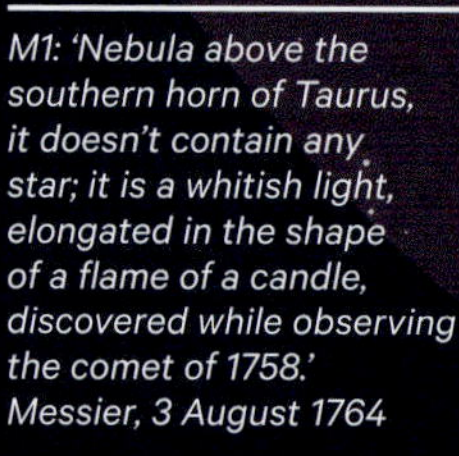

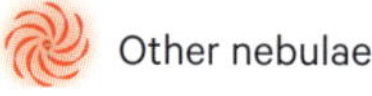 Other nebulae

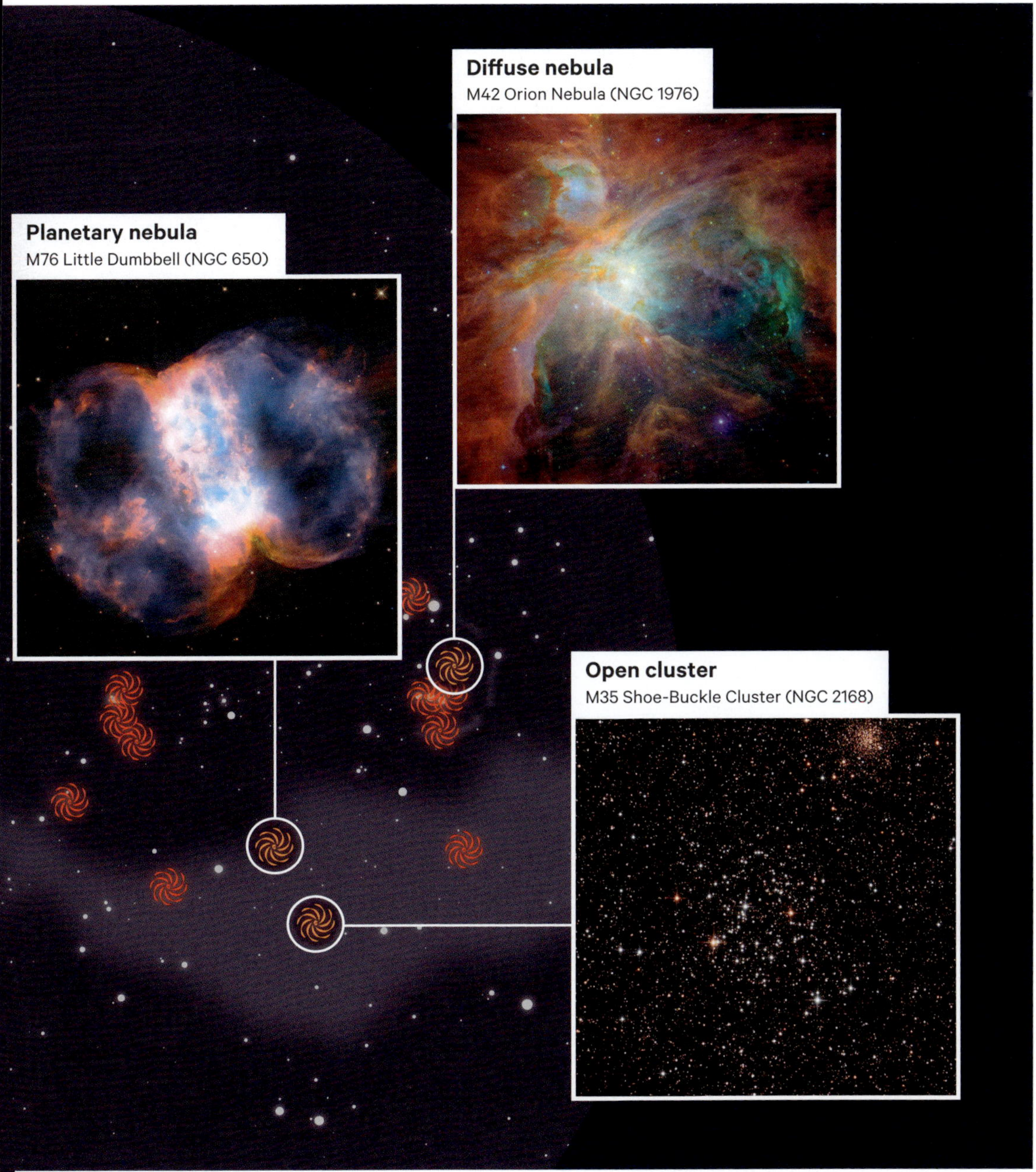

Diffuse nebula
M42 Orion Nebula (NGC 1976)

Planetary nebula
M76 Little Dumbbell (NGC 650)

Open cluster
M35 Shoe-Buckle Cluster (NGC 2168)

Messier's catalogue

Renowned comet-seeker Charles Messier (1730–1817) mistakenly thought the Crab Nebula was Halley's Comet, which he expected to return to that area of the sky (the constellation Taurus). In order to avoid any further mistakes, he decided to draw up a list of non-comet objects, known as the Messier Catalogue (1781). Unmoving relative to the stars, the nature of these nebulous objects, which appear as bright blurs of light, was unknown. As the first object in the catalogue, the Crab Nebula was named M1. Messier did not draw these nebulae, but described them very precisely (see the examples quoted opposite). Later, more complete catalogues were compiled. The Andromeda galaxy, the closest to our own Milky Way, was named M31 in the Messier Catalogue and NGC 224 in the New General Catalogue (1888). The six modern photographs of nebulae reproduced here show just how diverse they are, from interstellar clouds in the Milky Way to the Andromeda galaxy, much further away.

Inventors and artisans

Advances in mechanics, acoustics and optics, driven by the work of skilled craftspeople, were used to develop automata, music, scientific instruments (telescopes, microscopes, clocks, barometers, etc.), electric motors and explosive devices. This knowledge was spread through academic debate and publications such as Diderot's *Encyclopedia* in the 18th century. These advances also had an impact on our ability to observe the sky more closely and make new discoveries.

The global spread of technology

Scientific advances would not have been possible without the technical skills of craftspeople, which improved the instruments used for observing the sky and taking measurements in a variety of ways. In Italy and the Netherlands, local expertise in glassmaking and polishing was vital to the construction and improvement of optical instruments, such as refracting telescopes and mirrors. Similarly, clockmaking techniques played a key role in measuring longitude. This knowledge was spread, across Europe and then to the USA, through science museums – which replaced cabinets of curiosities – world's fairs, royal courts, academies, and more. Voyages by physicists and inventors, such as Denis Papin and Christiaan Huygens, as well as travellers who were amateur scientists, such as French diplomat Balthasar de Monconys, also played a part. Monconys met many scholars and philosophers on his travels, including Blaise Pascal in Paris, Huygens in the Netherlands, the Jesuits Honoré Fabri and Niccolò Zucchi in Italy and Otto von Guericke in Germany.

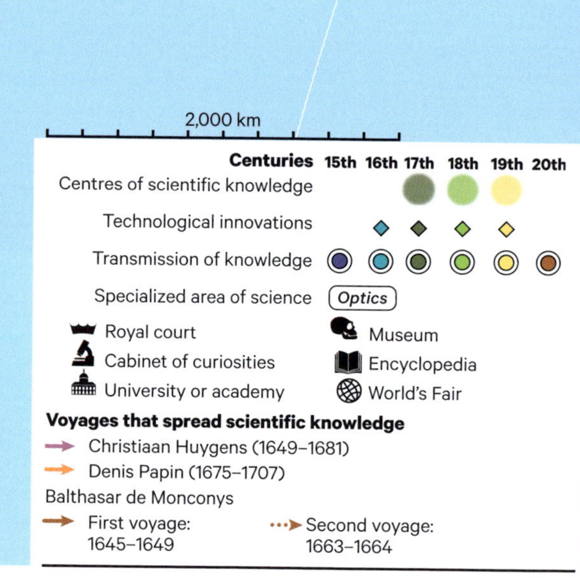

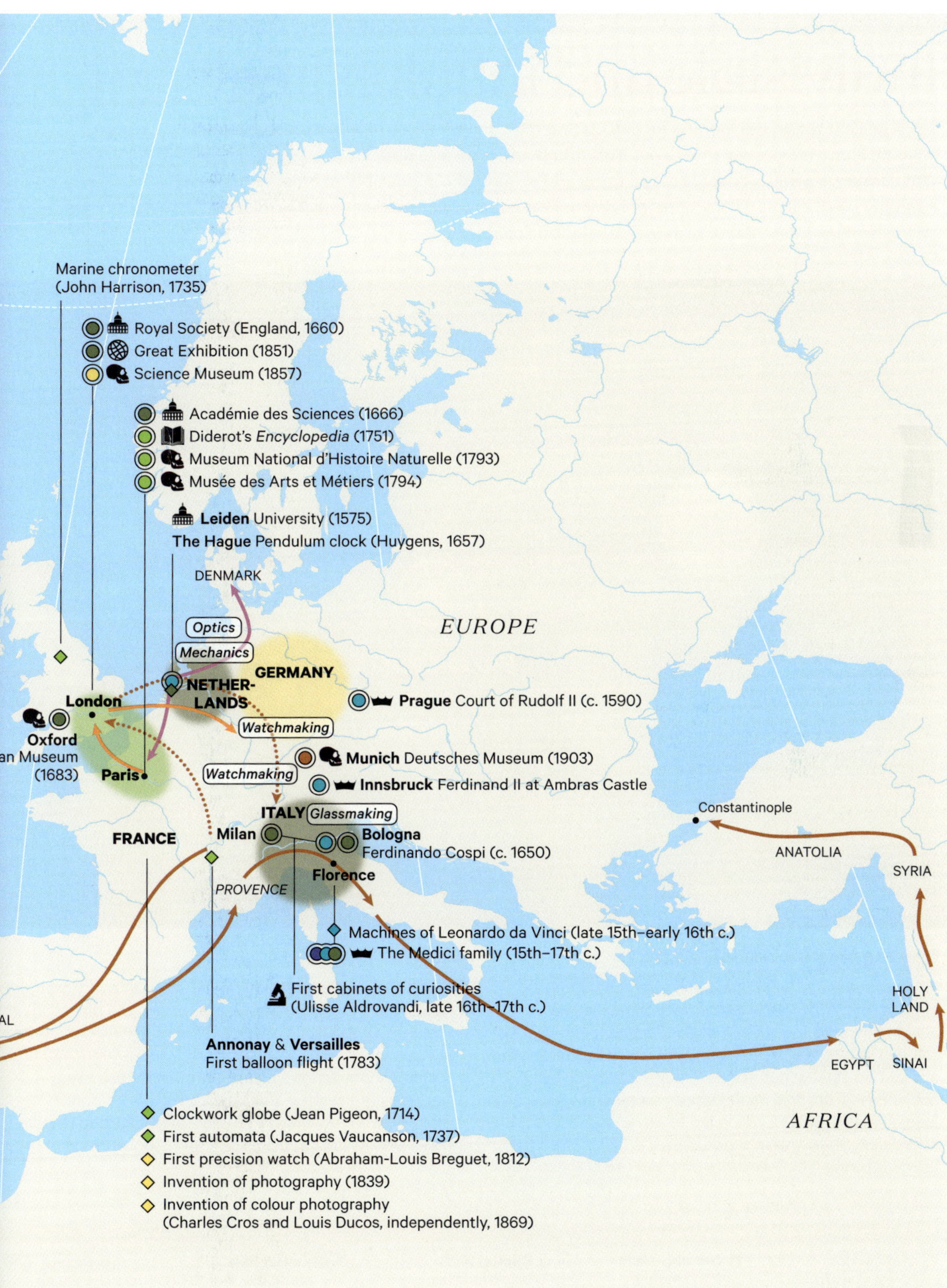
Marine chronometer
(John Harrison, 1735)
Royal Society (England, 1660)
Great Exhibition (1851)
Science Museum (1857)
Académie des Sciences (1666)
Diderot's Encyclopedia (1751)
Museum National d'Histoire Naturelle (1793)
Musée des Arts et Métiers (1794)
Leiden University (1575)
The Hague Pendulum clock (Huygens, 1657)
DENMARK
EUROPE
Optics
Mechanics
GERMANY
NETHER-
LANDS
Prague Court of Rudolf II (c. 1590)
London
Watchmaking
Oxford
an Museum
(1683)
Munich Deutsches Museum (1903)
Watchmaking
Paris
Innsbruck Ferdinand II at Ambras Castle
Constantinople
ITALY Glassmaking
FRANCE
Milan
Bologna
ANATOLIA
SYRIA
Ferdinando Cospi (c. 1650)
Florence
PROVENCE
Machines of Leonardo da Vinci (late 15th–early 16th c.)
The Medici family (15th–17th c.)
HOLY
LAND
First cabinets of curiosities
AL
(Ulisse Aldrovandi, late 16th–17th c.)
Annonay & Versailles
First balloon flight (1783)
EGYPT
SINAI
AFRICA
Clockwork globe (Jean Pigeon, 1714)
First automata (Jacques Vaucanson, 1737)
First precision watch (Abraham-Louis Breguet, 1812)
Invention of photography (1839)
Invention of colour photography
(Charles Cros and Louis Ducos, independently, 1869)

Indivisible and infinitesimal

Since ancient times, the method for calculating the circumference of a circle had involved dividing it into increasingly tiny arcs. This infinite division was difficult to imagine. In the 17th century, a series of people dared to try and achieve the impossible, in order to calculate infinitely small values with greater precision. This ushered in a mathematical revolution, giving humans a new ability to describe scientific concepts.

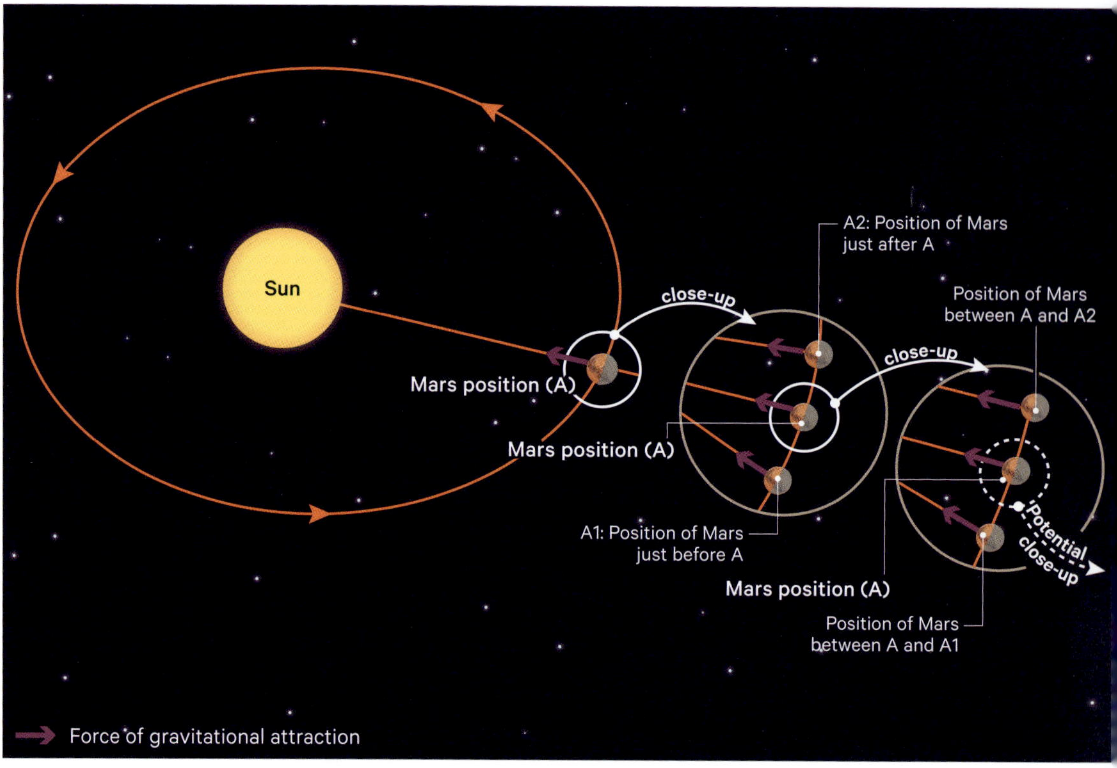

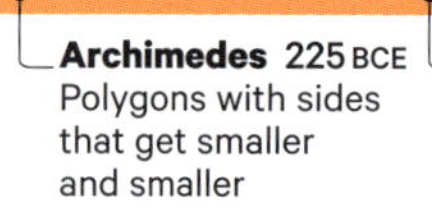

Dividing an orbit into infinite parts

To describe the motion of a planet, Johannes Kepler, in the early 17th century, came up with the idea of dividing its orbit into extremely thin slices. Although his calculations contained errors, these cancelled each other out and he established the 'law of areas' (see p. 98). Blaise Pascal studied a kind of curve known as a cycloid (see opposite above), which led to advances in this idea of division. Shortly afterwards, Newton discovered the law of attraction between the Sun and the planets.

When a planet travelled along its orbit, at every moment, however close to the previous moment, the direction and intensity of the force changed. Tracing the exact curve followed by the planet meant making the sequence of moments smaller and smaller, until eventually they became infinitesimal. But how could this be done? Isaac Newton invented a revolutionary method that he described in 1687 in his groundbreaking work on gravity.

Archimedes 225 BCE	**N. Oresme** 1482	**J. Kepler** 1609	**B. Cavalieri** 1635	**B. Pascal** 1658
Polygons with sides that get smaller and smaller	Uniform motion divided into extremely short stages	Orbit divided into infinitely small segments	The 'method of indivisibles', viewing forms as an indefinite series of lines or planes	Used 'indivisibles' to find the area within a cycloid

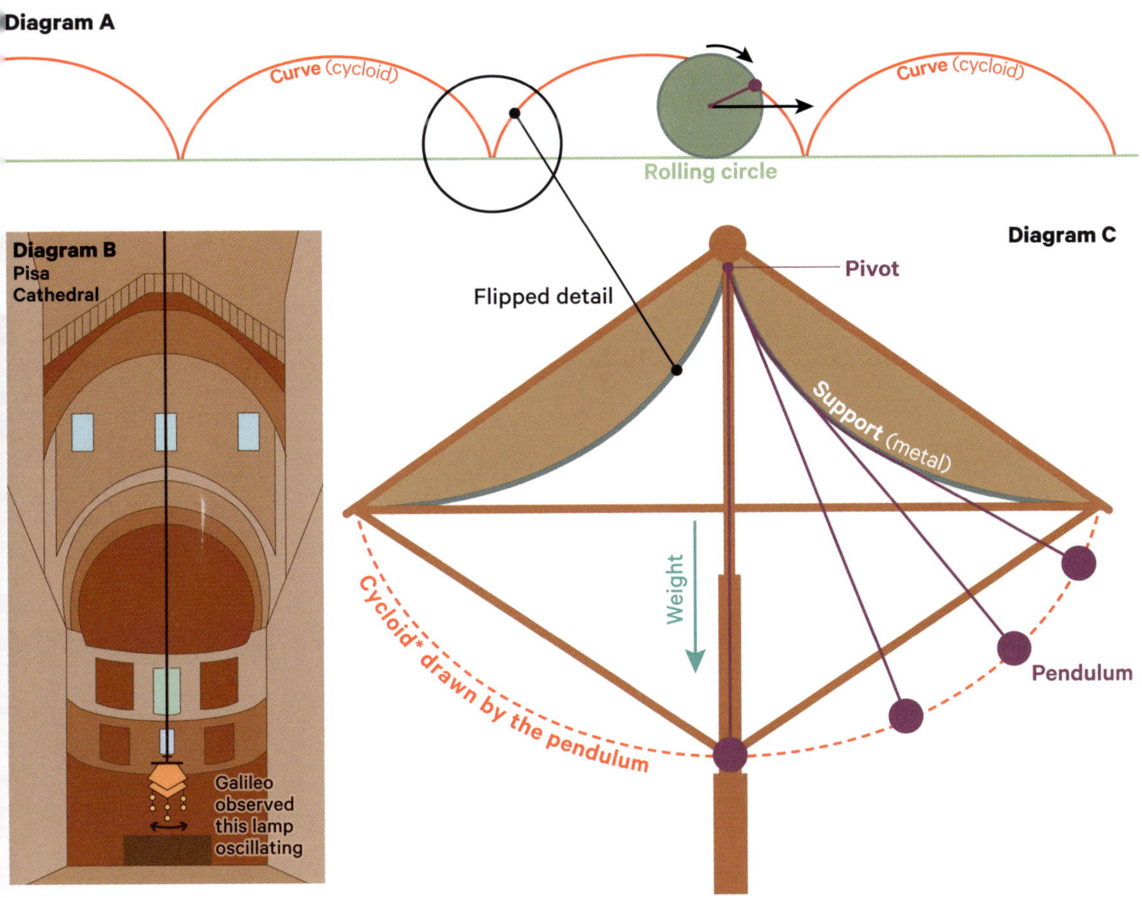

Diagram A

Diagram C

Diagram B
Pisa Cathedral

Measuring the oscillations of a pendulum

According to legend, Galileo observed the swinging motion of a hanging chandelier in Pisa Cathedral in 1602. Using the beating of his pulse to keep time, he noted the regularity of the chandelier's oscillations under the effect of gravity (diagram B). In this way, the idea of a relationship between a simple pendulum – a mass connected to a string – and the measurement of time was born. In 1658, Christiaan Huygens, a passionate horologist, realized that the period (the length of time taken to complete a single oscillation) depended on the width of the swing. To eliminate this source of imprecision, he set out to make a pendulum that would take precisely one second to swing, whatever its width. He discovered, through

'infinitesimal calculus', that the curve traced by the end of the pendulum was a section of a curve called a cycloid (diagram A).

Although this curve was not known in ancient times, it is easy to observe, because it is traced by a point situated on the edge of a wheel rolling along a plane. Huygens calculated that the shape that the string must trace to change its length in this way is the same cycloid. He developed an improved pendulum using this principle (diagram C) and built pendulum clocks that were more than fifty times as precise as any that had come before. He also invented the escapement mechanism, which regulates the mechanism of a clock and creates its characteristic tick-tock sound.

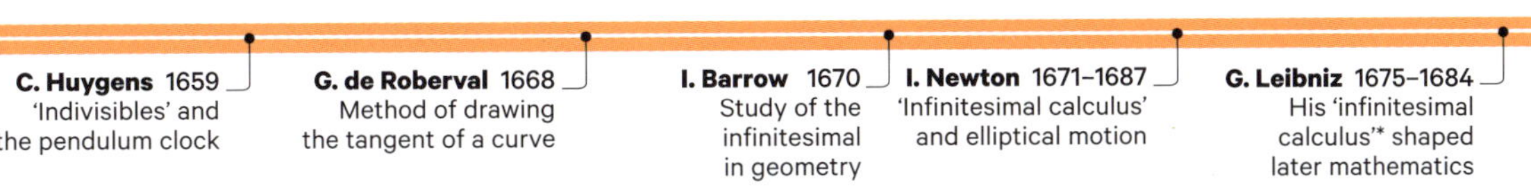

* Now split into differential calculus and integral calculus.

A timeline of light

The different stages in the history of light were each marked by a number of observations, hypotheses and contributions from at least one major figure. These developments were closely linked to advances in the understanding of the sky, particularly during the period that saw the emergence of modern science.

Ibn al-Haytham

In Cairo, this mathematician, astronomer and physicist described the role of the eye in forming an image that is then analysed by the brain. He established the laws of light rays being reflected and refracted and used a dark room to observe a solar eclipse.

1021

Ole Rømer

A Danish scientist based in Paris, he observed Io, a moon of Jupiter that was eclipsed by the planet once during each orbit of 1.77 days. This eclipse was sometimes regular, sometimes late, due to the changing distance between Earth and Jupiter, and the fact that the speed of light was finite; Rømer calculated it at 200,000 km/s. It was known that light travelled fast, but not infinitely so.

1676

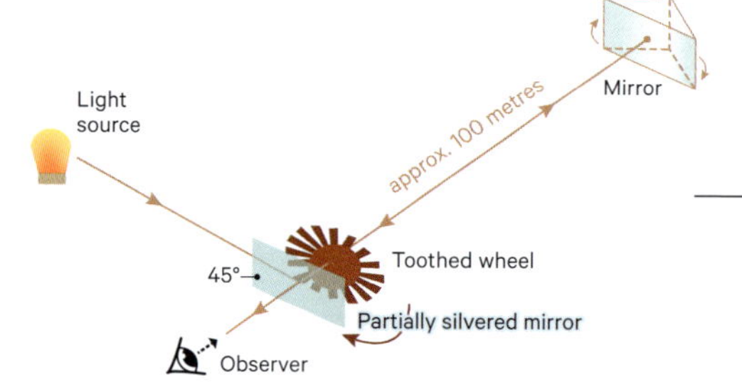

Hippolyte Fizeau

This French physicist and astronomer, who took the first photograph of the Sun, discovered and described the shift in the wavelength of light when its source was moving relative to the observer. He carried out an experiment to measure the speed of light in air, finding that it was 315,300 km/s, and proposed a method for measuring the diameters of stars using interference, which became very important in astronomy.

1850

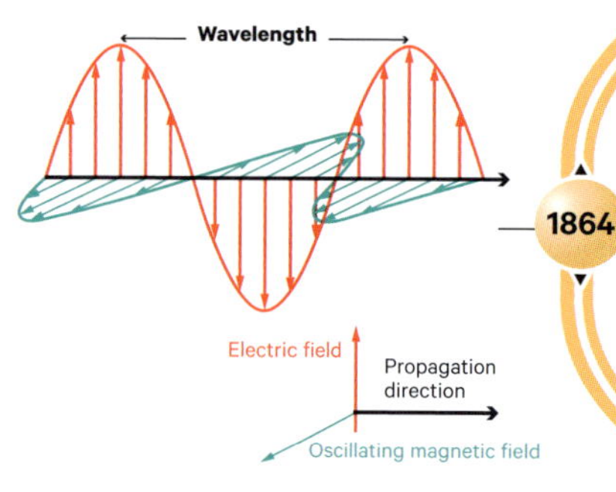

James Clerk Maxwell

This British physicist used mathematics to describe the relationship between electricity and magnetism, which experiments had already shown were connected. Light is an electromagnetic wave formed from oscillating electrical and magnetic fields, which moves through a vacuum at a speed that is directly related to the properties of electricity and magnetism. He predicted the existence of other kinds of light, which were then soon discovered (ultraviolet, X-rays, radio waves). These wavelengths of light came to revolutionize the understanding of the sky after 1940.

1864

1881

Albert Abraham Michelson

This remarkable experimental physicist, who moved from Prussia to the United States, was the first person to calculate the diameter of a star. He set out to show the motion of the Earth (30 km/s) relative to a light wave from a star, which it was believed required a medium to pass through. The negative result disproved the ancient idea that the interstellar void was filled with an invisible substance called aether.

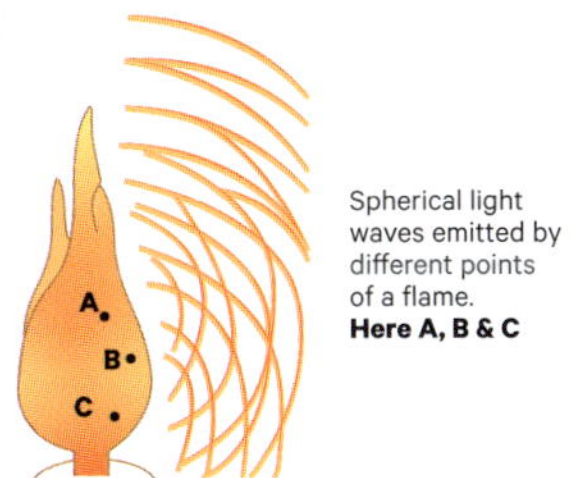

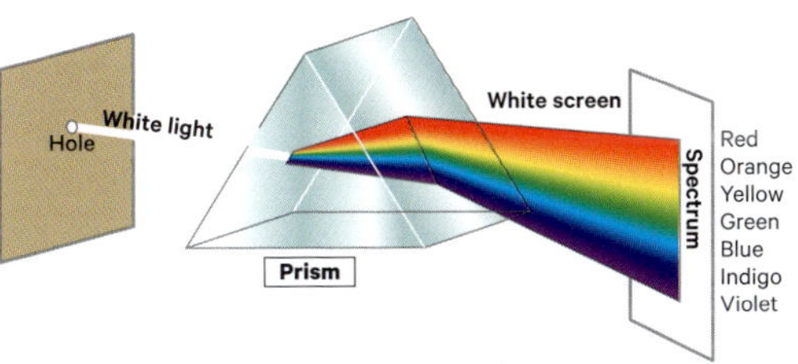

Christiaan Huygens

This Dutch astronomer drew on the laws of the refraction of light to make lenses. He observed the strange double refraction of light passing through a crystal of Iceland spar and found it could only be explained if light behaved like a wave. The first to suggest this hypothesis, he presented his *Treatise on Light* to the Académie des Sciences in Paris.

Isaac Newton

This English physicist set out a vision of the natural world that shaped both his own era and the centuries that followed. He separated the Sun's white light into a spectrum of colours. He thought of light as a flow of 'corpuscles' with different speeds and masses, and observed diffraction and interference colours, although he could not explain them. Although his work was groundbreaking, his conception of light was abandoned a century later, before making a comeback in the 20th century in a very different form.

1690   **1704**

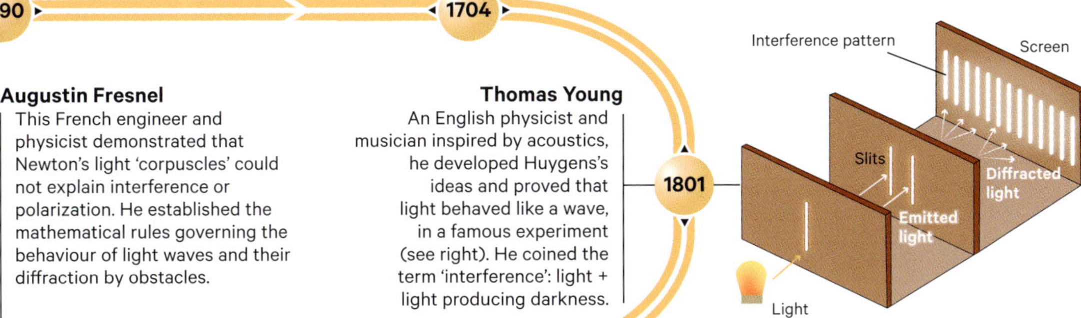

Augustin Fresnel

This French engineer and physicist demonstrated that Newton's light 'corpuscles' could not explain interference or polarization. He established the mathematical rules governing the behaviour of light waves and their diffraction by obstacles.

Thomas Young

An English physicist and musician inspired by acoustics, he developed Huygens's ideas and proved that light behaved like a wave, in a famous experiment (see right). He coined the term 'interference': light + light producing darkness.

1820   **1801**

Max Planck

This German physicist studied the way in which a hot object emits light, and established the law describing the distribution of wavelengths in this light, depending on the temperature of the object: the light emitted contains more short wavelengths at higher temperatures. This law made it possible to determine the surface temperature of a star from its spectrum. To explain this law, he could not avoid the idea that 'nature makes leaps' in exchanges of energy between hot objects and light. This was the start of quantum theory, which led to the birth of quantum physics in the 20th century.

Albert Einstein

In 1905, this German physicist interpreted the photoelectric emission of electrons by an object as being due to exchanges of quantum energy, and in 1926 he confirmed the existence of light 'particles', called photons, whose energy was directly proportional to their frequency. Aether was no longer required. In order to reconcile the principle of relativity of motion with Maxwell's description, he introduced a new kind of mechanics – special relativity – in which space and time are interconnected.

1900   **1905**

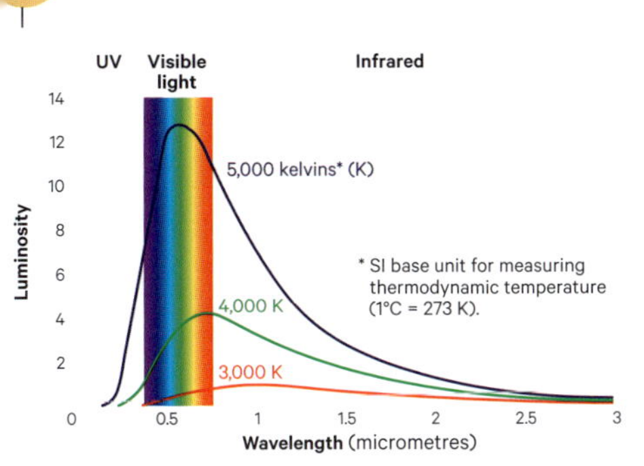

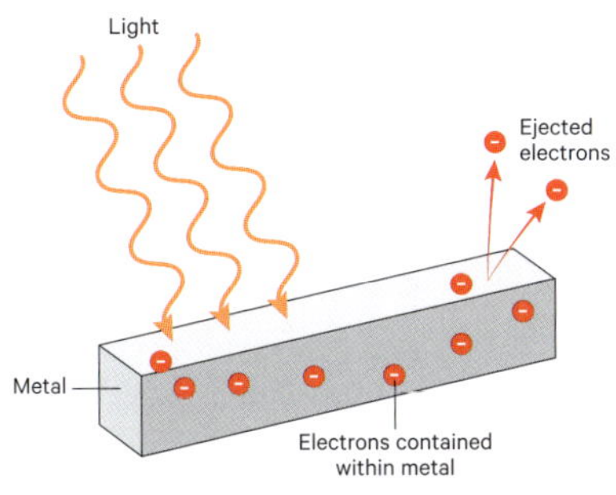

A better understanding of light

Visible light was the earliest known messenger from space. It was not understood, at the time, by Newton and Descartes, although Newton separated it into colours and Descartes analysed rainbows. Rømer established that it travelled at a fast but finite speed, which he measured. In the 19th century, Young and Fresnel proved that it was a wave. In 1864, Maxwell described it as an electromagnetic phenomenon.

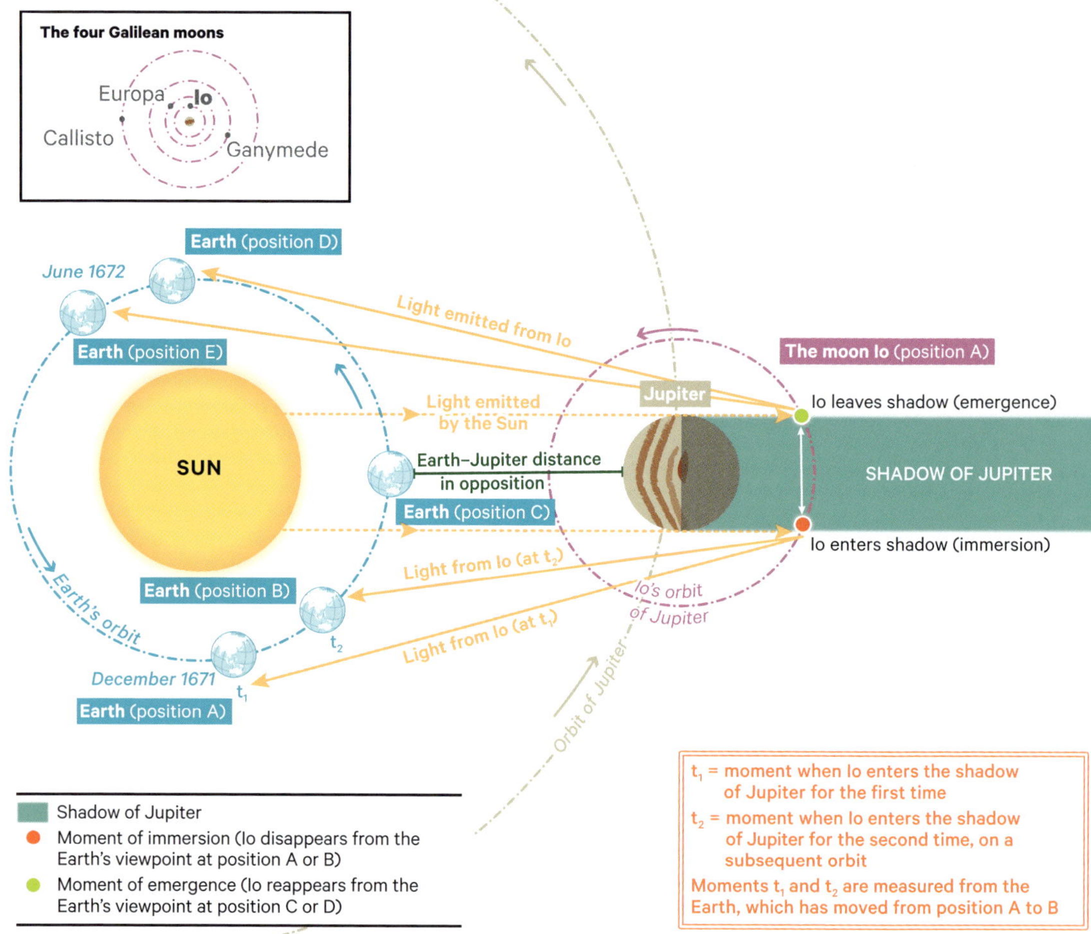

The speed of light

The earliest attempts to measure the speed of light were carried out by Galileo, but he only managed to conclude that it was too fast to be measured. In 1676, the Danish astronomer Ole Rømer observed an eclipse of Io, one of the moons of Jupiter, and noted differences in the time it took for Io's light to reach the Earth. When the Earth was moving away from Jupiter (positions D, E), the eclipse was late. Conversely, when the Earth was moving towards Jupiter (positions A, B), the eclipse was early. He realized that this time difference was caused by the difference in the distance travelled by the light. Therefore light must travel at a finite speed. This speed was not calculated until Huygens attempted it in 1678. He determined that it was 600,000 times greater than the speed of sound, at around 200,000 km/second.

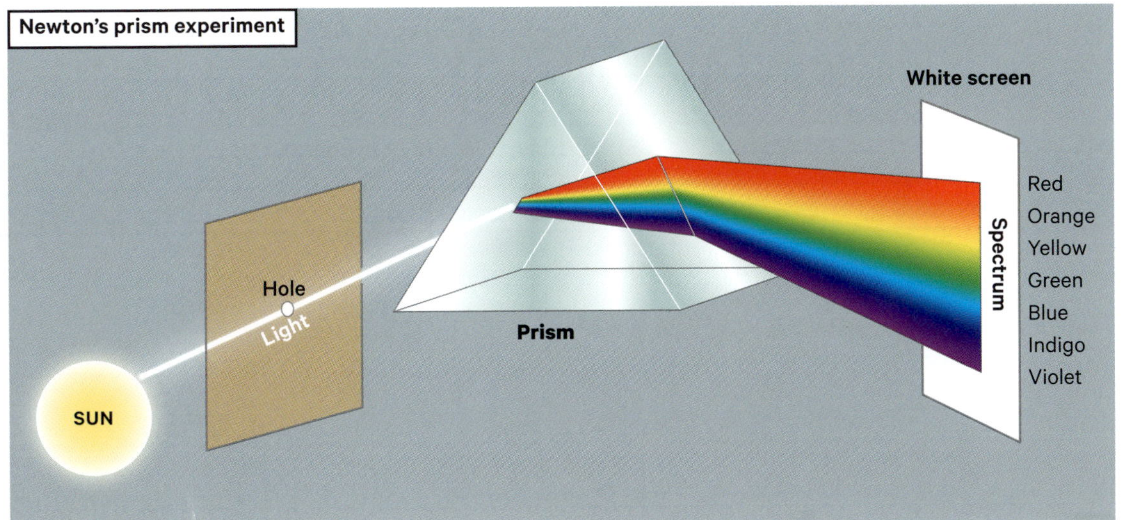

Light: a question of waves

Since ancient times, the trajectory of light had been described using the geometry of straight lines. Alongside this geometric description, Huygens developed a theory of light as a wave. It is based on the phenomenon of diffraction, which is also observed in waves on the surface of water: light 'scatters' when it encounters an obstacle. This contradicted Newton's theory of light particles. He theorized that light rays were made up of particles that he called 'corpuscles', with different masses and speeds. While Young and Fresnel's experiments supported Huygens, Newton had proposed in 1672 that white light was a combination of different colours, which he proved in his famous prism experiment (see above). The colours of the rainbow correspond to the light perceived by the human eye.

Other kinds of light were discovered in the early 19th century: infrared in 1800, then ultraviolet in 1801. Radio waves, X-rays and gamma rays were not detected until towards the end of the 19th century. Maxwell had predicted their existence when he stated that light was an electromagnetic wave. These different kinds of light each have their own wavelength (see below). The wavelengths of visible light range from 380 nanometres (nm) for violet to 780 nm for red.

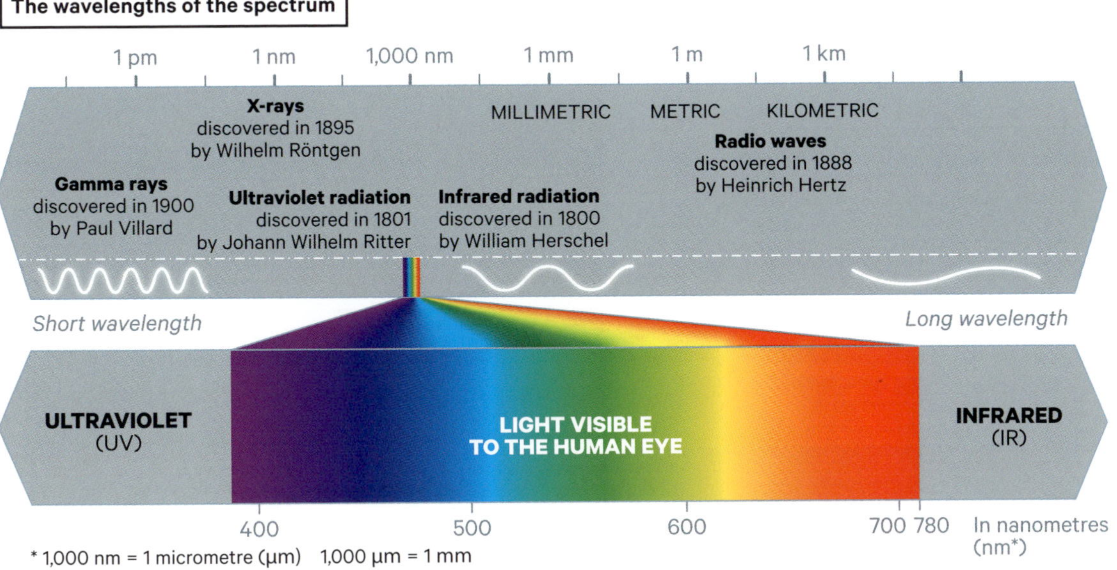

The physics of the Sun

The centre of the Earth's planetary system according to Aristarchus and Copernicus, and the source of gravitational attraction for the planets according to Newton, the physical nature of the Sun – which is 8 light-minutes away from the Earth – and the source of its energy remained a mystery until the mid-19th century. During a total eclipse, a vast corona can be seen around it. Black sunspots are sometimes seen on its surface, returning every eleven years.

Age
4.57 billion years

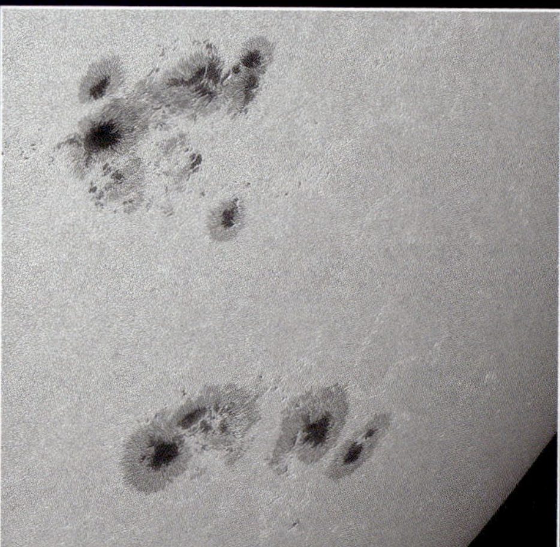

The gaseous nature of the Sun

It is a source of intense light, but is the Sun on fire? In 1811, shortly after the discovery that a light wave can oscillate in two directions (polarization), François Arago concluded from the non-polarization of light from the edge of the Sun that its surface could only be a gaseous body at a high temperature. In 1817, Joseph von Fraunhofer observed the spectrum of white light, dispersed by a prism, and saw black lines in it, which were then identified as being produced by chemical elements found on Earth (hydrogen, calcium, etc.). Later, Jules Janssen and Joseph Lockyer found an unknown element in the spectrum, which they called helium (from the Greek *helios*, Sun) and which was later discovered on Earth. In 1860, the relationship between the colour of the Sun and its surface temperature was established, with the latter being fixed at around 5,700 kelvin (K).

Surface temperature
around 5,700 °C

Mass
around 2×10^{30} kg = around 1,000 times the mass of Jupiter (mass of Jupiter = around 300 times the mass of Earth)

Average diameter
1,392,000 km
(around 110 times
the diameter of
the Earth)

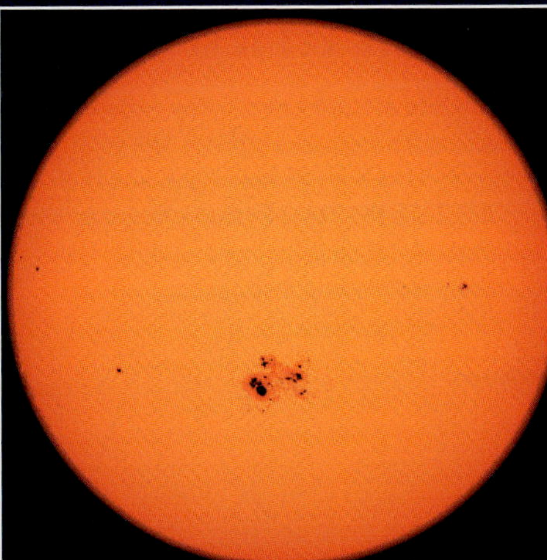

Sunspots

Changing spots on the surface of the Sun, seen by the naked eye through fog, were observed and recorded as early as 28 BCE in China, and later in some descriptions in the West, albeit rarely – most likely because of Aristotle's claim that the skies did not change. Galileo and Scheiner observed them in 1610, and after that there was no longer any doubt. The number of sunspots changes periodically over the course of a solar cycle of eleven years, as established by Wolf in 1849. A low number of sunspots called the Maunder minimum was recorded by Cassini between 1645 and 1715, during a period known as the 'Little Ice Age'. A sunspot is a colder region (4,200 K) on the Sun's surface, where energy is blocked by a magnetic field.

Solar corona

When the Moon blocks the Sun's face during a total eclipse, a faint halo of light appears around it: this is called the corona. It was first described by Plutarch in 98 CE. Edmond Halley (1715) and later François Arago (1842) believed it was the Moon's atmosphere. Observations made on Earth showed that the corona is the extended atmosphere of the Sun. This thin cloud is made up of hydrogen, split into protons and electrons by a temperature of a million kelvin, and cold dust particles, which extend a long way from the Sun, creating the 'false dawn' phenomenon known as zodiacal light. Coronal gas is 'blown', creating a wind that reaches the Earth and causes the aurora borealis. This image comes from the Institut d'Astrophysique de Paris (IAP).

Luminosity
around
4×10^{26} watts

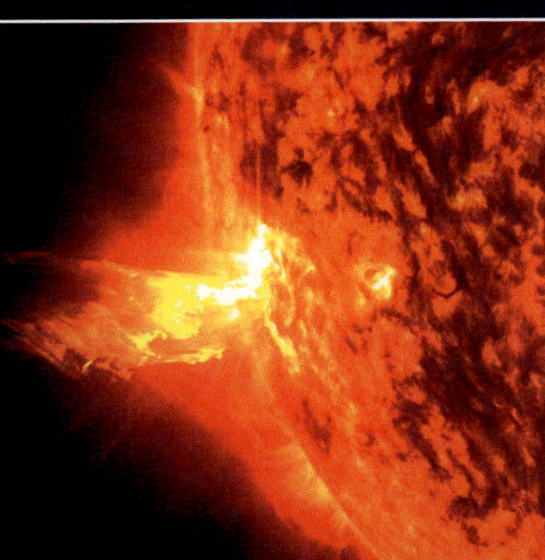

Solar prominences

Eruptions called prominences have been observed during total eclipses since the 19th century, extending beyond the edge of the Sun. The first photograph of prominences, taken by Angelo Secchi in 1860, showed they came from the Sun and not the Moon. They were later observed by balloon expeditions: by Jules Janssen in Algeria in 1870, and by Dmitri Mendeleev in Russia in 1887, at a height of 3,500 m. In 1937, Bernard Lyot invented the coronagraph, making it possible to observe these eruptions when there was no eclipse, without being dazzled by the Sun's light. In the 20th century, it became possible to take X-ray images (see left). The relationship between prominences, sunspots and the shape of the corona was established, showing that solar activity was connected to its magnetic field.

Stars and spectral analysis

Spectral analysis of light transformed astronomy into astrophysics. In the early 19th century, astronomers observed that some wavelengths were absent from the solar spectrum. This led to major discoveries: the Sun and stars are made of gas, and their chemical elements, classified by Dmitri Mendeleev, are the same as those found on Earth. The emission of light by a hot object was understood. Telescopes were equipped with spectroscopes, which measured the temperature, velocity and composition of stars.

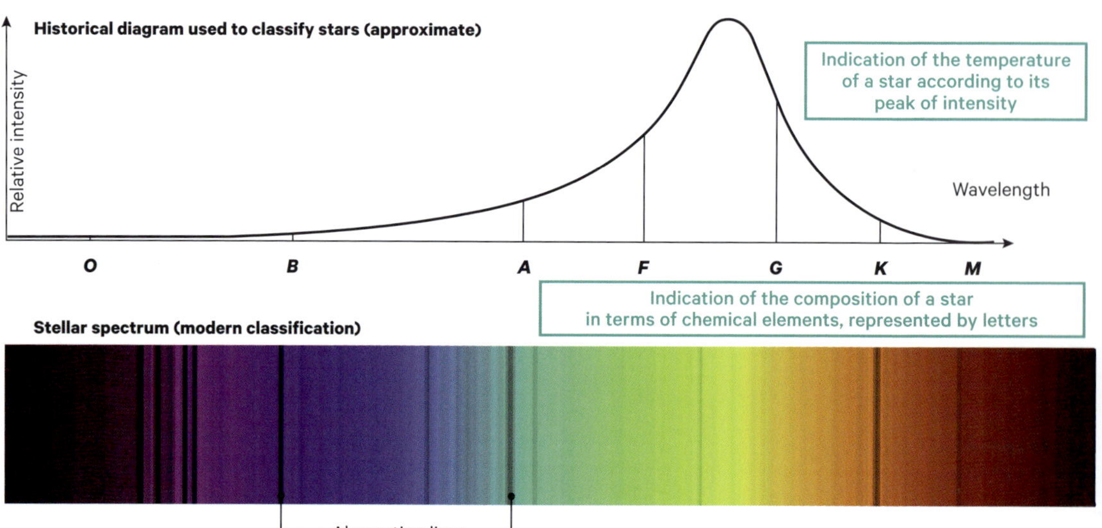

▲ The light spectrum of a star

This spectrum is a continuous distribution of light with black lines at certain wavelengths. In 1865, the chemist Robert Bunsen connected the positions of these lines, also observed in a laboratory, to certain elements (hydrogen, iron). These lines made it possible to identify those elements in the stars (spectroscopy). Close analysis of these black lines also led to other discoveries: the amount of the element present, the magnetic field, etc. Gustav Kirchhoff (1860) showed that there was a relationship between the most intense wavelength of light being emitted and the temperature of a hot object. Towards the end of the 19th century, photography, with its long exposure times, made it possible to see stars that were much less bright.

▼ The colour and temperature of stars

In a spectrum, Gustav Kirchhoff established the relationship between the wavelength being emitted at maximum intensity and the temperature of the emitting object: when an iron is heated up, first it turns reddish, then, when it is heated further, it turns white because it starts emitting more blue light. This was immediately applied to the stars, because their apparent colour shows their surface temperature, as can be seen from the diagram opposite, showing stars in Ursa Major. The spectrum also made it possible to understand velocity relative to the observer. In 1920, Albert Michelson directly measured the diameter of a star. However, its size can also be determined from its surface temperature, by measuring the amount of light emitted.

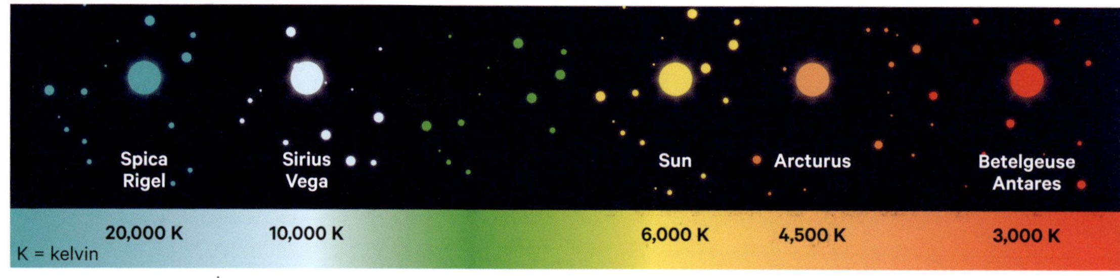

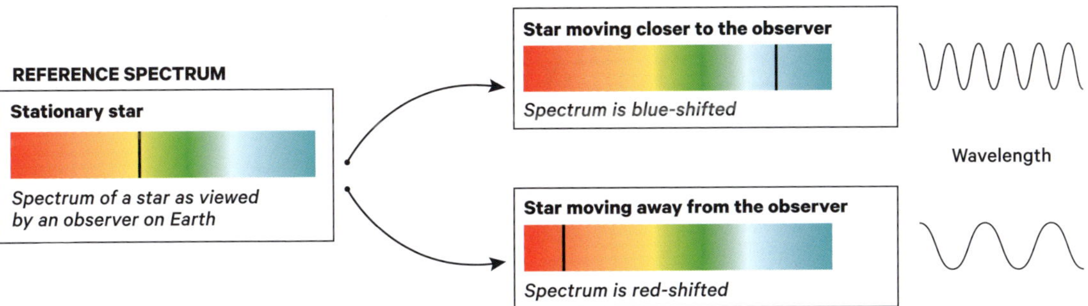

Spectral shift: the Doppler-Fizeau effect

In a spectrograph, the light of a star is dispersed by a prism or a grid (think of the rainbow reflections on the surface of a CD). The spectrum, which can be seen with the naked eye or photographed, is made up of bands of different colours. If a star is moving towards or away from the Earth (radial speed), this produces a shift in wavelength: towards the red end of the spectrum if it is moving away, or towards the blue end of the spectrum if it is moving closer. The size of this redshift or blueshift is proportional to the speed of movement. This method only gives the speed in the direction of sight.

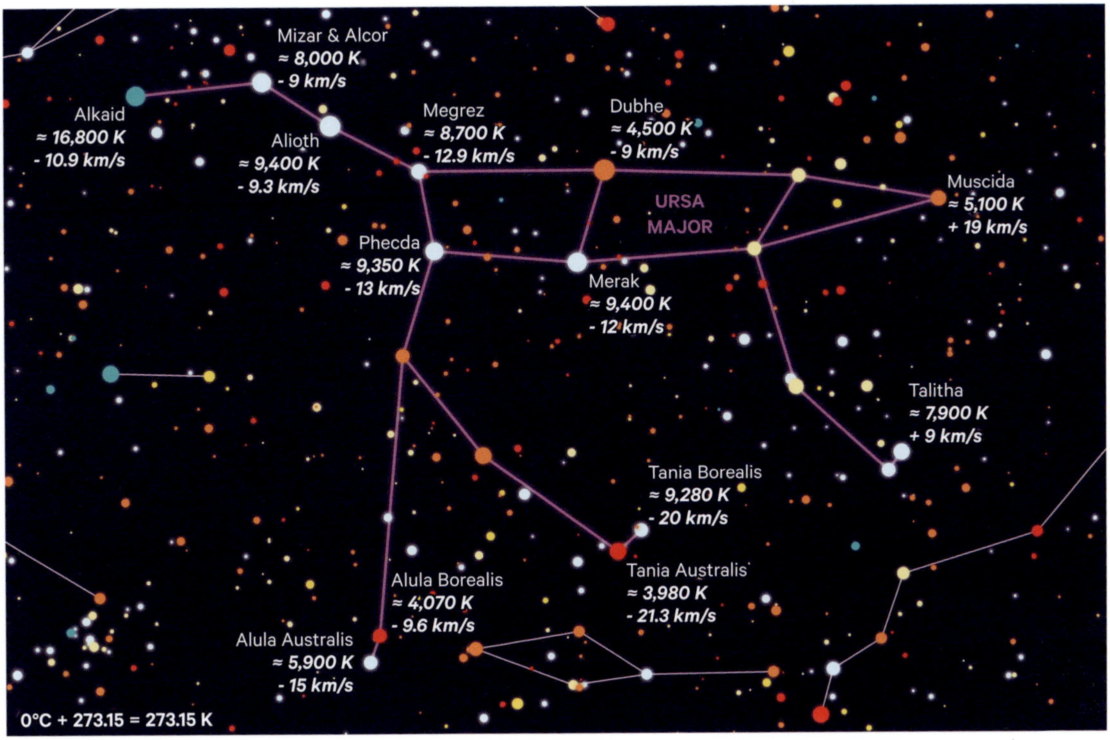

Astrophysics in Ursa Major

In this diagram of the constellation of Ursa Major, the brightest stars are identified by their name, the speed at which they are moving towards (-) or away from (+) the Earth, as measured using the Doppler-Fizeau effect and their surface temperature, expressed in kelvin (K). This unit of temperature starts at absolute zero and is therefore useful for calculations, and can easily be converted to degrees Celsius. Viewing this constellation, or even better, Orion, with binoculars is a simple way to see the differences in colour of the stars.

**Surface temperature
of stars** (in kelvin)

- Approx. 3,000
- Approx. 4,500
- Approx. 6,000
- Approx. 8,000
- Approx. 10,000
- More than 20,000

- 9 km/s Radial velocity of stars in comparison with the Sun

Measuring time without the sky

Longitude can be worked out from local time registered by a clock at sea, relative to a standard time. In 1675, Greenwich Mean Time (GMT) was created. Throughout the 18th century, competition between France and the United Kingdom eventually led to the invention of the marine chronometer, which had a tiny spiral balance spring and was accurate to within 3 seconds per day. James Cook took one with him on his voyages to the Pacific.

Longitude at sea

Galileo demonstrated the regularity of a pendulum's oscillation, and Huygens used this to synchronize a pendulum clock, which was sixty times more accurate than the clocks used since the Middle Ages. Local time, as determined using a sundial, was used everywhere. It could not be used to compare the time in two places, situated on two different meridians. In order to achieve this, a lunar eclipse in the Mediterranean was used as a 'clock'. Comparing two times was necessary to determine longitude, which was vital for navigation. London was used as the reference time zone, by which chronometers taken on ships were set. Long journeys required equipment that could tell the time accurately. This was not invented until the early 19th century.

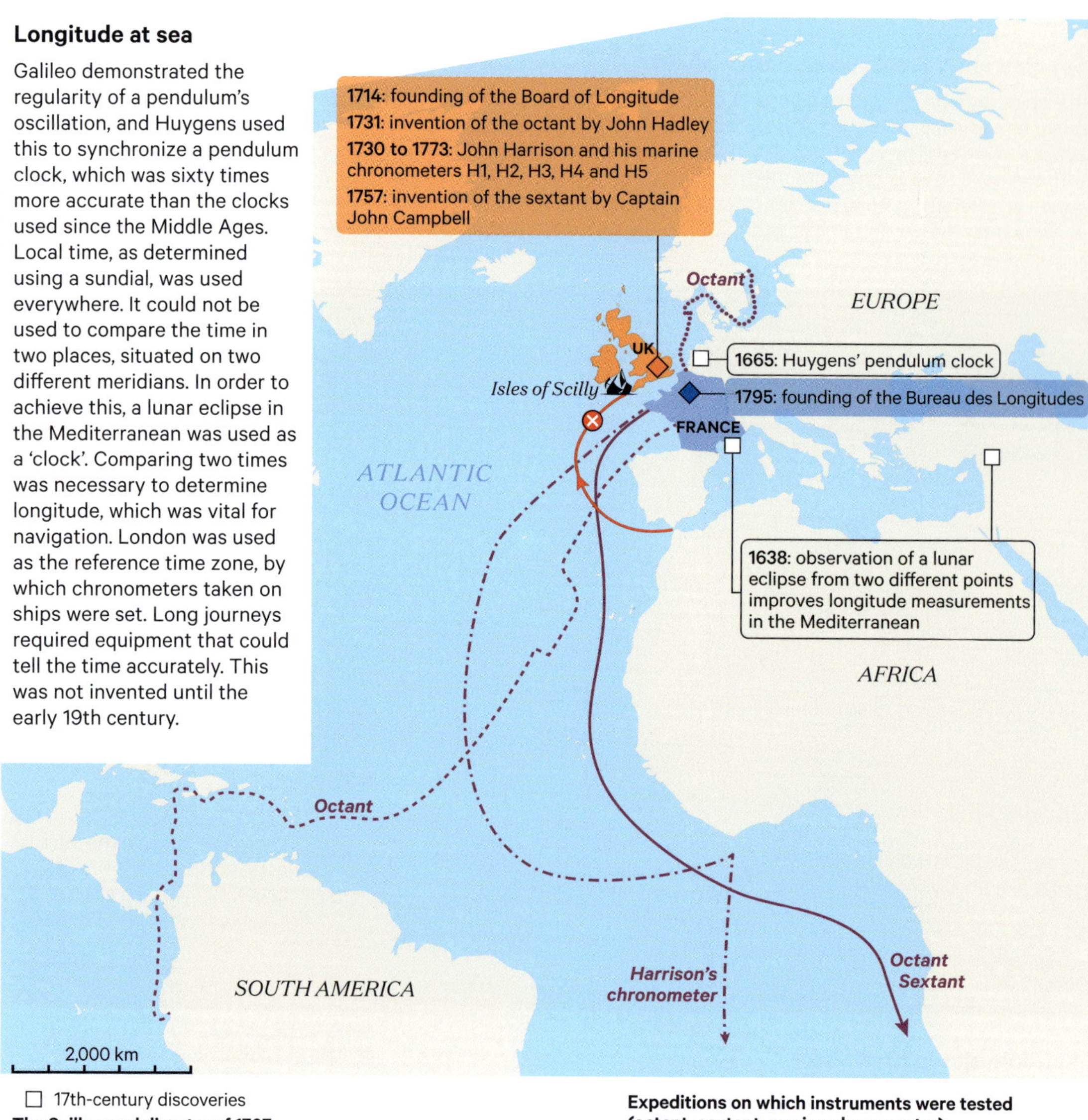

17th-century discoveries
The Scilly naval disaster of 1707
Route of English fleet
Estimated position of the ships by their pilots just before disaster
Shipwreck
18th-century developments in the establishment of longitude
In Britain   In France

Expeditions on which instruments were tested (octant, sextant, marine chronometer)
- - - Expedition to Peru (La Condamine, 1735)
······· Expedition to Lapland (Maupertuis, 1736)
—— Voyage to southern hemisphere (Lacaille, 1751)
-·-·- Cook's 2nd voyage to the Pacific (1772)
Octant Instrument tested

The sextant

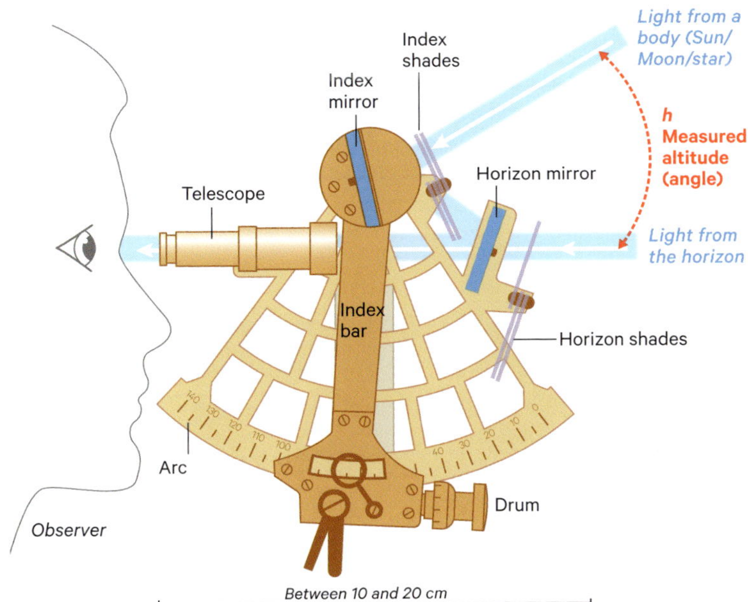

Measuring latitude

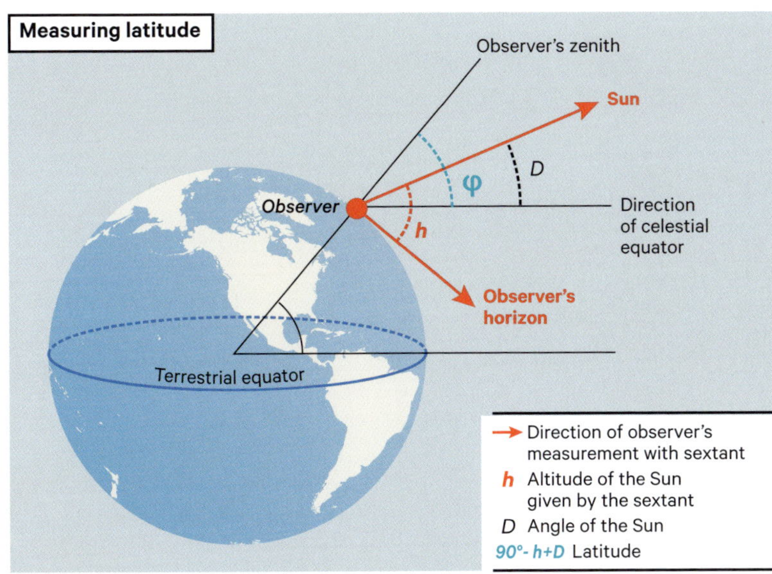

Time on land or at sea

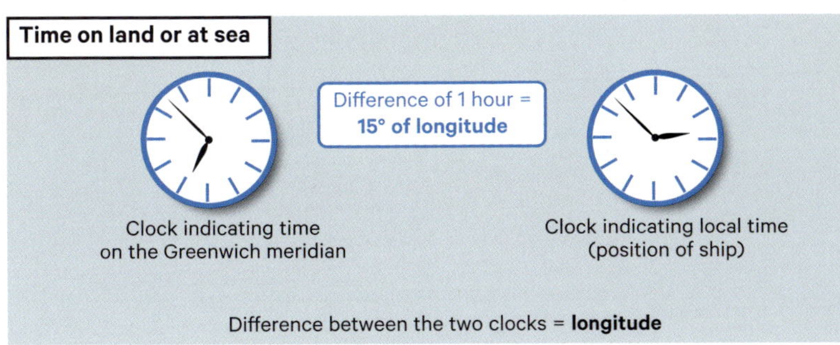

Navigating with the sky

Any traveller on land or sea needs to know their position in terms of latitude and longitude. Latitude is easy. At night, in the northern hemisphere, it can be calculated from the height of Polaris above the horizon and the time when certain bright stars appear over the southern horizon. In the southern hemisphere, it can be found using the Southern Cross near to the pole and the visible stars in the north. During the day, travellers can use local noon, the time when the Sun reaches its highest point. These measurements were originally taken with an astrolabe, which was replaced in the 17th century by the octant, then the sextant. This measures the angle between the Sun and the horizon at noon, and is accurate to within an arc-minute, a difference in latitude of one nautical mile (1,852 m). When combined with the date (the height of the sun depends on the season), this can be used to calculate latitude.

Finding the longitude of a place, relative to a reference meridian, is harder. If the same event – a solar or lunar eclipse, or the position of the Moon relative to the Sun – can be observed from this place and from the reference longitude, the difference between local times gives the longitude – as long as you have a good clock set to local noon. In the 18th century, expeditions took a Huygens pendulum clock and tables of the moons of Jupiter or the Moon, drawn according to local time in Greenwich or Paris. It is not possible to observe these moons at sea, from a ship. Advances in marine chronometers, which could remain accurate over a number of weeks, were made by clockmaker John Harrison. In 1761, he was able to keep time to within a few seconds, a difference of around a mile in longitude.

Universal time

In the 19th century, advances in international transport and trade required that countries should each have their own time zones, as well as a globally agreed reference time zone. Universal time, then regional time zones, were established, calculated using a measurement of time determined by the rotation of the Earth. Advances in the precision of mechanical clocks revealed irregularities in the Earth's daily rotation, with seconds being gained or lost. In 1967, this led to a new definition of the second.

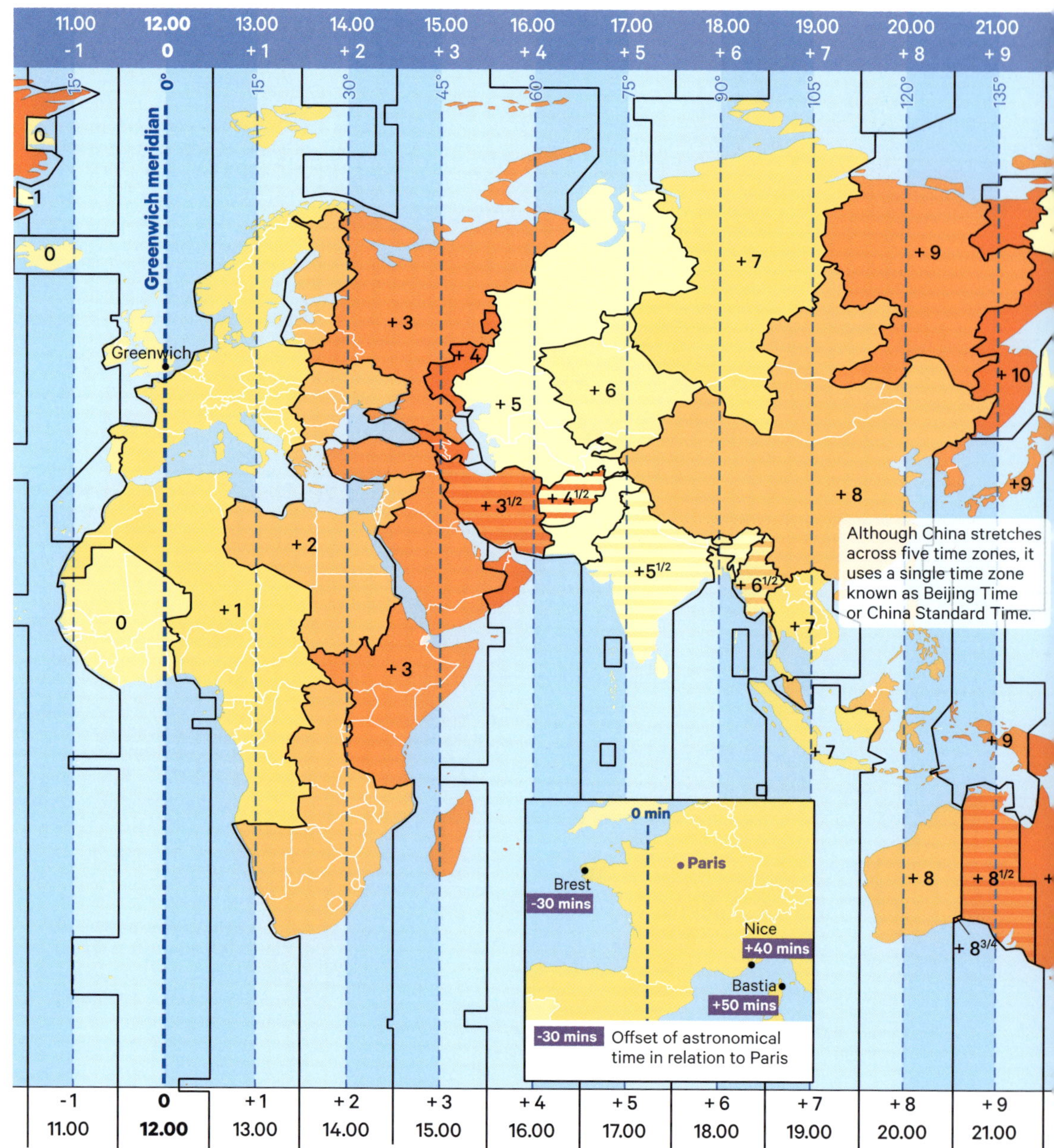

Universal time and local time

The invention of the telegraph and railways required standardized measurements of time. In 1847, GMT (Greenwich Mean Time), set by solar noon at Greenwich Observatory in London, was adopted across the UK. In 1884, at an international conference in Washington DC, the world was divided into 24 time zones, each covering 15° of longitude with the 0° meridian at Greenwich. Each country then adapted its time zone for its own territory. Because the rotation of the Earth is slightly irregular, GMT was no longer used as a baseline from 1972 onwards. International Atomic Time (TAI) was created, which is not dependent on the Earth's rotation. In order to keep this in line with the length of a day, Coordinated Universal Time (UTC) is used, and reset if necessary from TAI.

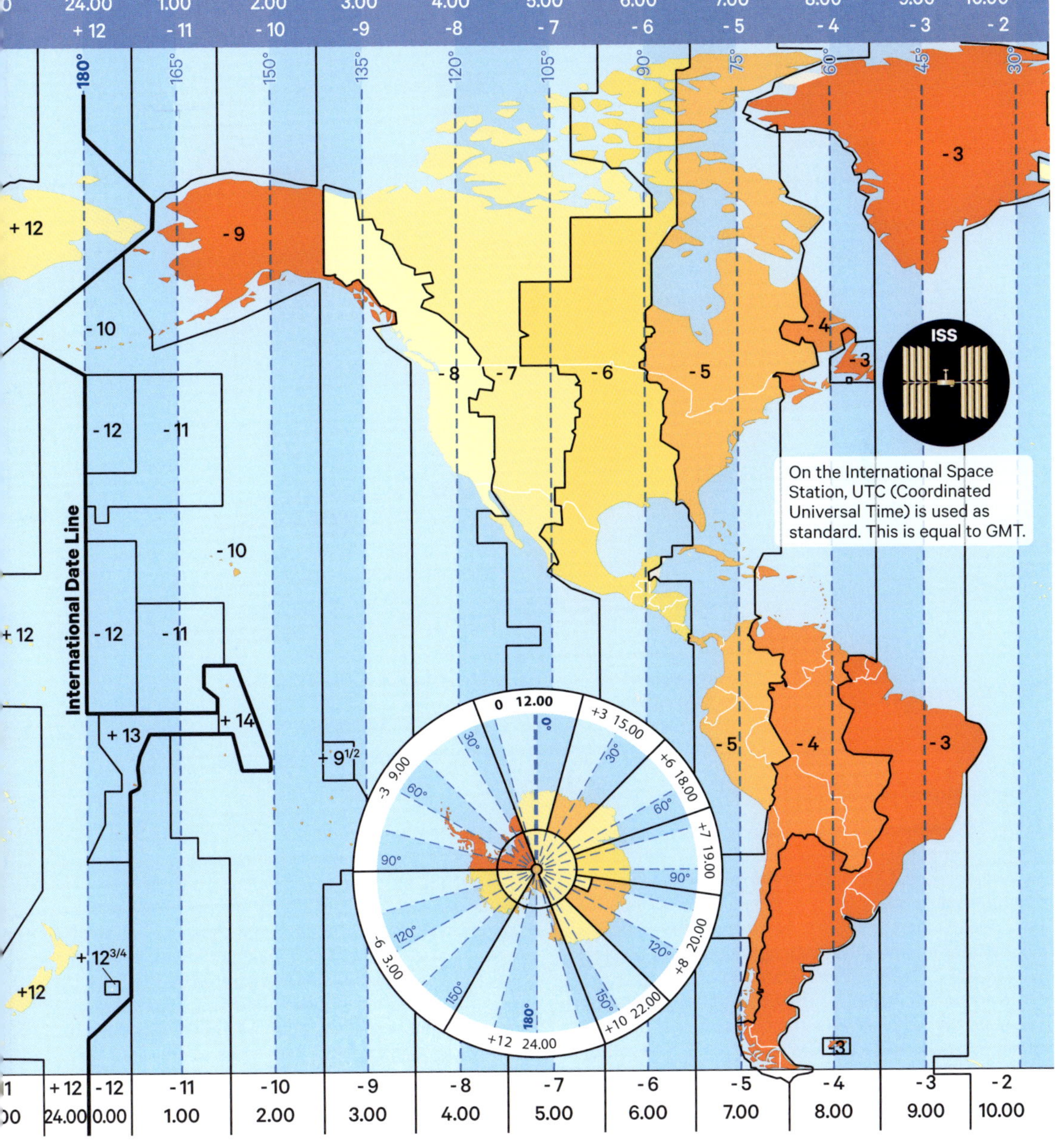

The Astrographic Catalogue

This global endeavour, which was never completed, was the first international scientific project. Launched at Paris Observatory in 1887, it included 18 countries. The aim was to photograph the whole of the sky using telescopes with a diameter of 30 centimetres, then to draw up complete maps of stars up to the 14th magnitude – two million stars. Their positions were accurate to the arc-second, as the precision of measurements was limited by atmospheric turbulence. In the 20th and 21st centuries, the satellites Hipparcos and Gaia continued this project, taking measurements that were more than a thousand times more accurate.

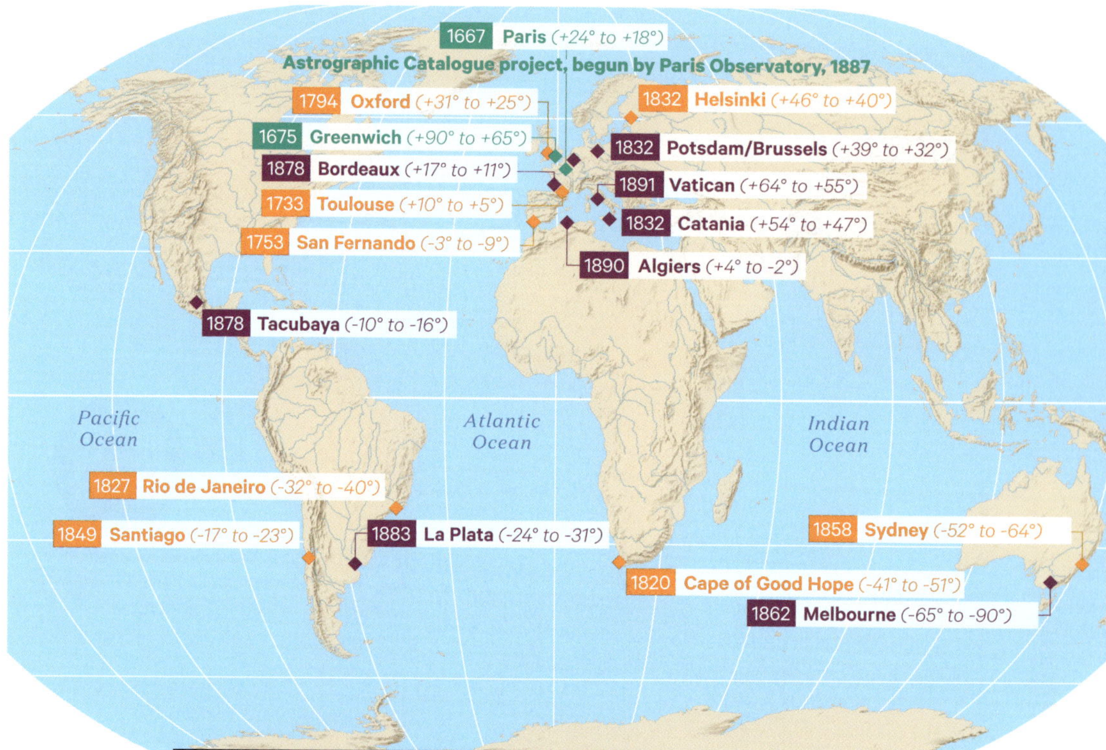

A huge global project

In 1887, the Astrographic Catalogue project, the first major international initiative of its kind, was launched by Admiral Ernest Mouchez at Paris Observatory. Photographing the entirety of the celestial sphere, the catalogue set out to accurately measure the equatorial coordinates and brightness (magnitude) of at least two million stars. The telescopes required had to have a large enough diameter for the photograph to show all the stars that were a million times less bright than the dimmest stars visible to the naked eye. Some twenty observatories were equipped with the same instrument, to guarantee precision, as well as a machine to analyse

the photographic plates. On these plates, the images of stars have a positional accuracy of around 0.5 arc-seconds due to the Earth's atmospheric turbulence, but the position of their centre was more precisely measured. Each image shows a section of the celestial sphere. This catalogue, which included 22,000 photographs, was interrupted by the First World War and remains incomplete. A comparison of the Catalogue's measurements of the positions of stars, published in 1958, with those established by the satellite observatory Hipparcos (1989–1993), demonstrates that a large number of stars have moved.

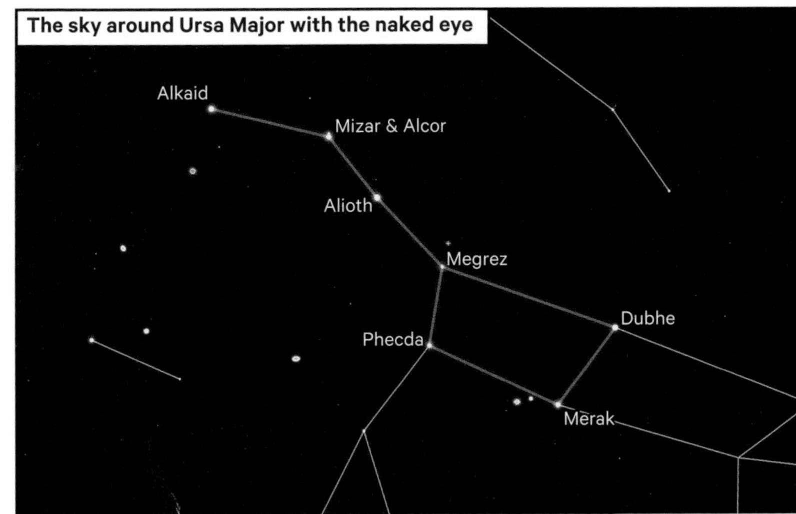

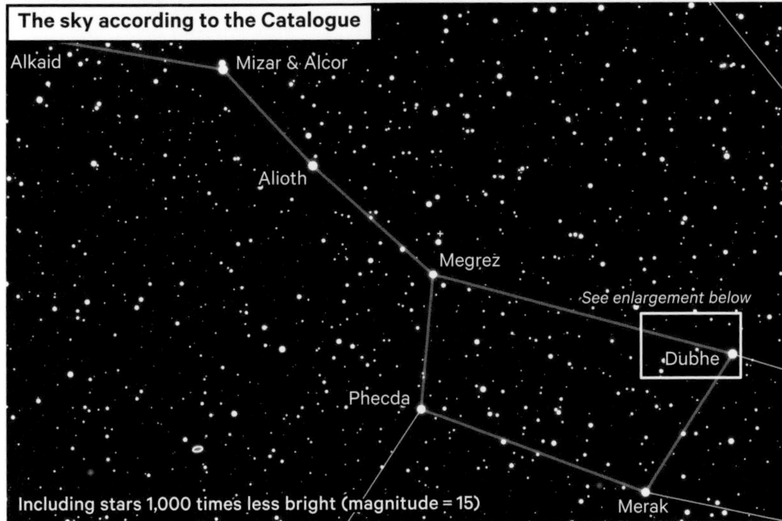

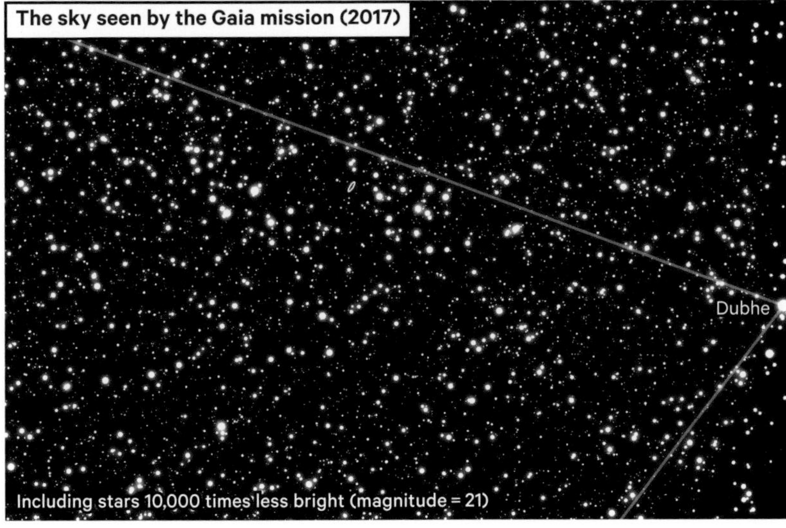

An abundance of stars

On a moonless night, the naked eye, with the small opening of its pupil (7 millimetres at night), can only make out a fairly small number of stars, as seen here in the constellation of Ursa Major. Binoculars or a telescope collect light through a much larger opening, and can therefore see less bright stars. As the eye only has an exposure time of around 0.1 seconds, a photographic exposure of many minutes increases the sensitivity even further, as long as the instrument is on a motorized 'equatorial' mount so it follows the Earth's diurnal rotation. This became possible after 1880 with the invention of photographic plates, which revealed ever more stars. The position of a few stars in the photographed area is determined independently (with a transit telescope) and serves as a reference for all the other stars, thanks to a 'measuring machine'. This explores the finished photographs star by star and determines the area's astrometry. These astrometric measurements are very important for the study of the stars, so much so that in 1989 the Hipparcos satellite mission was launched, its small telescope tasked with drawing up a new map of the sky. It was so successful that in 2013 Europe launched the Gaia space observatory, which could capture the stars in the Milky Way with greater sensitivity and accuracy, a project that would take twelve years. Of the hundred billion stars in the galaxy, Gaia measured around 1.7 billion. In the centre diagram, the highlighted region shows where the stars in the Gaia catalogue are found. It allows us to see stars that are ten thousand times less bright than those in the image at the top of the page.

The Westernization of the sky

European science spread throughout the 19th century, particularly to the USA and the colonies of the British empire. Precolonial scientific knowledge almost always disappeared or became invisible (Africa, America, Oceania), except when intellectual similarities made it easy to amalgamate (India, Japan). Although China had a strong scientific tradition, it did not often collaborate with Western science. In the late 20th century, the globalization of science, still dominated by the US and Europe, left many less wealthy countries behind.

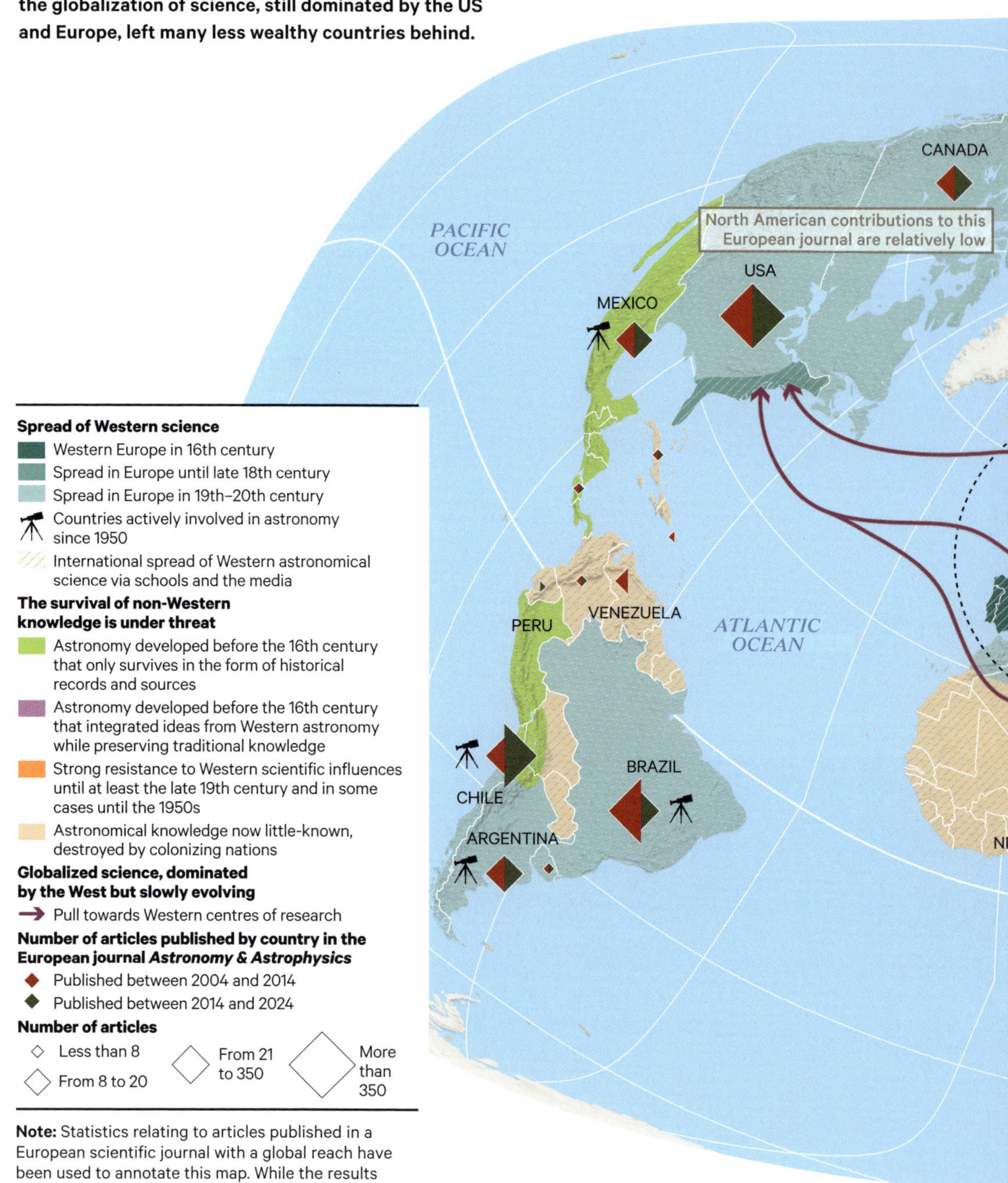

Note: Statistics relating to articles published in a European scientific journal with a global reach have been used to annotate this map. While the results are interesting, they are necessarily biased.

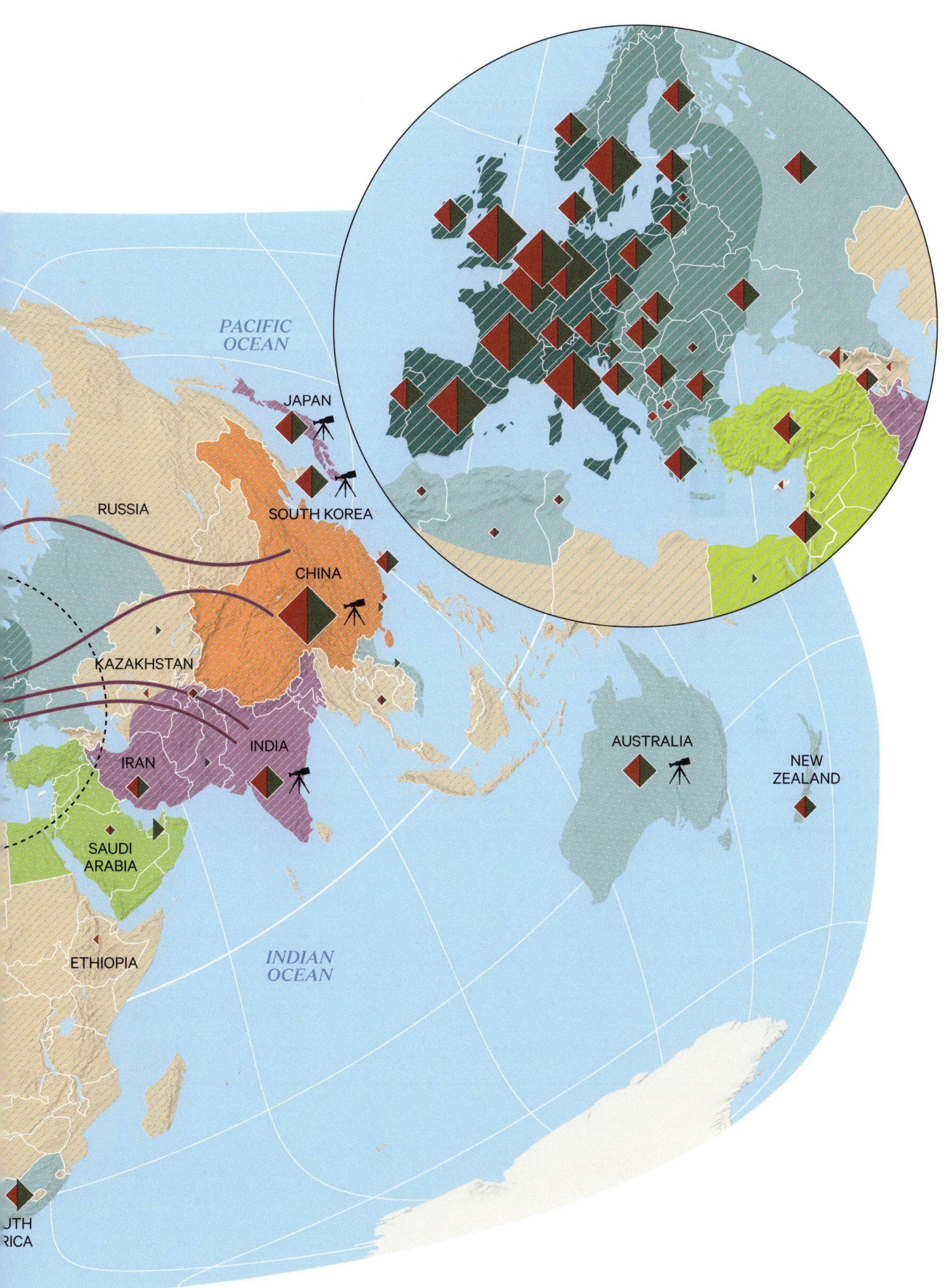

PACIFIC
OCEAN
JAPAN
RUSSIA
SOUTH KOREA
CHINA
KAZAKHSTAN
INDIA
IRAN
SAUDI
ARABIA
ETHIOPIA
INDIAN
OCEAN
AUSTRALIA
NEW
ZEALAND
SOUTH
AFRICA

4

Understanding the universe (20th–21st century)

Advances in scientific instruments and a new understanding of physics led to a radical transformation. The universe as we understand it expanded when a multitude of new galaxies were discovered. Questions about the origin and nature of time, both physical and philosophical, were studied and new concepts were introduced to the realm of physics. The nature of light, almost our only means of receiving information from space, was better understood. Other forms of information were detected, such as gravitational waves, paving the way for future revolutions. The discovery of countless planetary systems in the galaxy rekindled the question of whether there is life elsewhere in the universe. The digital revolution allows this new knowledge to be translated into images that are accessible around the world.

A new kind of physics is born

Albert Einstein [°]
(1879–1955)
This German physicist resolved the contradictions between classical mechanics and electromagnetic waves using a revolutionary approach. Time and space are relative; mass and energy are exchanged; nothing can travel faster than the speed of light. He also established the physical laws governing the energy of light quanta.

1900
German physicist Max Planck [°], studying radiation from hot objects, discovered that energy was exchanged between light and matter through 'jumps' or 'leaps'.

Georges Lemaître
(1894–1966)
In 1927, this Belgian cosmologist and priest's studies of relativity led him to independently conclude that the universe was expanding. After Hubble's observations in 1929, he hypothesized a 'primeval atom' from which the universe expanded: an early formulation of the Big Bang theory.

1922
Russian cosmologist Alexander Friedmann used general relativity to describe the theory that the universe was expanding.

Edwin Hubble
(1889–1953)
This American astronomer observed nebulae and proved they were distant galaxies by studying their variable stars. He classified them according to their shapes. In 1929, he studied the redshift of nebulae and established what is now known as the Hubble-Lemaître constant.

1925–1950
Erwin Schrödinger [°] (German) and Paul Dirac [°] (British) established the laws of quantum physics and their fundamentally random nature.

1931
Karl Jansky invented radio astronomy. This US physicist and radiocommunications engineer discovered radio waves emitted by the Milky Way.

Hendrik van de Hulst
(1918–2000)
This Dutch astrophysic[ist] predicted the existence [of] the hydrogen line, a thi[n] layer of neutral interste[llar] hydrogen. He later help[ed] to map it in our galaxy using radio astronomy.

1956
British astronomer Martin Ryle [°] combined multiple radio telescopes to create the first interferometer, producing higher-quality images of astronomical phenomena.

1953
The laser and maser (producing microwaves) were invented. Both use the stimulated emission of electromagnetic radiation to create a focused beam.

Margaret Burbidge
(1911–2020)
British astrophysicist. In 1957, working with Geoffrey Burbidge and William Fowler, she discovered that chemical elements were created by nucleosynthesis in stars. She actively campaigned against discrimination towards women in astronomy.

1958
The use of silicon as a semiconductor and advances in microtechnology began to revolutionize electronics. Light receptors in telescopes became more powerful.

1964
US radio astronome[r] Arno Allan Penzias and Robert Woodro[w] Wilson [°] identified radiation coming from beyond the Milky Way galaxy, in every direction.

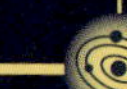

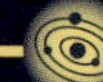

Nobel prizewinners:
o one prize
oo two prizes

1903
Marie Skłodowska-Curie oo was a Polish physicist. Working in Paris, where radioactivity had recently been observed for the first time, she discovered radioactive substances such as radium.

Henrietta Swan Leavitt
(1868–1921)
This American astronomer studied Cepheid variable stars, whose luminosity varies periodically, and discovered a fundamental relationship between their period and their luminosity.

1912
French scientist Jean Perrin o proved the existence of atoms through experiments studying the random motion of particles suspended in gas or liquid.

1913
Danish physicist Niels Bohr o, the founder of quantum physics, discovered the structure of atoms.

1916
A telescope with a mirror with a diameter of 2.5 m was installed on Mount Wilson in California, making major astronomical discoveries possible.

Karl Schwarzschild
(1873–1916)
This German astrophysicist advanced the study of general relativity, demonstrating the curvature of space, the possibility of black holes, and how gravity and light would interact there.

1916
Einstein published his general theory of relativity, following his special theory of relativity of 1905. Relativity defines the relationship between curved space, time and the presence of mass–energy.

Subrahmanyan Chandrasekhar o
(1910–1995)
Born in India, this astrophysicist used relativity to describe the structure and evolution of stars. At the University of Chicago, he discovered the Chandrasekhar limit, a maximum mass beyond which a star will collapse to form a neutron star.

1933
Irène Joliot-Curie o and her husband Frédéric o, discovered artificial radioactivity. Their studies prefigured Otto Hahn's discovery of nuclear fission.

1939
Working in the US, German physicist Hans Bethe o established that the light from a star is created by nuclear reactions inside its core.

1945
Birth of computer science: the British mathematician and cryptologist Alan Turing established its basic foundations.

George Gamow o
(1904–1968)
Working with Ralph Alpher, this Soviet physicist, who had moved to the US, proved that hydrogen and helium were created during the Big Bang. He also predicted the existence of cosmic microwave background radiation.

1940
In the US, Enrico Fermi o discovered a formula to describe quantum systems, studied radioactivity, and built the first nuclear reactor.

1989
The Hubble Space Telescope, with a mirror diameter of 2.4 m, was launched by NASA and went on to play a key role in many astronomical discoveries.

Anne-Marie Lagrange
(born 1962)
Building on the work of astronomers Michel Mayor o and Didier Queloz o, who developed a spectroscopic method and discovered the first exoplanet, French astrophysicist Anne-Marie Lagrange created the first direct image of an exoplanet.

2023
The successor to the Hubble telescope, the James Webb Space Telescope, with its 6.4 m mirror, is able to detect infrared light from distant galaxies.

A universe of galaxies

In 1915, Einstein proposed a new physics of gravity, which applied to the entire universe. In 1920, the universe expanded even further when it was discovered that there were many galaxies outside of the Milky Way and that the universe changed over time. Little by little, observations confirmed that galaxies were moving away from Earth, and more powerful telescopes revealed billions of galaxies.

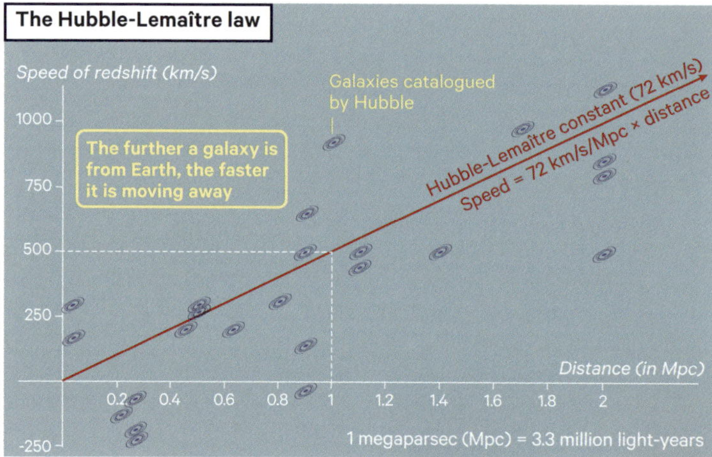

The expanding universe

In 1910, Henrietta Swan Leavitt measured the distance of stars known as Cepheid variables, whose luminosity changed periodically. Was the Milky Way the entire universe? In 1920, this 'Great Debate' was resolved by observing Cepheids in the Andromeda nebula, which was revealed to be a galaxy far beyond the Milky Way. Using Mount Wilson Observatory, Edwin Hubble studied dozens of nebulae. Their redshift showed that they were moving away from Earth, at a speed proportional to their distance: this is now called the Hubble-Lemaître constant. The expansion of the universe had been proven. Being able to see further also made it possible to see into the the universe's past.

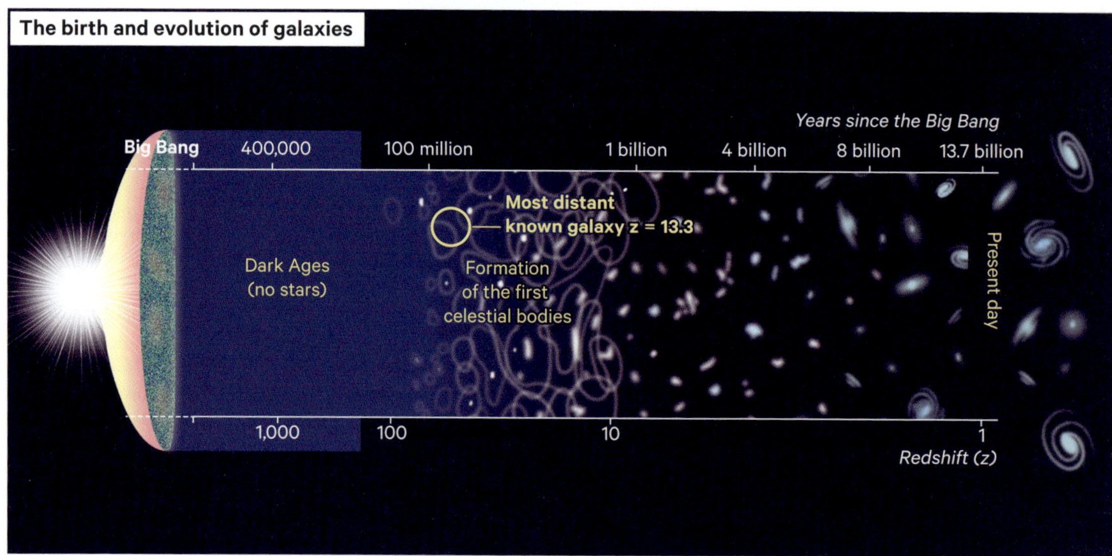

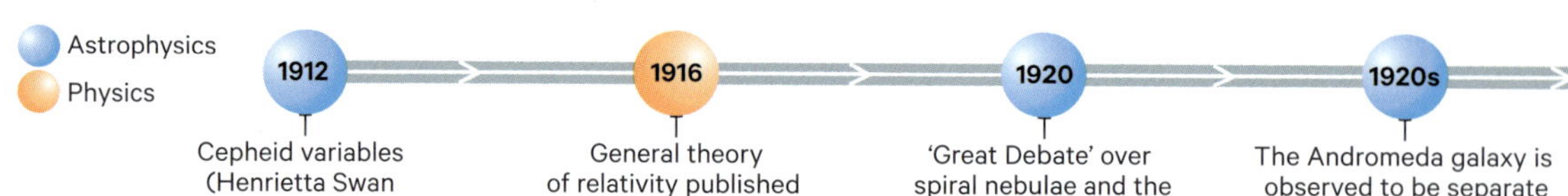

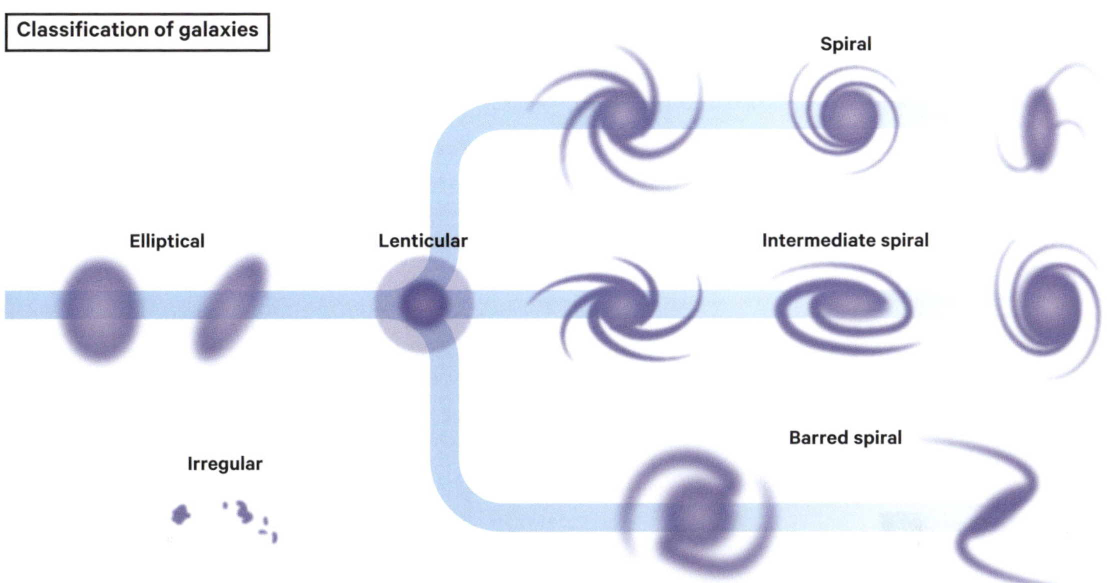

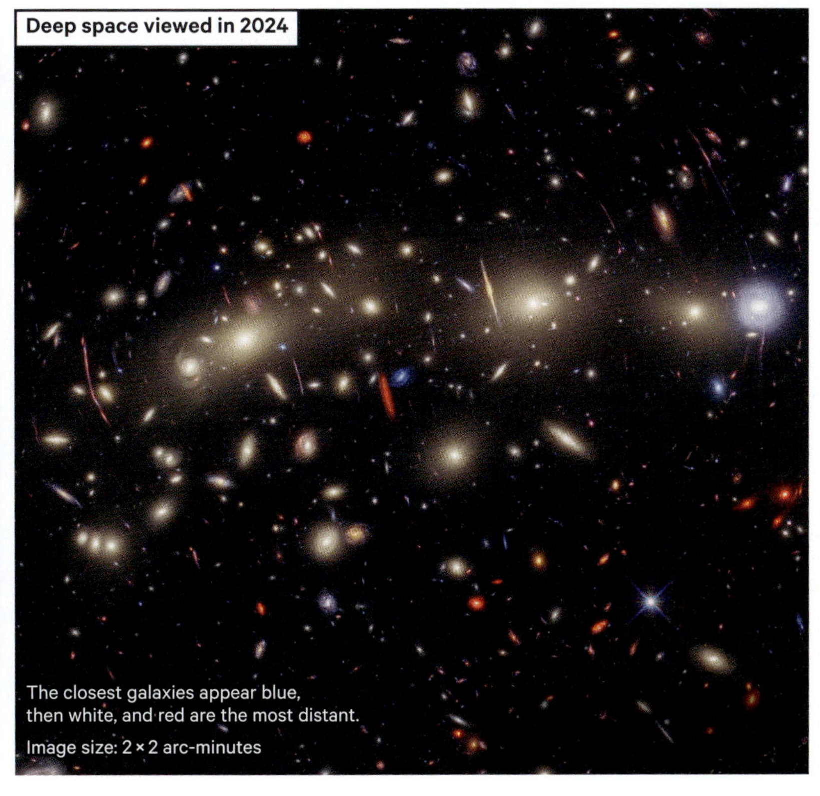

Galaxy types

From 1926 onwards, Hubble was able to see much dimmer galaxies filling the near universe, which he classified according to their shape. Spiral galaxies, with a shape like a flattened disc, are full of young gaseous stars. Elliptical galaxies contain old stars, often billions of them gathered into globular clusters, with very little interstellar matter. Galaxies move through space. Sometimes they collide and combine to form clusters made up of hundreds of galaxies. Many galaxies have a black hole at their centre, with a mass that may be billions of times that of the Sun. In 2023, the James Webb Space Telescope took pictures from orbit of some of the most distant galaxies ever observed, up to 13.3 billion light-years away from Earth.

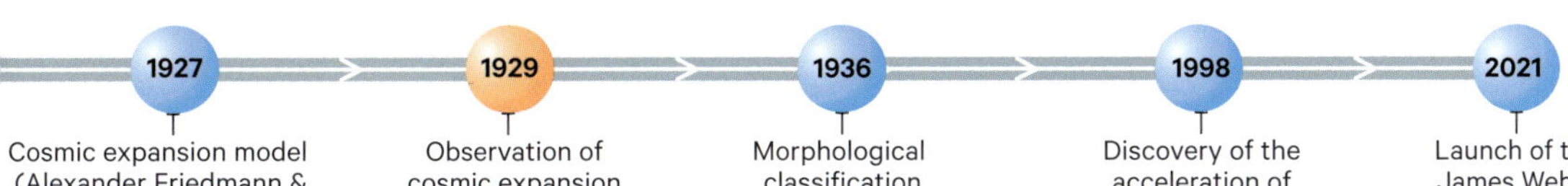

The life cycle of stars

Because they are so diverse, stars are classified according to their brightness and their mass. Their brightness remains stable for millions, sometimes billions of years, depending on their mass. When the fuel for their nuclear reactions is exhausted, they transform or explode, dispersing matter throughout space.

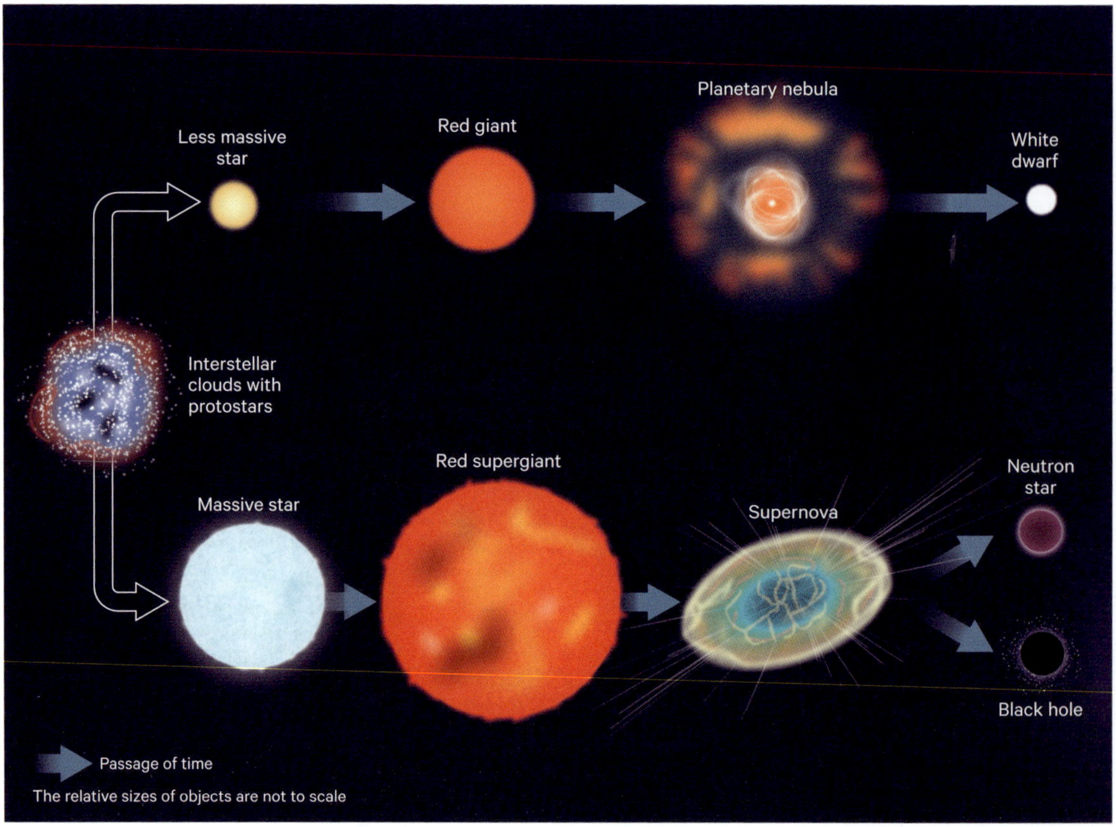

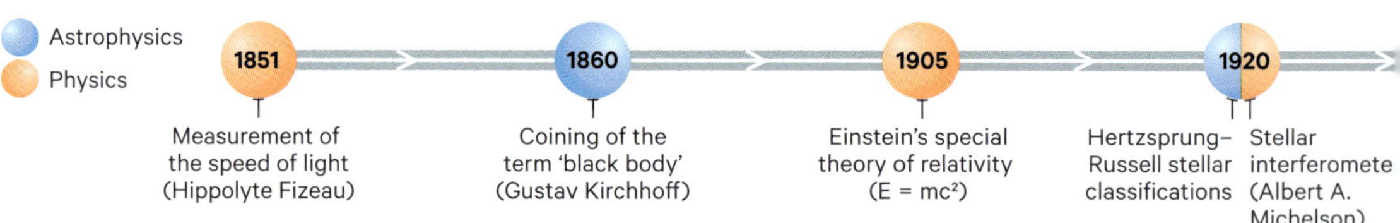

The formation and evolution of stars

In a galaxy, gravity condenses the matter of interstellar clouds (gas and dust) into 'clumps' of variable mass. These clumps collapse in on themselves, forming one or two stars, often with an associated planetary system. Less massive stars, such as the Sun, remain at a stable luminosity for many billions of years, until the fuel for nuclear reactions in their core is exhausted. They then expand, transforming into red giants, before becoming less luminous white dwarves. Stars with greater mass have a shorter period of stability (millions of years) before the nuclear reactions at their core become out of control. They expand to form a supergiant, then explode in a few seconds (supernova), briefly emitting a very intense light and dispersing matter into space.

Astrophysics
Physics

- **1851** Measurement of the speed of light (Hippolyte Fizeau)
- **1860** Coining of the term 'black body' (Gustav Kirchhoff)
- **1905** Einstein's special theory of relativity ($E = mc^2$)
- **1920** Hertzsprung–Russell stellar classifications
- **1920** Stellar interferomete (Albert A. Michelson)

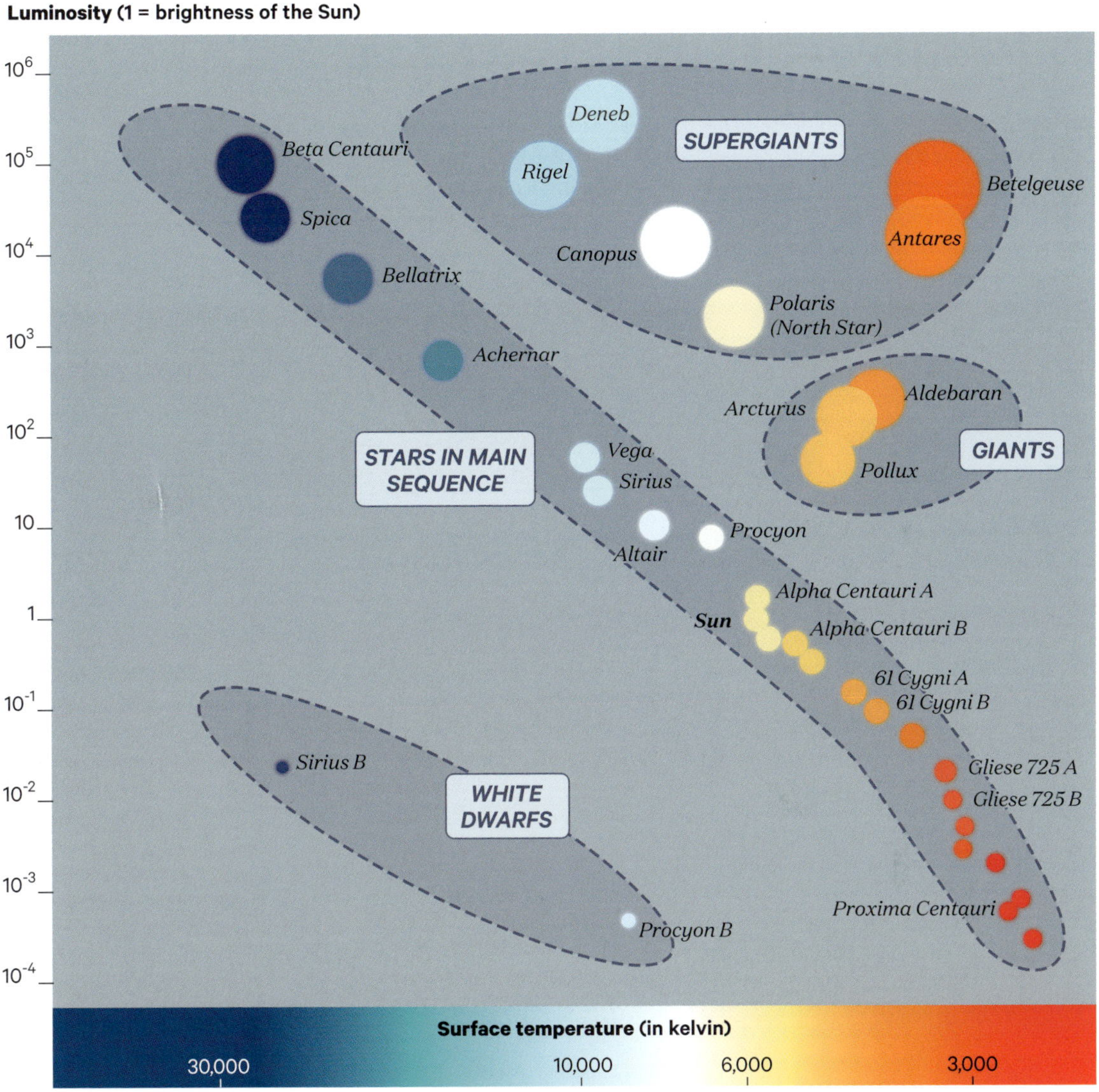

Luminosity (1 = brightness of the Sun)
10^6
10^5
10^4
10^3
10^2
10
1
10^-1
10^-2
10^-3
10^-4
SUPERGIANTS
Deneb
Rigel
Canopus
Betelgeuse
Antares
Polaris (North Star)
Beta Centauri
Spica
Bellatrix
Achernar
STARS IN MAIN SEQUENCE
Vega
Sirius
Altair
Procyon
Alpha Centauri A
Sun
Alpha Centauri B
61 Cygni A
61 Cygni B
Gliese 725 A
Gliese 725 B
Proxima Centauri
GIANTS
Aldebaran
Arcturus
Pollux
WHITE DWARFS
Sirius B
Procyon B
Surface temperature (in kelvin)
30,000
10,000
6,000
3,000

The lifespan of a star

The criteria for classifying a star are relatively simple: its temperature, its luminosity and its mass. Every star can be positioned within a so-called H–R diagram, created by Ejnar Hertzsprung and Henry Russell in 1910. This includes four groups: main sequence stars, white dwarfs, giants and supergiants. As the temperature of a star goes up, its colour changes from red to white, then to blue. Main sequence stars are stable in their luminosity and nuclear combustion. They remain in this phase for different lengths of time, depending on their mass, before transforming or expanding into one of the other three categories.

- **1934** Coining of the term 'supernova' (Walter Baade & Fritz Zwicky)
- **1938** Explanation of stellar energy generation (Hans Bethe)
- **1939** Prediction of the existence of black holes (R. Oppenheimer)
- **1965** First black hole discovered
- **1967** Existence of neutron stars (F. Pacini, J. Bell & A. Hewish)

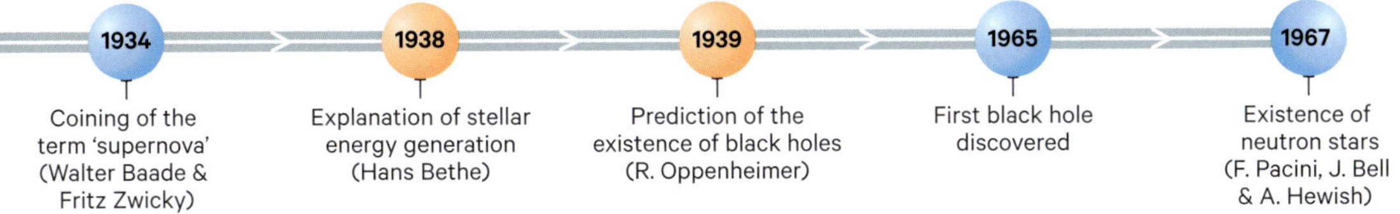

The interstellar medium

Within a galaxy, the interstellar medium (ISM) refers both to the vast space between the stars and the matter that it contains. This takes the form of gas and dust clouds, either frozen or heated by the stars. The interaction of matter and light can be explained and calculated using quantum physics. It is a constantly evolving cycle: the matter within the interstellar medium creates stars, which, in turn, eject matter at the end of their cycle.

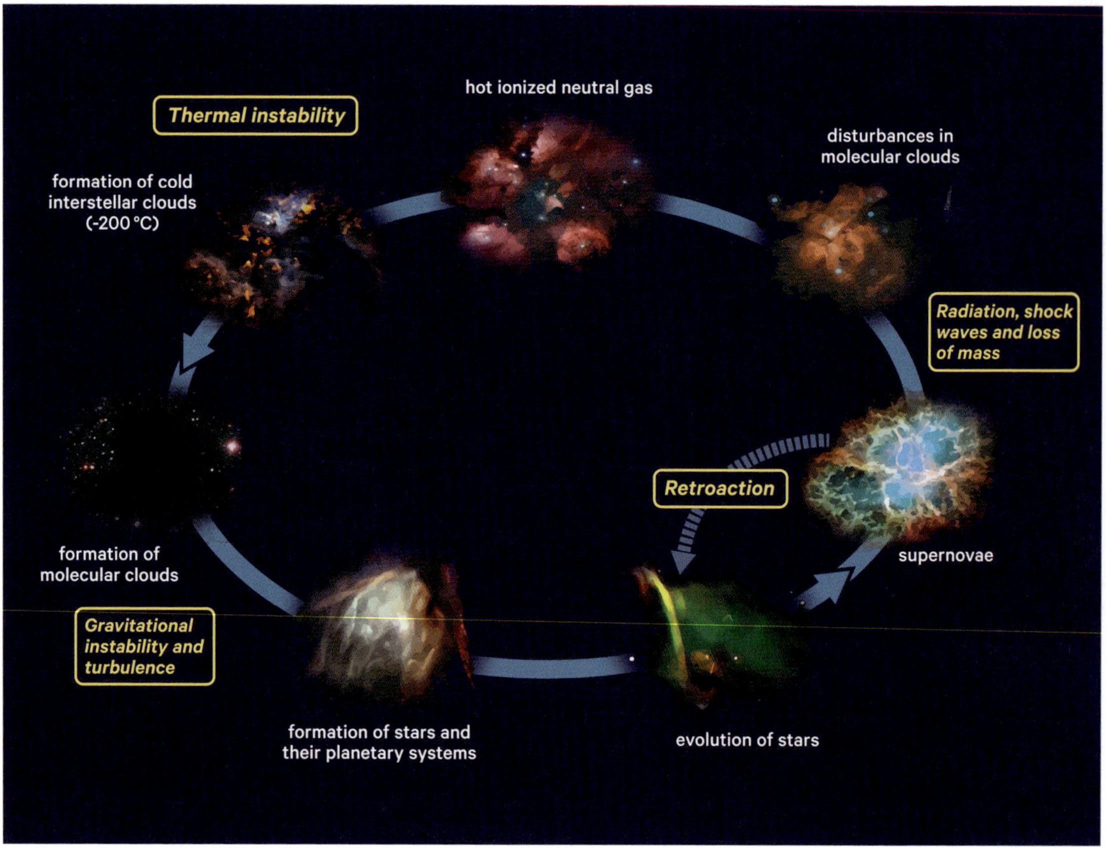

The constant recycling of matter

Within a galaxy, matter is either condensed into stars or dispersed between them in the interstellar medium. It forms vast clouds, either gaseous (hydrogen and other elements) or solid, in the form of minuscule grains of rocky dust (silicates). The gravitational attraction of each particle draws these dust particles together, forming stars and planets. During or at the end of its lifespan (the supernova stage), a star ejects some of its elements, transformed by nuclear reactions in its core, into space. The cycle can then start again, leading to changes in the chemical composition of the galaxy, and therefore the universe.

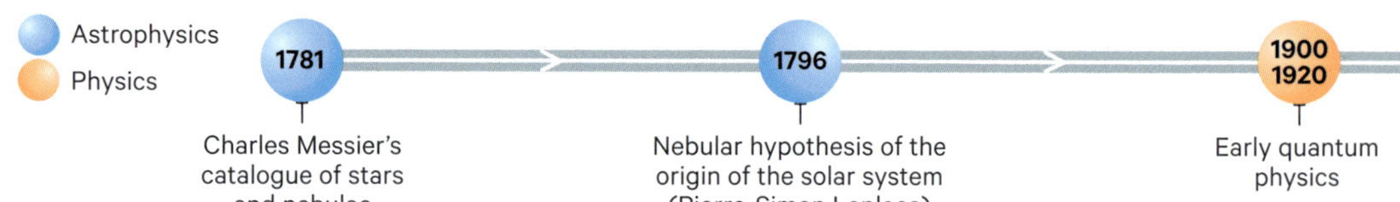

▲ This image shows the dust cloud of the Andromeda galaxy, captured in infrared by the Spitzer Space Telescope. Interstellar dust is a component of the interstellar medium.

Composition of the interstellar medium

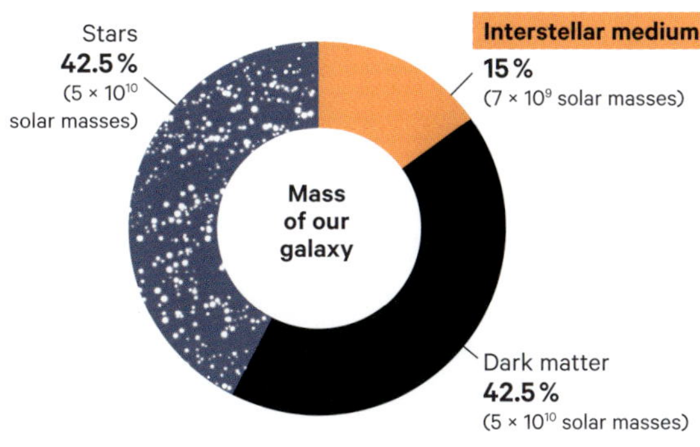

A significant part of the mass of the universe

An unstable medium

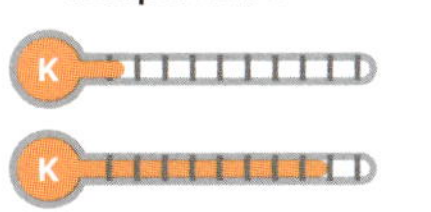

From over 10 kelvins to millions of kelvins

From less than 1 to more than 10^8 particles per cm^3

Elements and molecules

Interstellar clouds are mostly composed of hydrogen and helium, but other atoms are also present (carbon, oxygen, nitrogen, silicon, etc.). The interstellar medium is unstable and constantly changing. Depending on temperature, the matter may be in the form of gas, at varying temperatures, or cold dust particles. In clouds of colder gas, a wide variety of chemical compounds form molecules, ranging from simple to complex. Stars eject tiny solid crystals (silicates), around which some molecules (CO_2, H_2O, NH_3) form as ice.

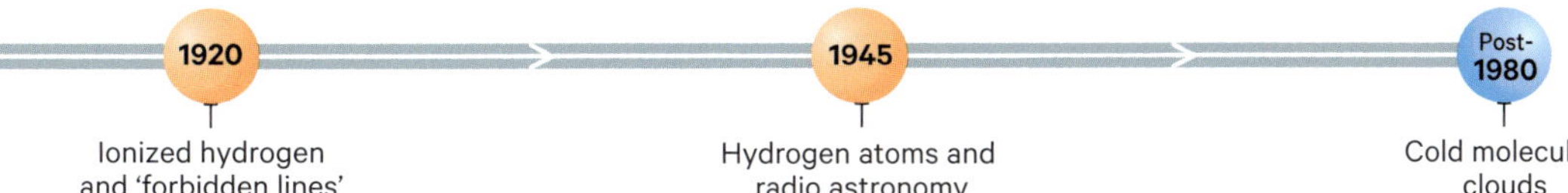

From hydrogen to uranium

Stars are nuclear furnaces that fuse hydrogen atoms together to create elements that are greater in mass. When massive stars explode, temperatures reaching billions of degrees produce very heavy elements, which are ejected into the interstellar medium. These will form the next generation of stars. Back on Earth, advances in nuclear physics have made it possible to build nuclear reactors that generate electricity but the same developments have also led to the creation of weapons of mass destruction.

Chemical elements in the universe

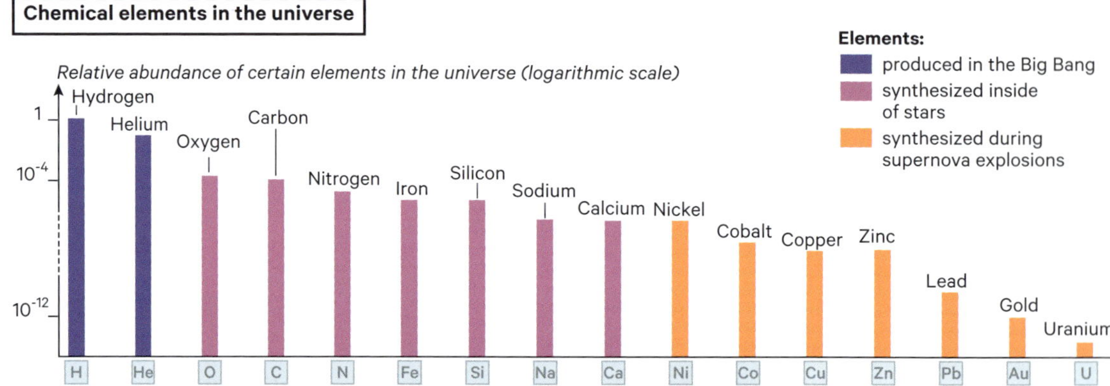

Inside a massive star

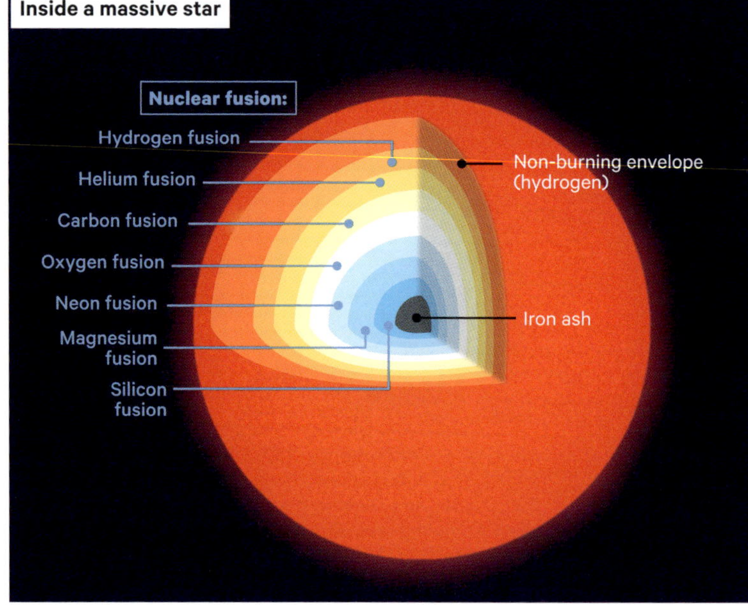

A nuclear reactor

Spectral analysis of the light of stars can be used to determine their chemical composition and identify the atoms present. The relative proportion of various atoms is almost the same across the universe (above), with hydrogen (H) and helium (He) being the most prevalent. Between 1935 and 1958, physics described the process of nuclear fusion that takes place in a star. If the star is low in mass, such as the Sun, the hydrogen nuclei in its core are transformed into helium, then into carbon and oxygen. Producing heavier elements (lead, gold, uranium) requires very high temperatures, which are reached inside the layers of a massive red supergiant shortly before it explodes into a supernova.

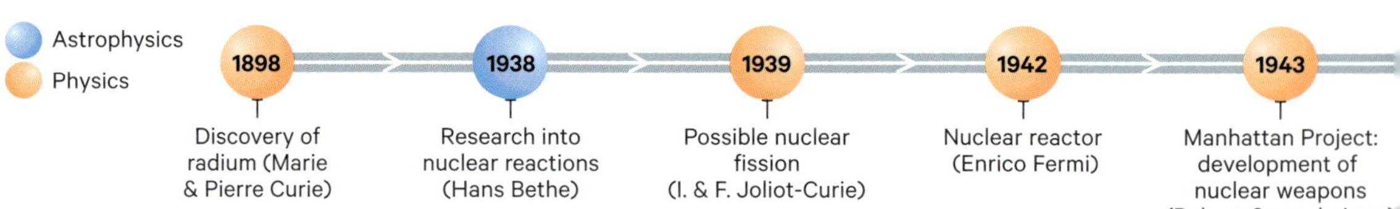

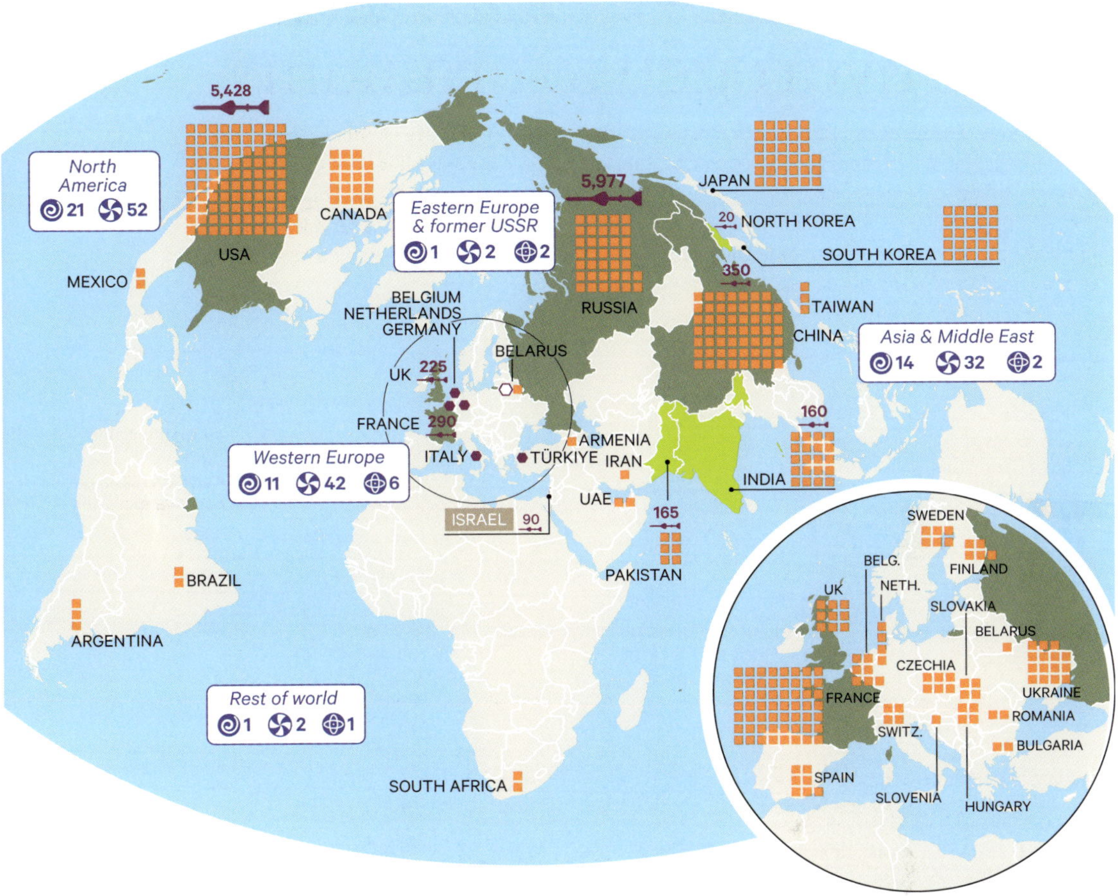

Map labels

North America
- 21 cyclotron (isotopes)
- 52 cyclotron (imaging)

5,428 — USA

CANADA

MEXICO

Eastern Europe & former USSR
- 1 cyclotron (isotopes)
- 2 cyclotron (imaging)
- 2 medium-energy accelerator

5,977 — RUSSIA

BELARUS

BELGIUM
NETHERLANDS
GERMANY

UK — 225

FRANCE — 290

ITALY

Western Europe
- 11 cyclotron (isotopes)
- 42 cyclotron (imaging)
- 6 medium-energy accelerator

ARMENIA

TÜRKIYE

IRAN

UAE

ISRAEL — 90

JAPAN

NORTH KOREA — 20

SOUTH KOREA

TAIWAN

CHINA — 350

Asia & Middle East
- 14 cyclotron (isotopes)
- 32 cyclotron (imaging)
- 2 medium-energy accelerator

INDIA — 160

PAKISTAN — 165

BRAZIL

ARGENTINA

Rest of world
- 1 cyclotron (isotopes)
- 2 cyclotron (imaging)
- 1 medium-energy accelerator

SOUTH AFRICA

Inset (Europe):
SWEDEN, FINLAND, UK, BELG., NETH., SLOVAKIA, BELARUS, CZECHIA, UKRAINE, FRANCE, ROMANIA, SWITZ., BULGARIA, SPAIN, SLOVENIA, HUNGARY

Legend

Nuclear weapons (in 2022)
- State with nuclear weapons, according to the NPT*
- State that admits possession of nuclear weapons
- State that possesses nuclear weapons but without official recognition
- State that hosts nuclear weapons via NATO**
- State that hosts nuclear weapons outside of NATO
- Stock of nuclear weapons

Civil nuclear reactors (electricity production)
- 1 reactor

Nuclear medicine
- Cyclotron producing medical isotopes
- Cyclotron used in medical imaging (positron emission tomography)
- Medium-energy accelerator producing medical isotopes

*Treaty on the Non-Proliferation of Nuclear Weapons
**North Atlantic Treaty Organization

Nuclear power around the world

Between 1935 and 1940, advances in the physics of the nuclei of atoms explained where the Sun and stars got their energy. When the Second World War began, this knowledge was used to create the nuclear fission bomb (A bomb), eventually used on Hiroshima and Nagasaki. H bombs were also developed, using the explosive fusion of hydrogen. The use of nuclear fission reactors to produce energy for the world also became more widespread. The energy released is transformed into heat, which drives turbines that produce electricity. Uranium became a strategic resource, whose use is controlled by the International Atomic Energy Agency (IAEA). Particle accelerators are used to create radioactive material for medical use.

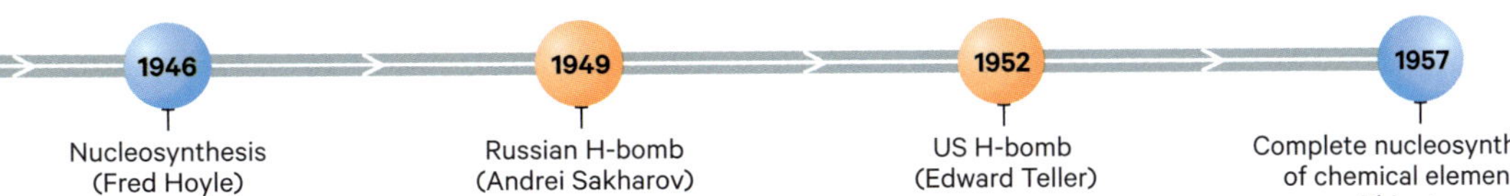

The Solar System

Our knowledge of the Solar System has been increased by exploration, using probes that fly past or land on planets and moons. Beyond the eight known planets, a vast domain is still being explored.

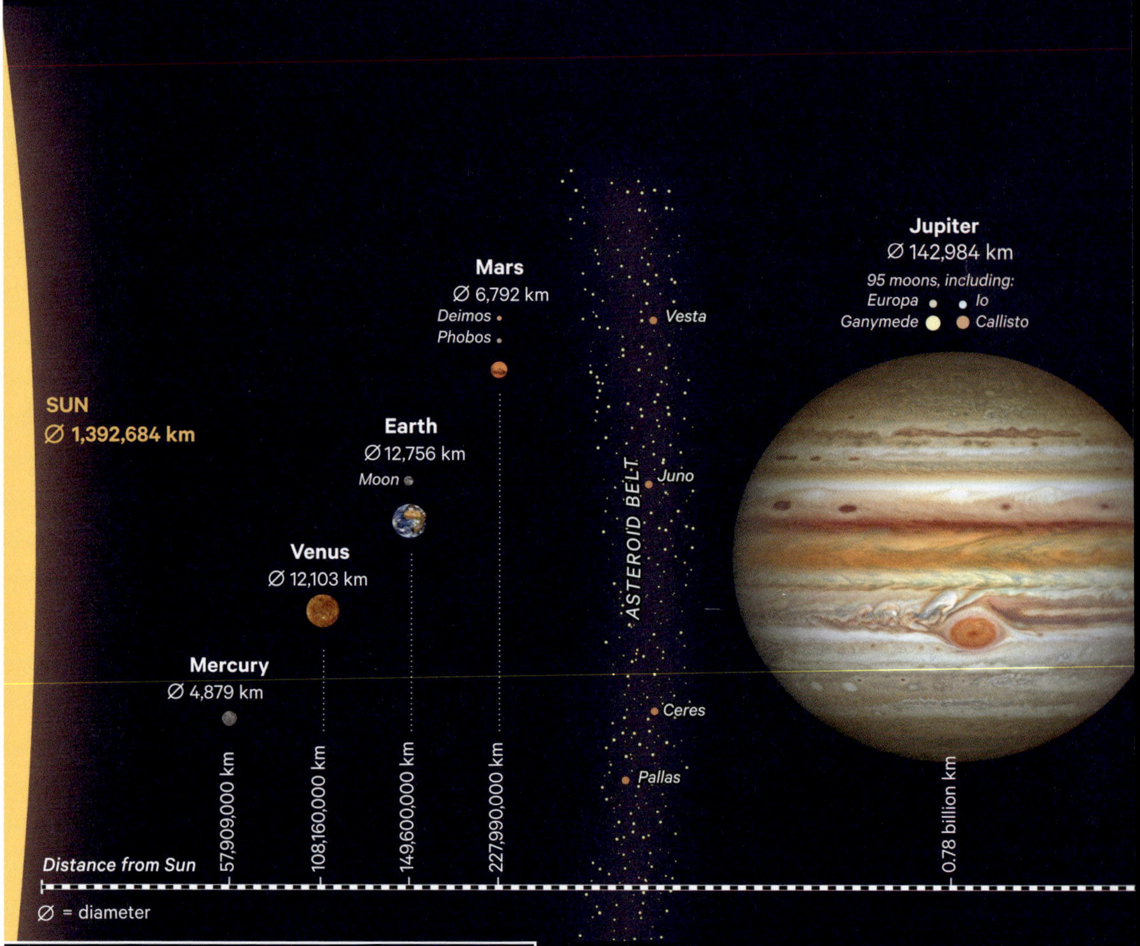

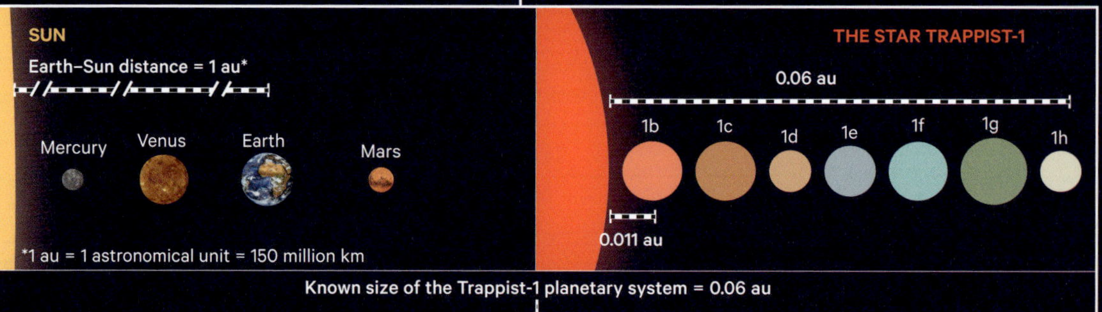

The Solar System compared to the Trappist-1 system

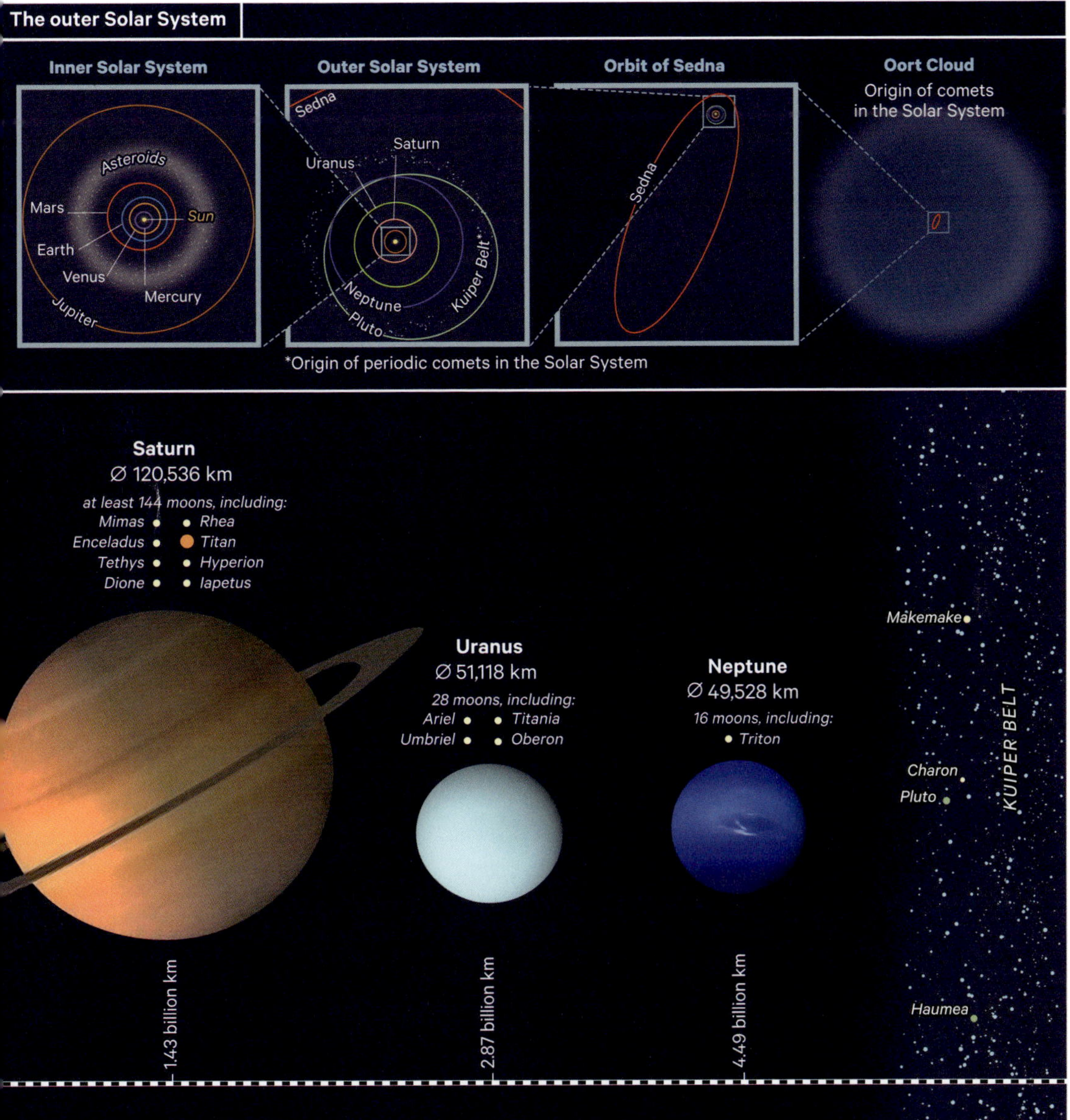

The unity of the Solar System

This unity derives from the power of gravity, which the huge mass of the Sun exerts on everything around it. It also stems from its history, when the Sun was formed 4.7 billion years ago, along with the solid or gaseous bodies that became planets, satellites, asteroids, as well as clouds of dust and gas in the interstellar medium where comets are formed (top). Exploration of the Solar System has been ongoing since the 1970s, with uncrewed probes transmitting ever more precise observations back to Earth. The presence of life elsewhere in our Solar System remains an open question, and is sure to be a driving force in future exploration missions. Beyond the dwarf planet Pluto and its moon Charon lies a vast region of space that remains virtually unexplored. Up until the late 20th century, the Solar System was the only known planetary system. Since then, thousands of others have been discovered, which are very diverse and allow us to draw interesting comparisons. Around the star Trappist-1, for example, there is a system of seven planets (opposite below). Their distances from their star and their sizes are comparable to the characteristics of the rocky terrestrial planets in our Solar System, from Mercury to Mars.

Exoplanets

Until the discovery of the first exoplanet in 1995, the Solar System was the only known planetary system in the universe. Exoplanets are objects that are similar to the planets of the Solar System, but they orbit other stars. It is now known that there are thousands of exoplanetary systems in the galaxy and scientists are searching for exoplanets that are similar to Earth. Instruments are able to measure the temperature and atmosphere of some of these planets. Astrobiology studies the issue of potential life in space.

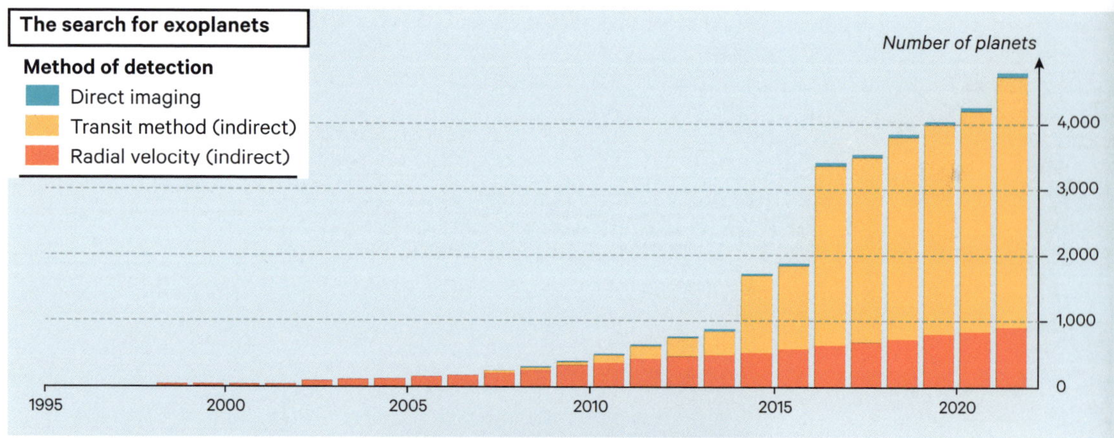

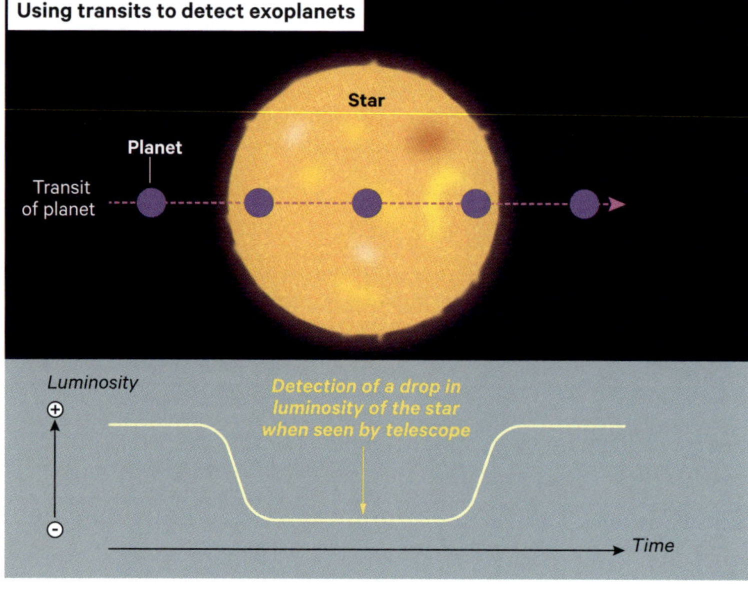

How are exoplanets detected?

The mass of known exoplanets ranges from that of the Earth to a thousand times greater (Jupiter). The more massive the planet is and the closer it is to its star, the easier it is to detect. This is achieved through three main methods: 1) images, which identify the planet in the form of a faint point of light next to a star; 2) the transit method, which detects a decrease in the star's luminosity each time that a planet passes between the telescope and the star (see left); 3) radial velocity, which measures a change in the star's velocity, under the gravitational effect of the planet (spectroscopy).

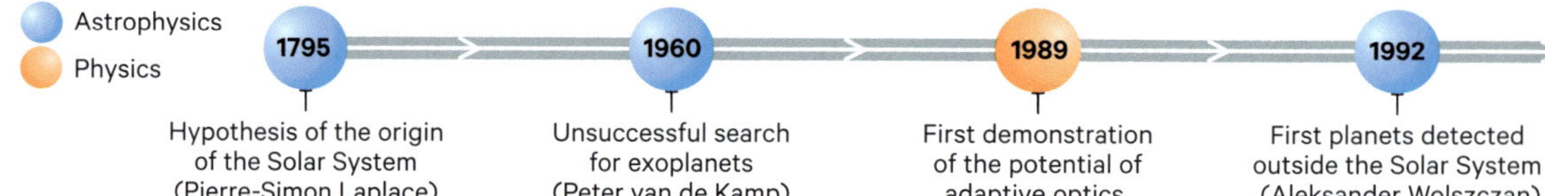

The closest exoplanets (southern hemisphere)

Sky over Santiago, Chile, 24 January 2024 at 6 am

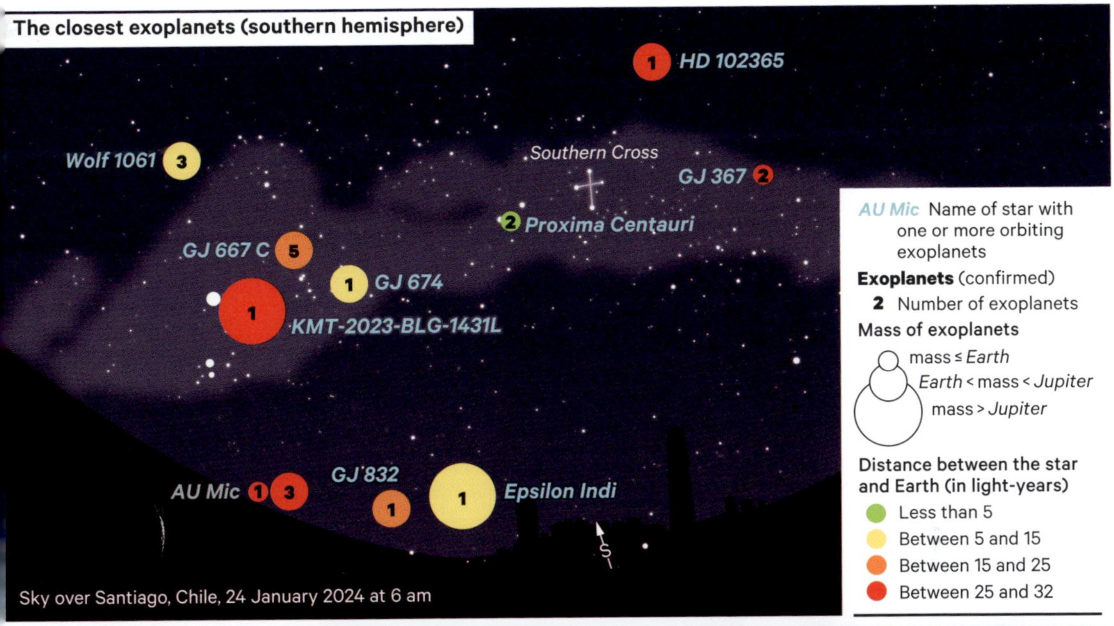

The Proxima Centauri system

*1 kelvin (K) = -272.15 °C

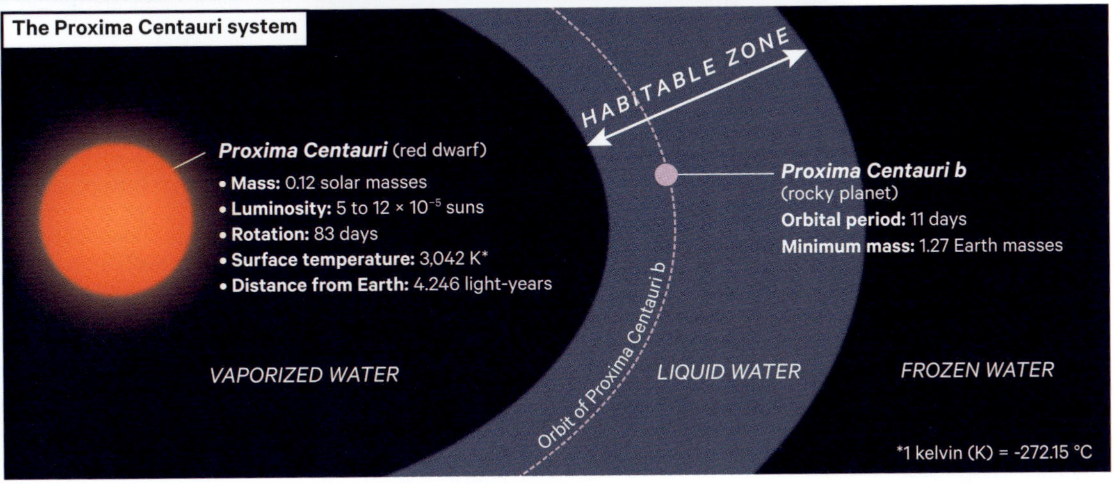

The habitable zone

The top diagram shows the stars with planetary systems that are closest to Earth. The three stars of the southern constellation Centaurus, which form a triple system, are 4.5 light-years away from the Sun. One of them, the red dwarf Proxima Centauri, is the closest star to the Solar System. It has between two and four exoplanets (b, c, d, e – the first planet discovered takes the name of its star followed by the letter b, the second by c, etc.).

The surface temperature of a planet depends on how far it is from its star. Too close, and it may be above 100°C; too far away, and it falls below 0°C. Between those two points, water can exist in liquid form. Living things like those on Earth would need liquid water, so the area around a star where the temperature is between 0°C and 100°C is called the 'habitable zone'. This does not mean that the planets in this zone are inhabited.

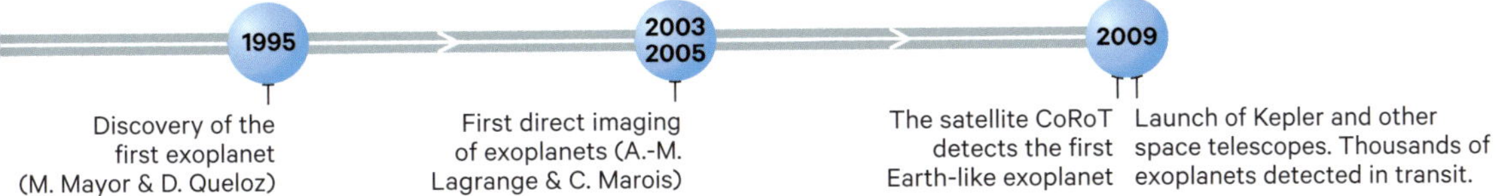

The curvature of spacetime

Since Einstein published the general theory of relativity in 1916, it has been known that the path of light in a space containing mass is not a straight line, but a curve. Time itself is dependent on it. German physicist Karl Schwarzschild proposed the existence of black holes, which are so dense that no light can escape. Subsequently, the British scientist Arthur Eddington (1919) proved the curvature of space close to the Earth during a solar eclipse. The curvature affects the trajectory of GPS signals and the light from other galaxies in the universe.

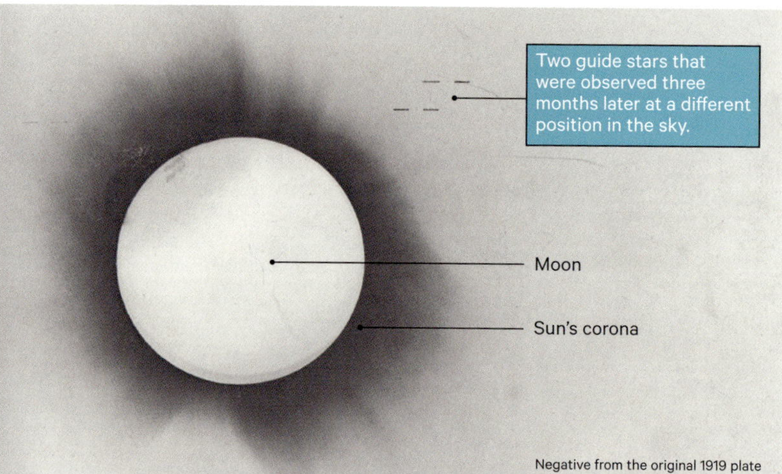

Einstein's relativity and the solar eclipse of 1919

Classical mechanics state that a photon (a light particle) passing through a gravitational field would be thrown off course, like a projectile on Earth. In 1916, Einstein published his general theory of relativity, reconceptualizing Newtonian gravity. To test this new concept of gravity, the total solar eclipse of 1919 was photographed, from Brazil and Africa (the island of Príncipe), by the British scientist Arthur Eddington and his team. Some stars, those closest to the edge of the solar disc, were visible in the photographs, their positions having shifted due to the deviation of light caused by the Sun's mass. Photographing the same area of the sky a few months later made it possible to measure the shift in the position of the stars between the two dates. Although it was difficult to carry out these measurements, they provided the first confirmation of this new physics of spacetime.

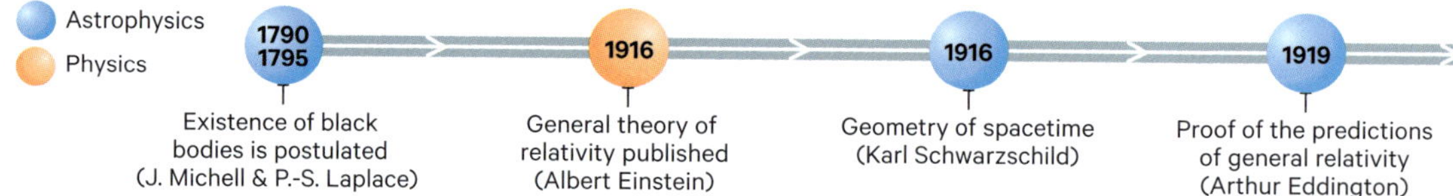

▲ An Einstein ring

In 1936, Einstein predicted a phenomenon that he thought would be impossible to observe, but which can be observed today. In the constellation Leo, two galaxies appear aligned for an observer on Earth. The light of the more distant galaxy, LRG 3-757 (10 billion light years away), is diverted and amplified when it passes by the other galaxy, forming a 'gravitational lens' situated between us and LRG 3-757.

▼ How gravitational lensing works

Seen from the Earth, star B, a moving star within our galaxy, is precisely aligned with the more distant star A. The light from star A is diverted by the gravity of the closer star B, which may have an exoplanet orbiting around it. Star B becomes a gravitational lens, acting like a magnifying glass that produces a distorted image of the distant star A.

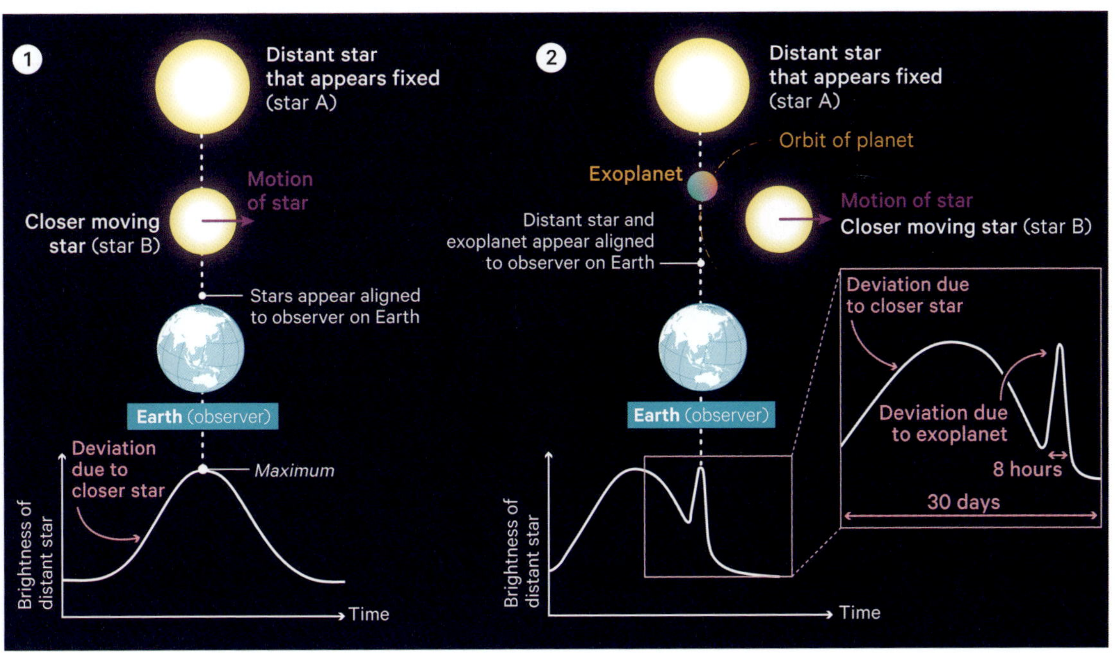

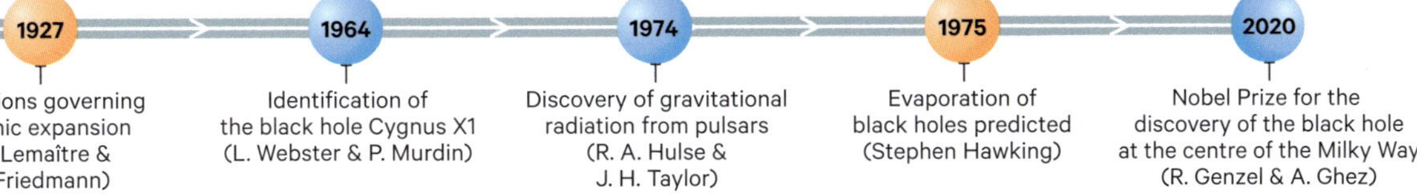

Where do galaxies come from?

In the 1930s, models of the universe predicted a very distant past when there were no galaxies, but there was a source of light. This light was detected in 1962 and is known as cosmic background radiation. This radiation suggests there was once a universe that was almost homogenous, where today's galaxies were formed.

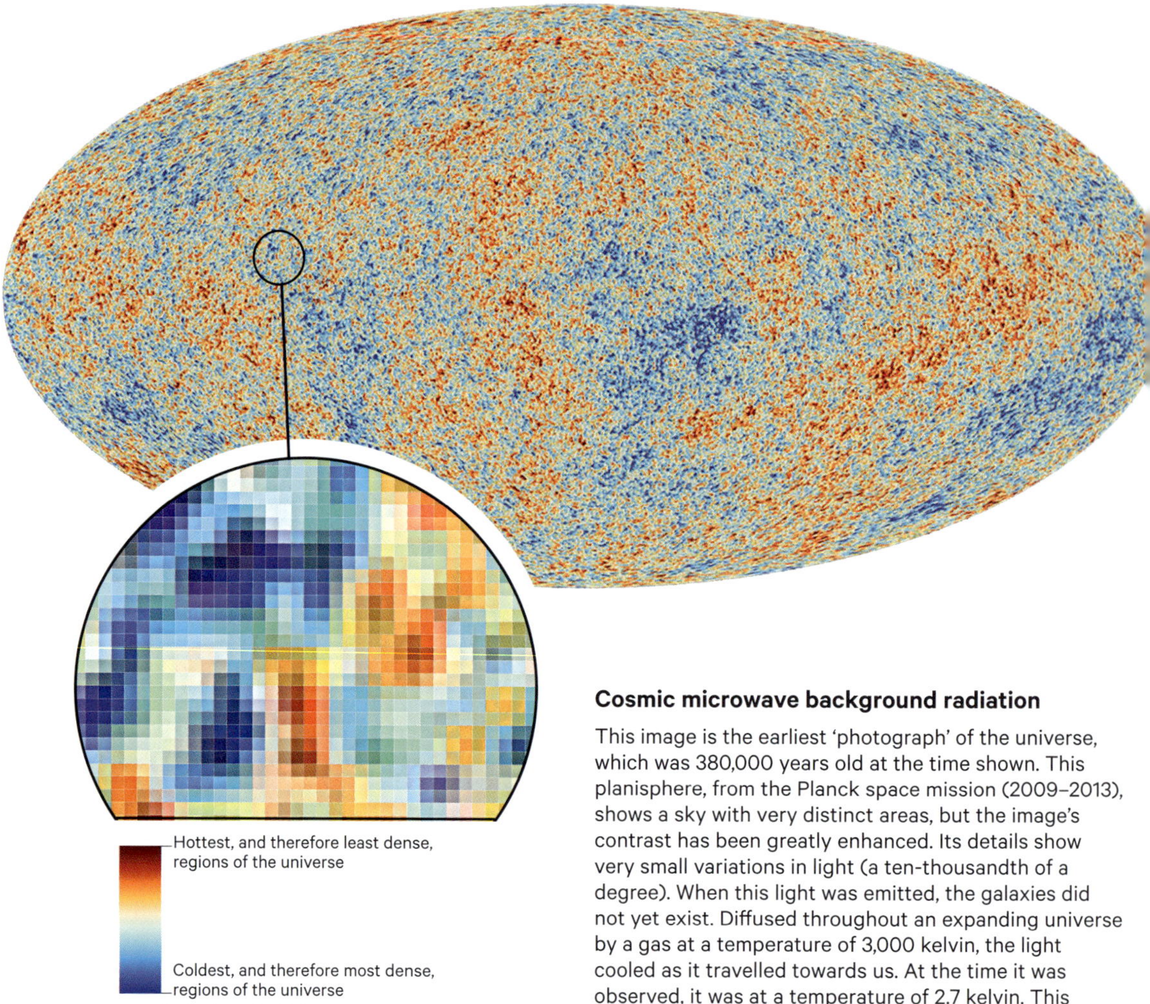

Hottest, and therefore least dense, regions of the universe
Coldest, and therefore most dense, regions of the universe

The pixels show the highest resolution possible with the equipment

Cosmic microwave background radiation

This image is the earliest 'photograph' of the universe, which was 380,000 years old at the time shown. This planisphere, from the Planck space mission (2009–2013), shows a sky with very distinct areas, but the image's contrast has been greatly enhanced. Its details show very small variations in light (a ten-thousandth of a degree). When this light was emitted, the galaxies did not yet exist. Diffused throughout an expanding universe by a gas at a temperature of 3,000 kelvin, the light cooled as it travelled towards us. At the time it was observed, it was at a temperature of 2.7 kelvin. This considerable variation in wavelength means that today its observed spectrum is at millimetre wavelengths.

Astrophysics
Physics

1795 Hypothesis of the origin of the Solar System (Pierre-Simon Laplace)

1960 Unsuccessful search for exoplanets (Peter van de Kamp)

1989 First demonstration of the potential of adaptive optics

1992 First planets detected outside the Solar System (Aleksander Wolszczan)

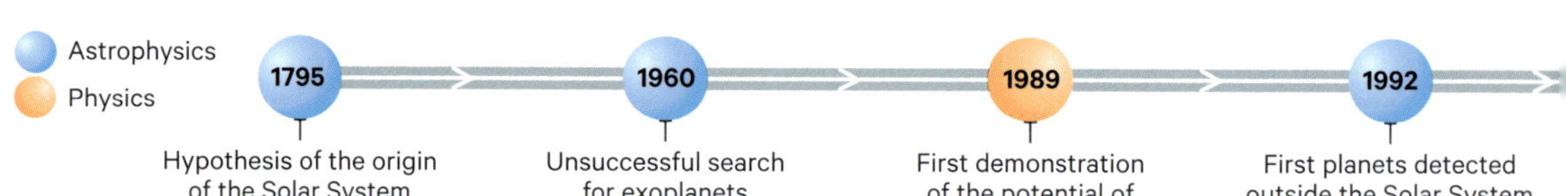

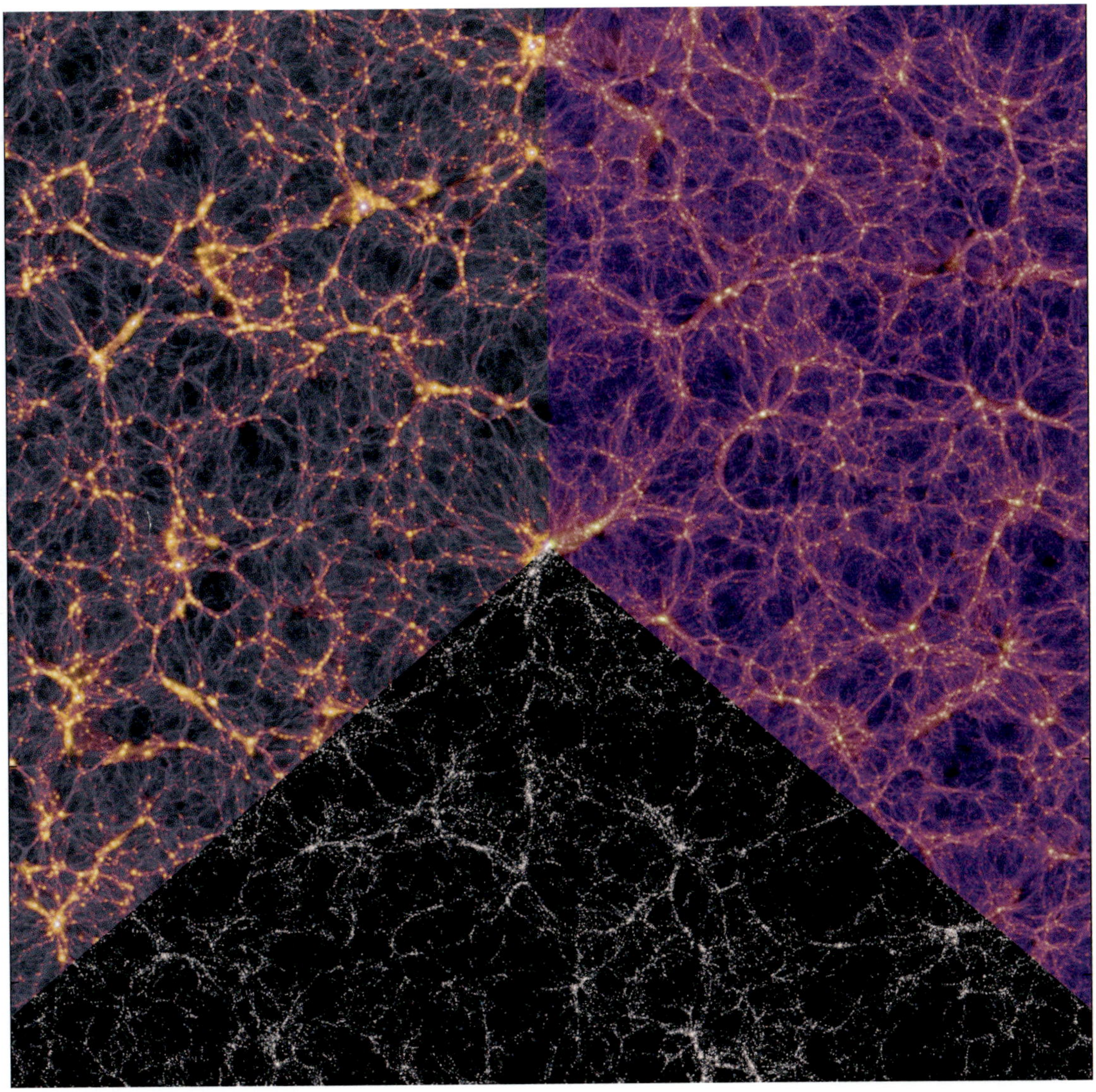

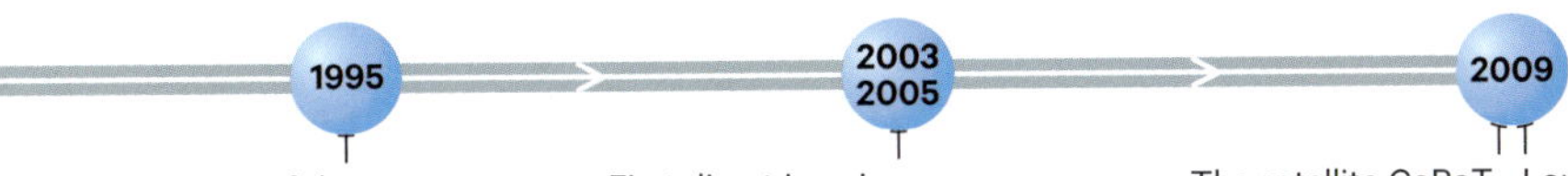

Protogalaxies and protostars

Advances in computer processing have made it possible to calculate the evolution of clumps or clusters of hydrogen from the time when the galaxies were formed and over the hundreds of millions of years that followed. This calculation can be compared with what we know about the most distant galaxies observed in 2024. The simulation must take into account not only the hydrogen, but also the mass of dark matter, which is even more abundant. The three sections of this diagram, taken from the Millennium Simulation Project (Germany), show the results of calculations: the density of hydrogen gas, the first stars in the protogalaxies and dark matter, in a past that our instruments no longer have access to.

1995 – Discovery of the first exoplanet (M. Mayor & D. Queloz)

2003 2005 – First direct imaging of exoplanets (A.-M. Lagrange & C. Marois)

2009 – The satellite CoRoT detects the first Earth-like exoplanet

Launch of Kepler and other space telescopes. Thousands of exoplanets detected in transit.

The Big Bang: further than science can reach

In 1933, Georges Lemaître proposed that the universe may have begun its existence in the form of a primeval atom, within which all the matter of the universe was condensed. This hypothesis was hotly debated. Today, background radiation and observed matter confirm that the primeval state of the universe was hot and dense. Cosmology can provide a model of this past, but physics reaches its limits, because it is not yet able to come up with a unified theory of gravity and quantum physics.

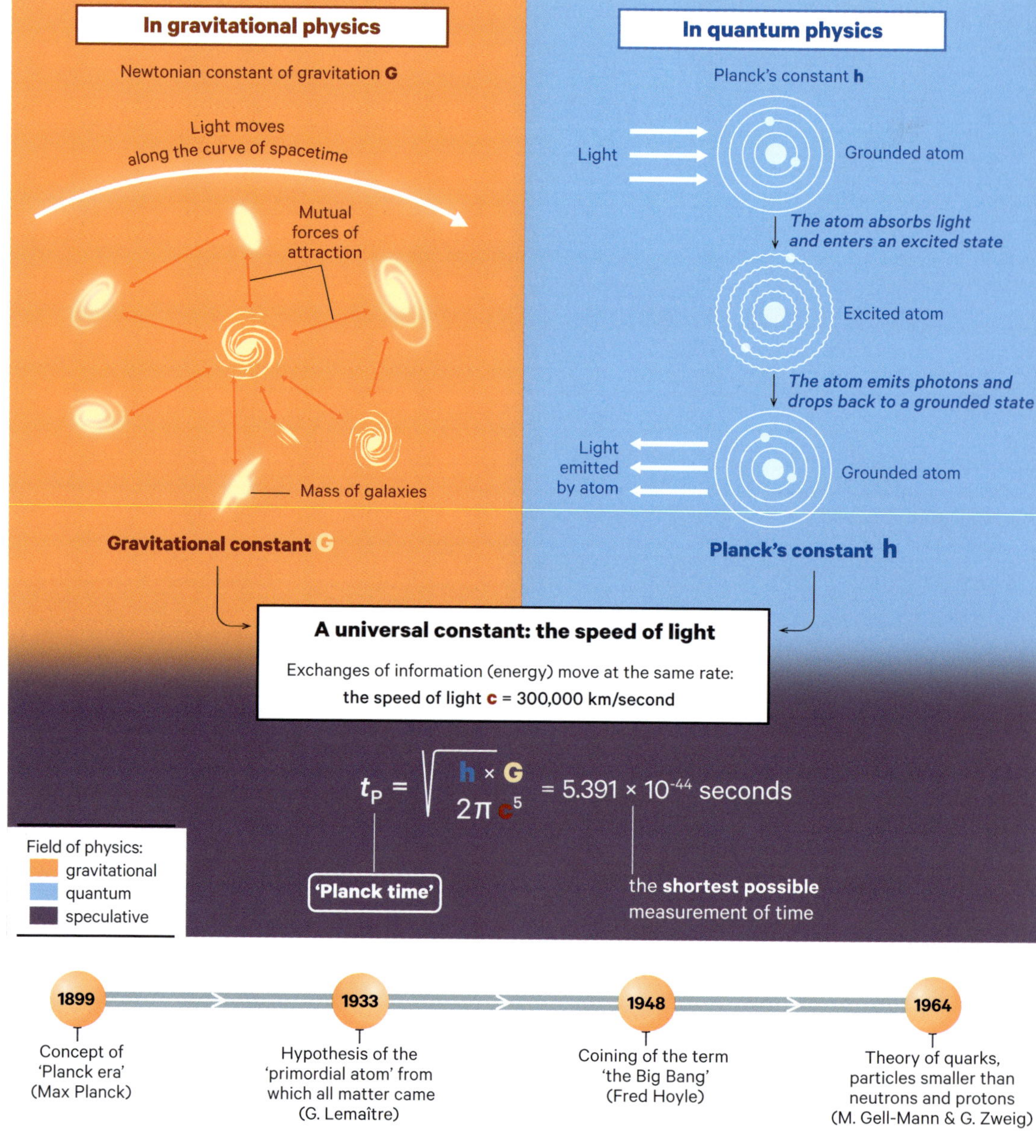

$$t_\mathrm{P} = \sqrt{\frac{h \times G}{2\pi\, c^5}} = 5.391 \times 10^{-44} \text{ seconds}$$

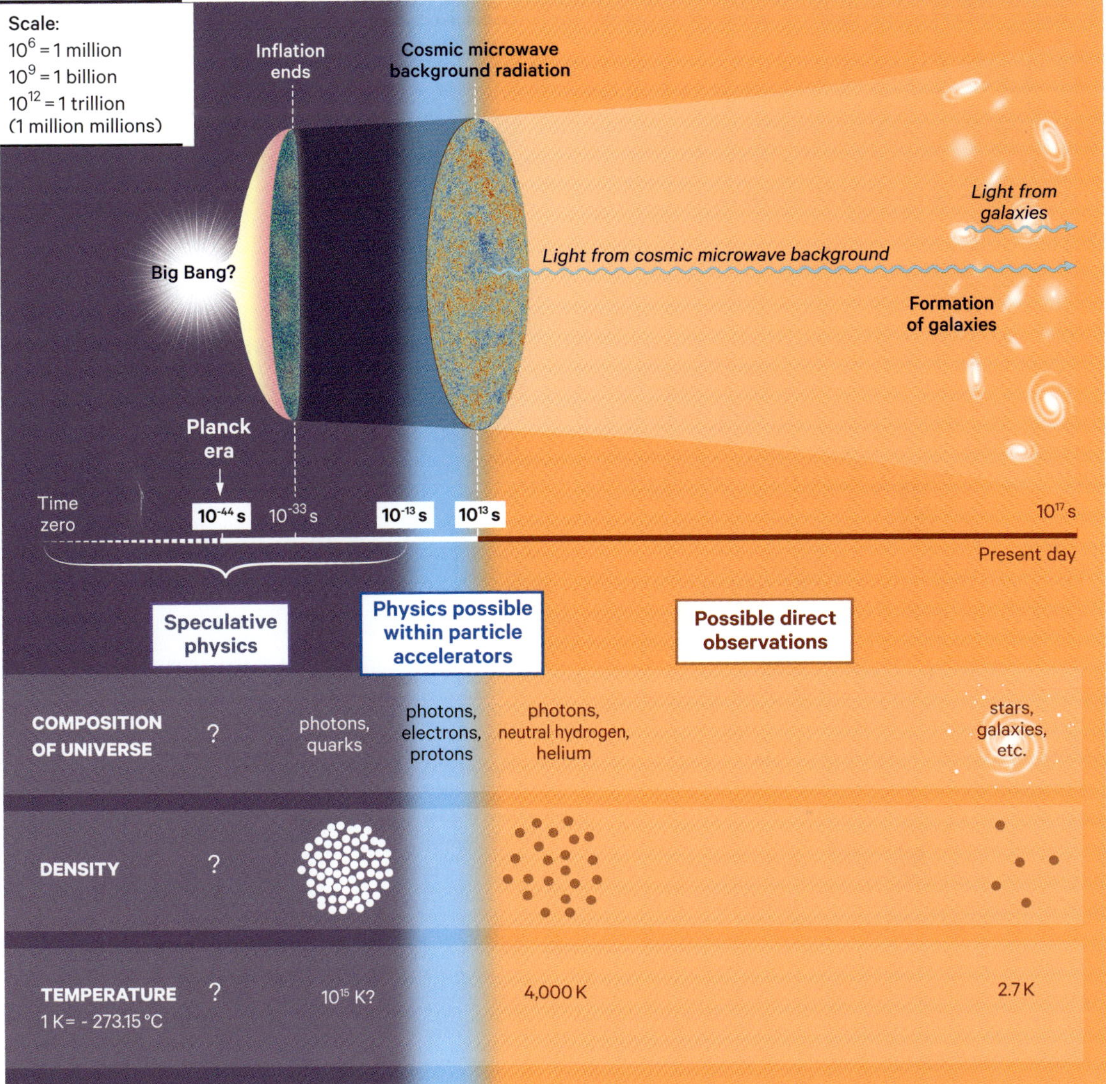

◀ Planck time

There were two major breakthroughs in physics in the 20th century. Through gravity and relativity, physics was able to describe the universe on the macroscopic level (galaxies, stars, etc.). Through quantum physics, it was possible to describe it at the microscopic level (atoms, molecules, gas). Gravitational waves and light carry information at the same speed (c).

The universe is structured around three universal numbers: 'c' for all exchanges of information (light, gravity), the gravitational constant 'G', and Planck's constant 'h', which defines energy leaps in the quantum realm of atoms. In 1899, physicist Max Planck realized that a simple combination of these three numbers would create a length of time, perhaps the smallest length of time that physics can conceive of, which is now known as 'Planck time'.

▲ As distant as the past can get

Describing the evolution of the primordial universe, before the formation of galaxies and the emission of cosmic background radiation, is a challenge for physics and astrophysics. There are a few surviving traces of this period, but it is not yet possible to interpret them fully. A different scale must be used to describe the passing of time than the one used on Earth today. Going back to a time situated infinitely far in the past, this scale reframes the question of the origin of the universe. Moving 'backwards' through history, physics describes a universe that grows progressively denser and hotter. Reaching Planck time, which combines the constants 'G' (gravity) and 'h' (quantum physics), requires a new kind of physics, which is capable of describing a universe where the quantum world and gravity can both be applied.

Optical telescopes

As their mirrors have become larger, telescopes have become able to collect a broader spectrum of light, both visible and infrared. Advances in photoreceptor technology have also considerably increased the sensitivity of telescopes.

The largest optical telescopes

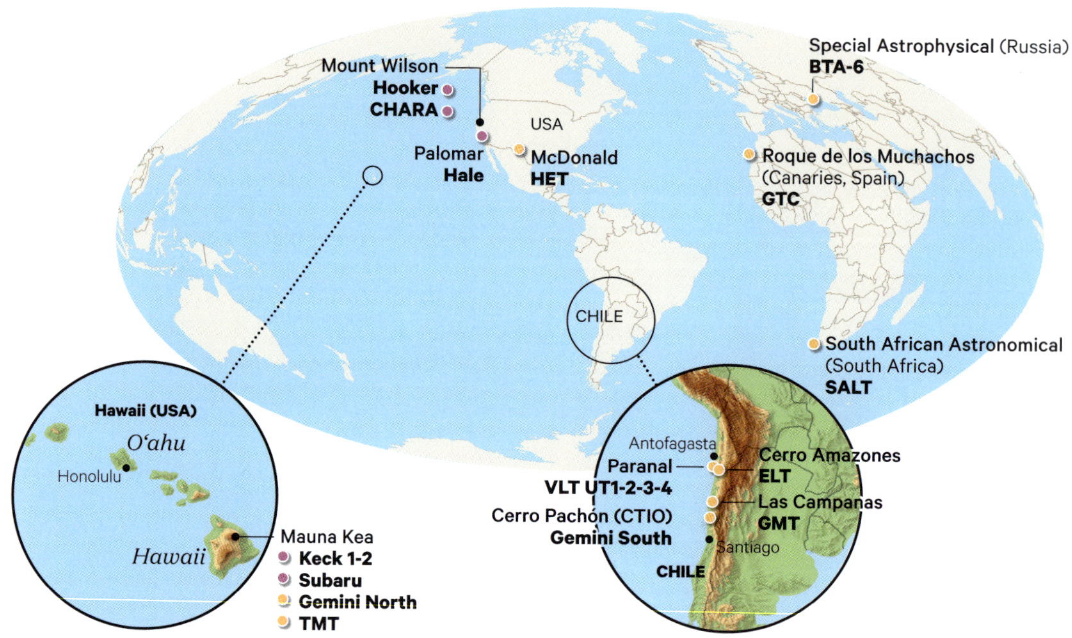

Growing ever larger

Primary mirror type
- Monolithic mirror
- Segmented mirror
- Multiple mirrors

NB: Timeline not to scale

Financing and running
- Nationally funded and run
- Funded and run by international cooperation

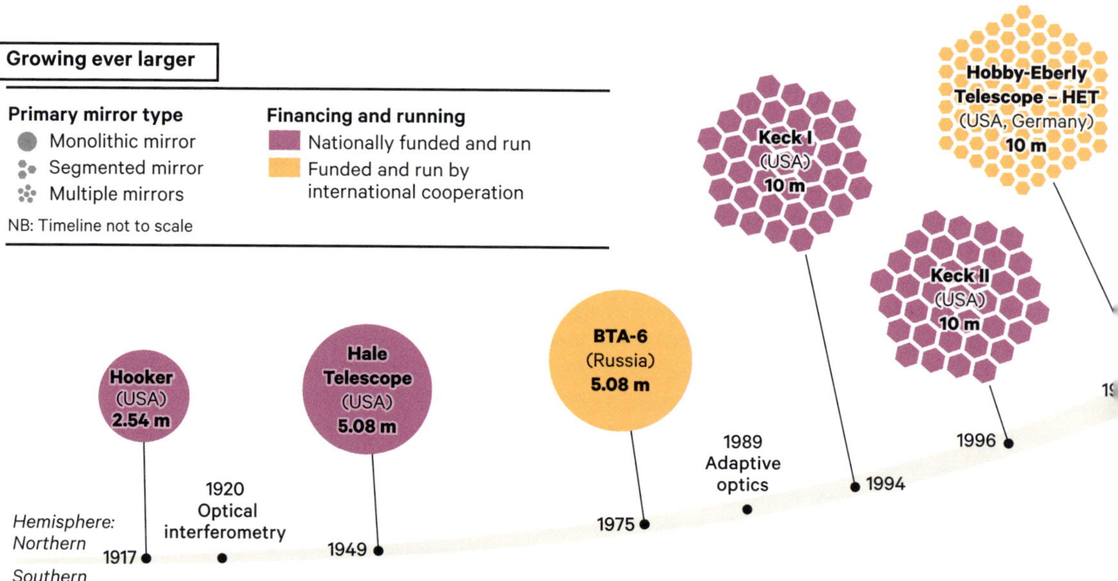

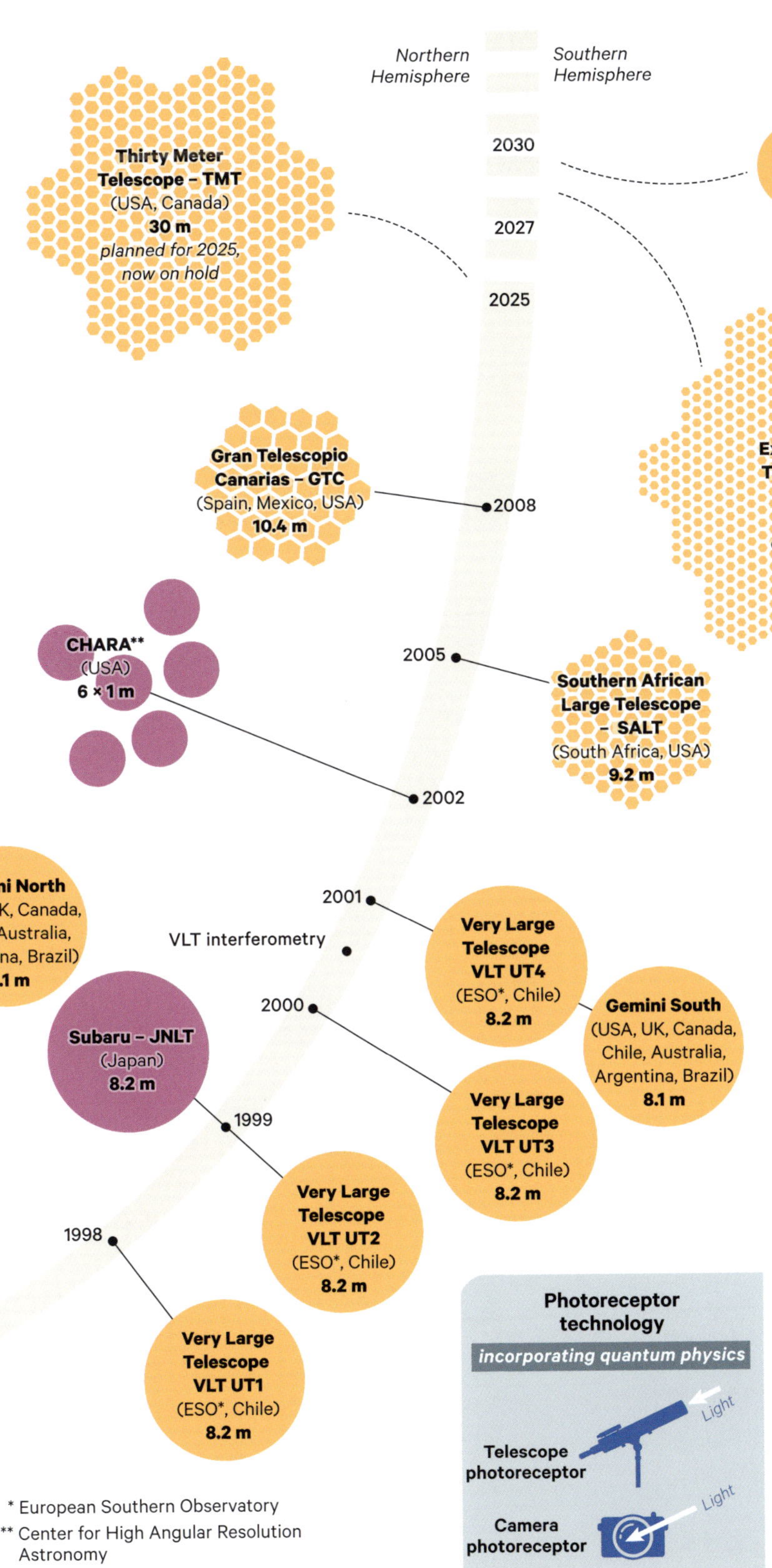

* European Southern Observatory
** Center for High Angular Resolution Astronomy

Seeing further and better

The field of optics deals with visible and infrared light, most of which is not absorbed by the Earth's atmosphere up to a wavelength of 12 micrometres. The amount of light captured depends on the diameter of the primary mirror, which forms an image of the celestial objects inside the telescope, which is equipped with cameras and spectroscopes. After 1990, these mirrors became so large that it was no longer possible for them to be a single unit, so they were made up of smaller segments tessellated together. To improve the images and mitigate the distortions caused by the Earth's atmosphere, telescopes use adaptive optics, which significantly improve image resolution. Today, they are digitally adjusted to compensate for the Earth's rotation. Combining images from a number of telescopes in distant locations has led to significant advances in the level of detail captured.

Radio telescopes

Advances in electronics and telecommunications led to the birth and development of radio astronomy from 1933 onwards. This deals with wavelengths of more than a millimetre, making it possible to discover new phenomena and perceive much smaller details.

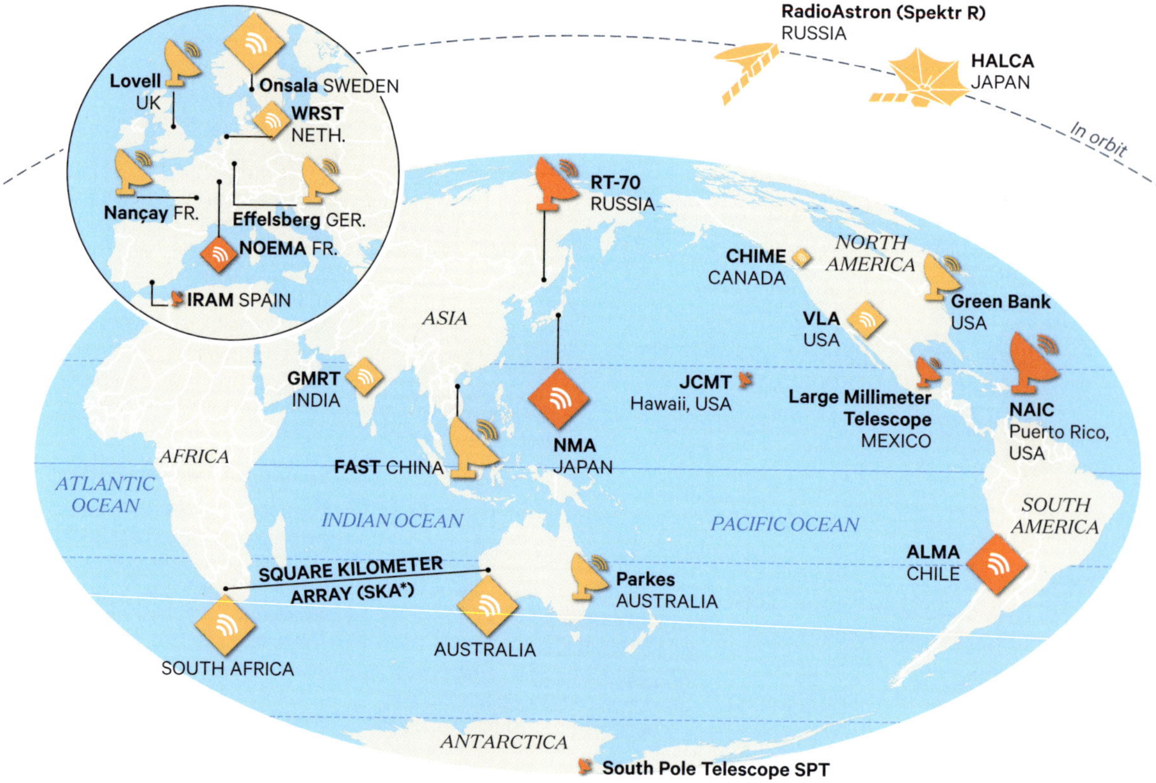

A huge range of wavelengths

To cover the entirety of the sky, radio telescopes are spread around the Earth's surface. Some are even located in space, and are capable of combining with instruments on Earth. The wavelengths of light that they collect cover a huge range, from a millimetre to a kilometre and beyond. To improve the level of detail in the images, radio telescopes form local networks that can then be combined into global networks. The larger the radio telescope's mirror (up to hundreds of metres in diameter), the more its bulk can make it difficult to orientate in order to follow the Earth's rotation. These giant instruments are therefore fixed in place and 'capture' light from the object as it passes into their field.

* International project with telescopes at two different sites

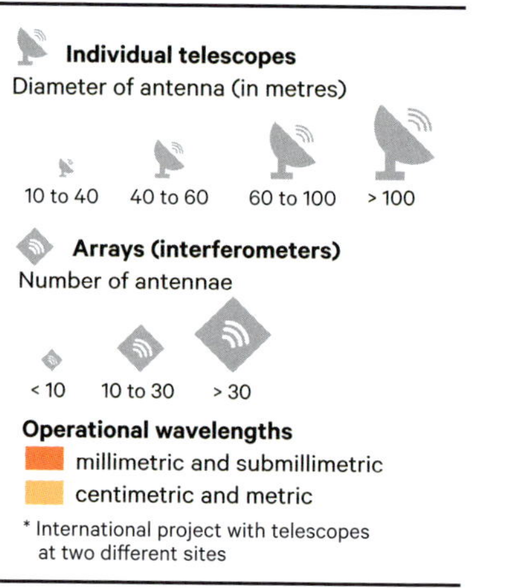

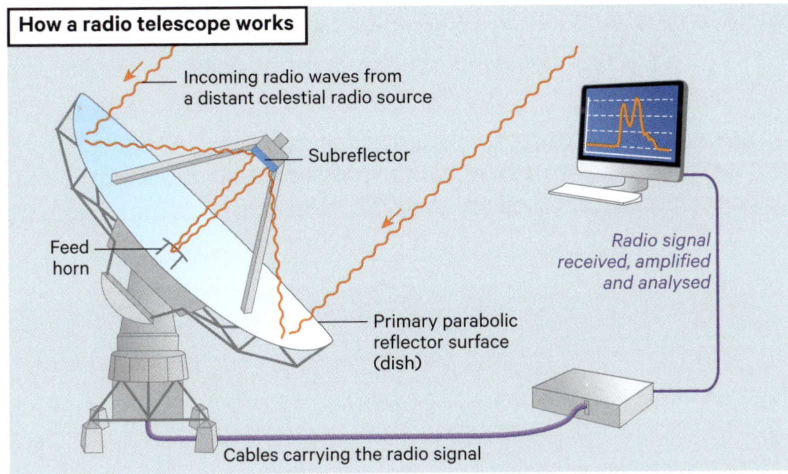

Radio astronomy instruments

Like an optical telescope, a radio telescope has a primary mirror, which reflects and focuses the light from celestial objects. Inside it, receptors transform the light wave into an electrical signal. As in optics, the details of the image are limited by the diffraction of light by the mirror, but not greatly distorted by the atmosphere.

When a number of radio telescopes are combined into a network, their radio signals are all transmitted to a single site. They are then combined using an electronic device called a 'correlator', which measures their similarity and creates an image of the object. The further apart the telescopes are, the more detailed the image. Combining the data in one place makes it possible to measure the similarity and construct an image of the object being observed. The global reach of the EHT (Event Horizon Telescope) network makes it possible to perceive details as small as a millionth of an arc-second.

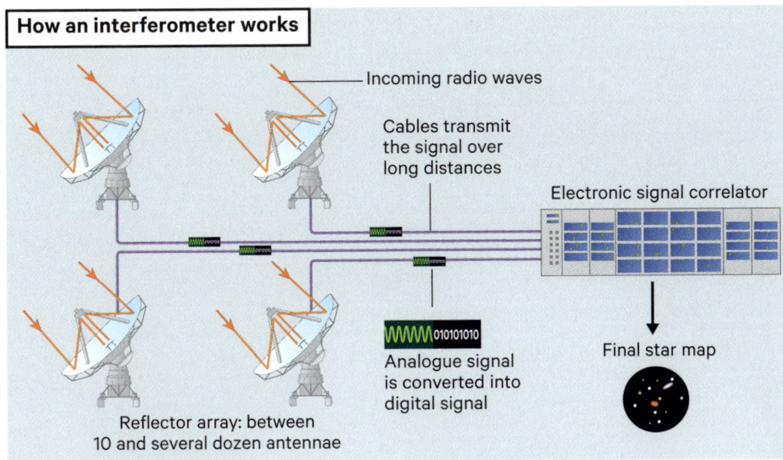

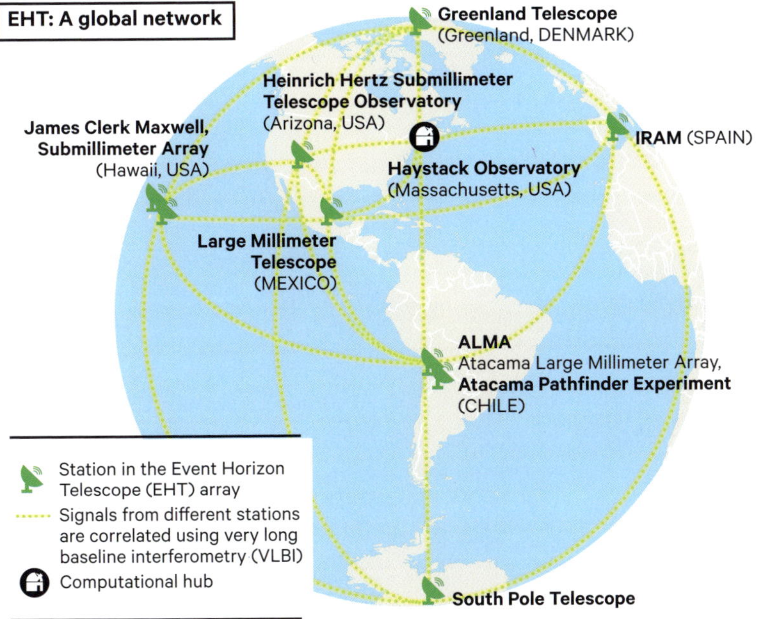

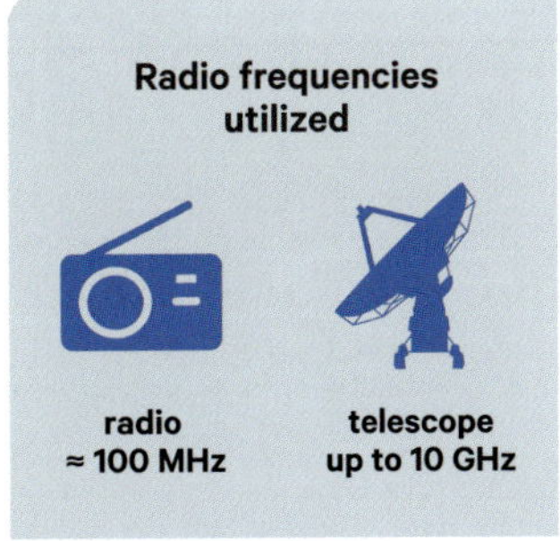

Space telescopes

The ability to place telescopes in orbit transformed our understanding of the sky. By leaving the Earth's atmosphere, these telescopes were able to observe all wavelengths of light. Others could capture particles of high-energy matter emitted by the Sun (solar winds) or by the distant universe (cosmic background radiation).

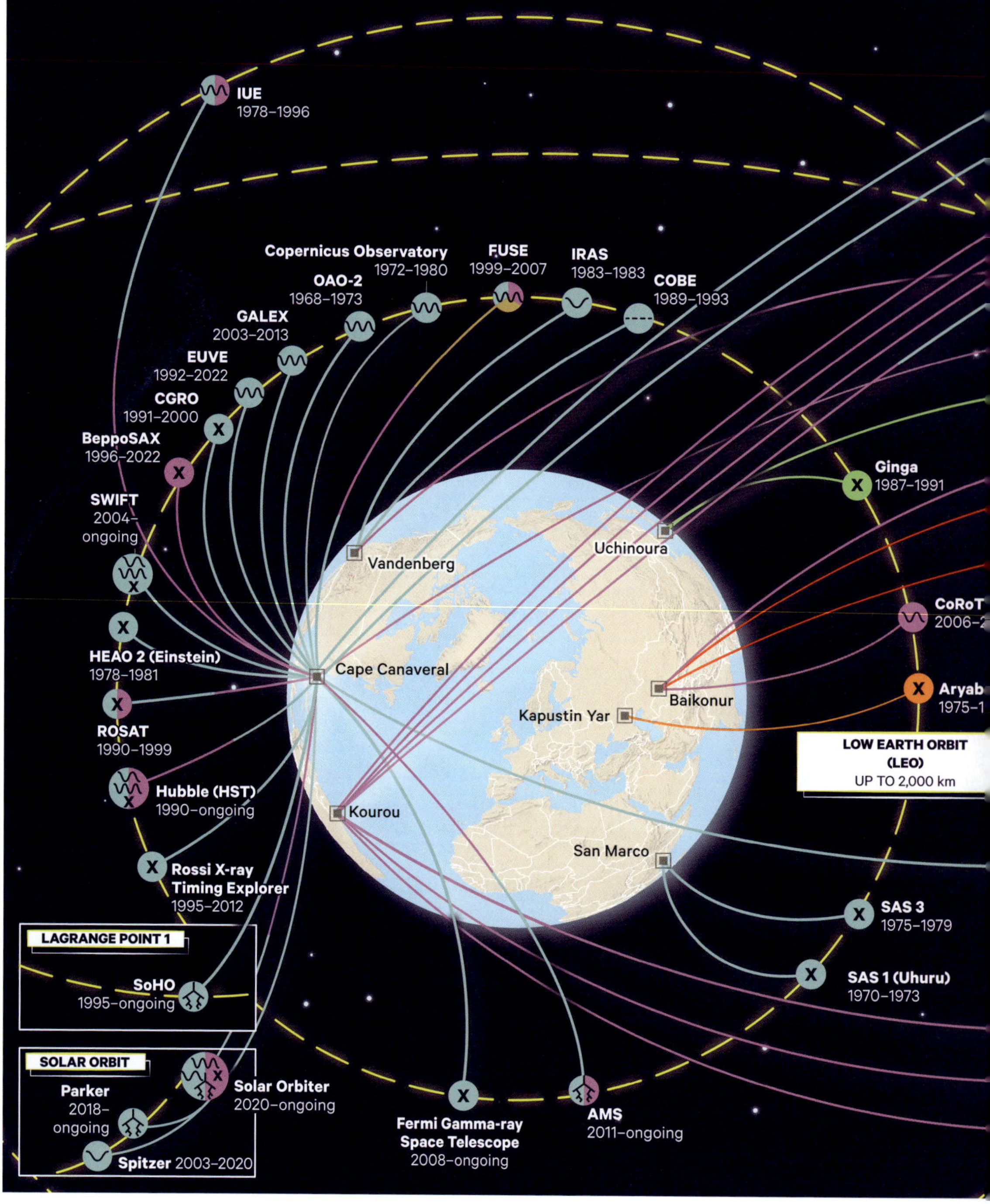

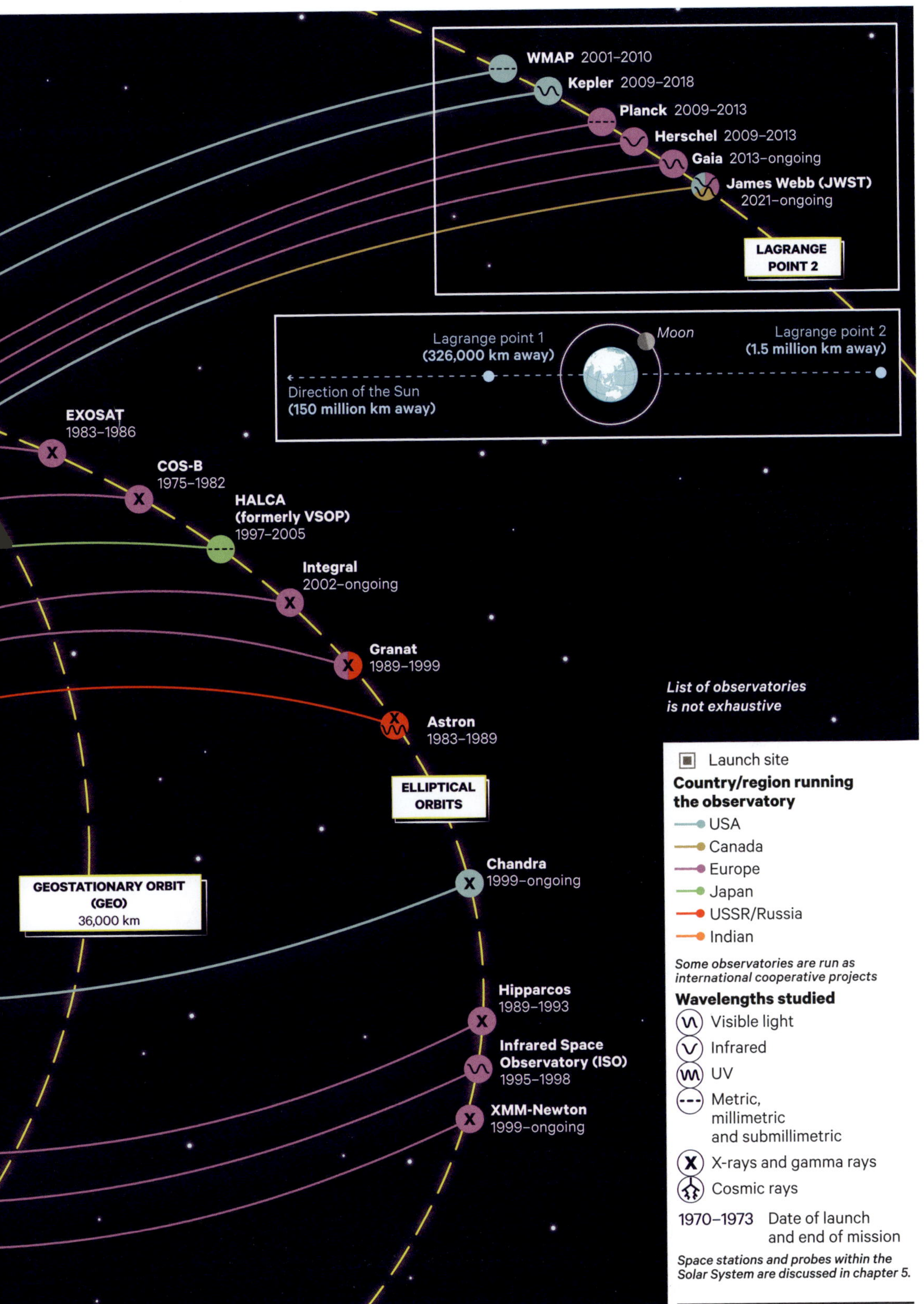
WMAP 2001–2010
Kepler 2009–2018
Planck 2009–2013
Herschel 2009–2013
Gaia 2013–ongoing
James Webb (JWST) 2021–ongoing
LAGRANGE POINT 2
Lagrange point 1 (326,000 km away)
Moon
Lagrange point 2 (1.5 million km away)
Direction of the Sun (150 million km away)
EXOSAT 1983–1986
COS-B 1975–1982
HALCA (formerly VSOP) 1997–2005
Integral 2002–ongoing
Granat 1989–1999
Astron 1983–1989
ELLIPTICAL ORBITS
List of observatories is not exhaustive
Chandra 1999–ongoing
GEOSTATIONARY ORBIT (GEO) 36,000 km
Hipparcos 1989–1993
Infrared Space Observatory (ISO) 1995–1998
XMM-Newton 1999–ongoing
Launch site
Country/region running the observatory
USA
Canada
Europe
Japan
USSR/Russia
Indian
Some observatories are run as international cooperative projects
Wavelengths studied
Visible light
Infrared
UV
Metric, millimetric and submillimetric
X-rays and gamma rays
Cosmic rays
1970–1973 Date of launch and end of mission
Space stations and probes within the Solar System are discussed in chapter 5.

Solar and cosmic particles

Photons are not the only particles that reach the Earth; others include the nuclei of atoms, protons, electrons and neutrinos. In 1920, particles of celestial origin were detected using stratospheric balloons. These particles are emitted and accelerated by the Sun, stars and other extraterrestrial sources. Detectors have been placed under the sea, underground or in space in order to measure them and help us to understand where they come from.

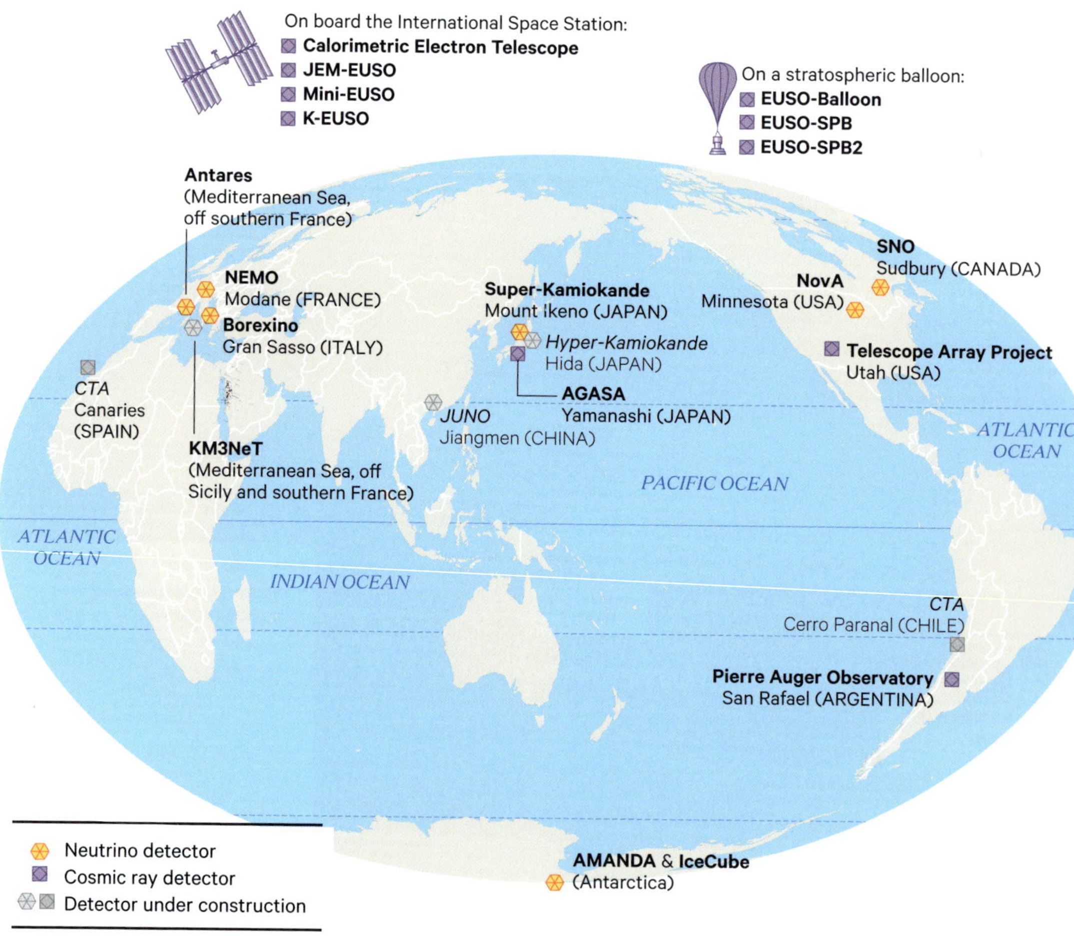

Cosmic rays and neutrinos

The term 'cosmic ray' dates back to the 1900s, when it was difficult for physicists to distinguish between photons (X-rays, gamma rays) and other elementary particles, which were also classified as rays. These particles come from the Sun, as well as exploding stars in our galaxy and beyond. They travel very quickly and possess a huge amount of energy: a single particle can almost have the same amount of energy as a gunshot. Neutrinos are another category of elementary particles. These do not interact strongly with matter, which makes them difficult to detect. Because the information about the Sun and stars that these particles carry is so important to our understanding of the universe, scientists began to build observatories dedicated to them half a century ago, both on Earth and in space.

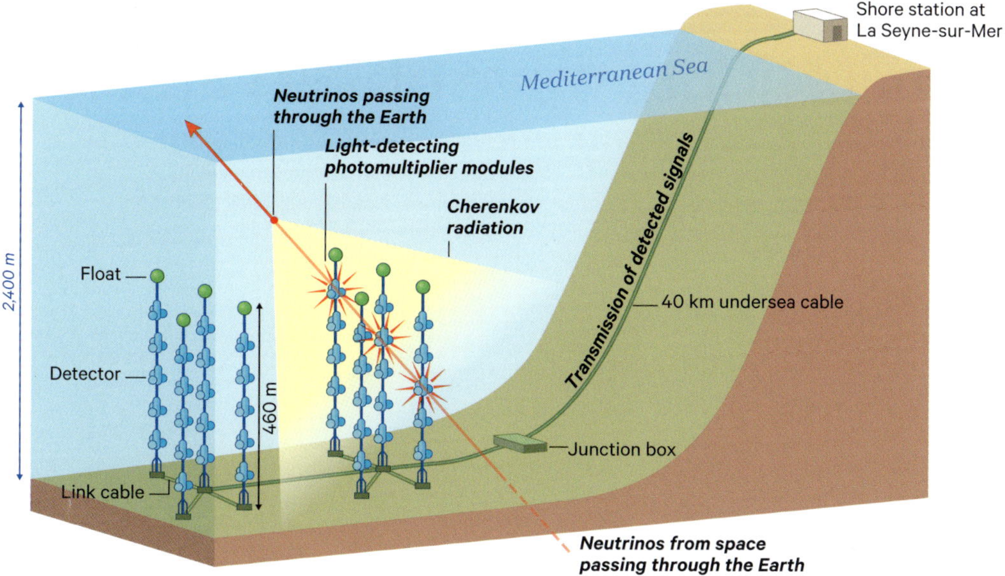

▲ Undersea neutrino detection

A neutrino, coming from a source in the Milky Way or beyond, can travel through the Earth without being blocked. Sometimes, however, it collides with an atom on the seabed, producing another particle known as a muon, which travels in the same direction through the water, creating a cone of visible light around itself called Cherenkov radiation. A network of underwater detectors record this radiation as it passes by and transmit it to a shore station, which can calculate the neutrino's energy and the direction in space that it is travelling from.

▼ Detecting cosmic particles

The atoms present in the Earth's atmosphere serve as a target for cosmic particles that contain an amount of energy that no accelerator on Earth is able to produce. This shock produces showers of secondary particles. The showers also create fluorescence in the atmosphere, which can be detected by the light receptors in fluorescence telescopes. Fluorescence and Cherenkov radiation can also be observed from the International Space Station (ISS), looking back towards the Earth's surface.

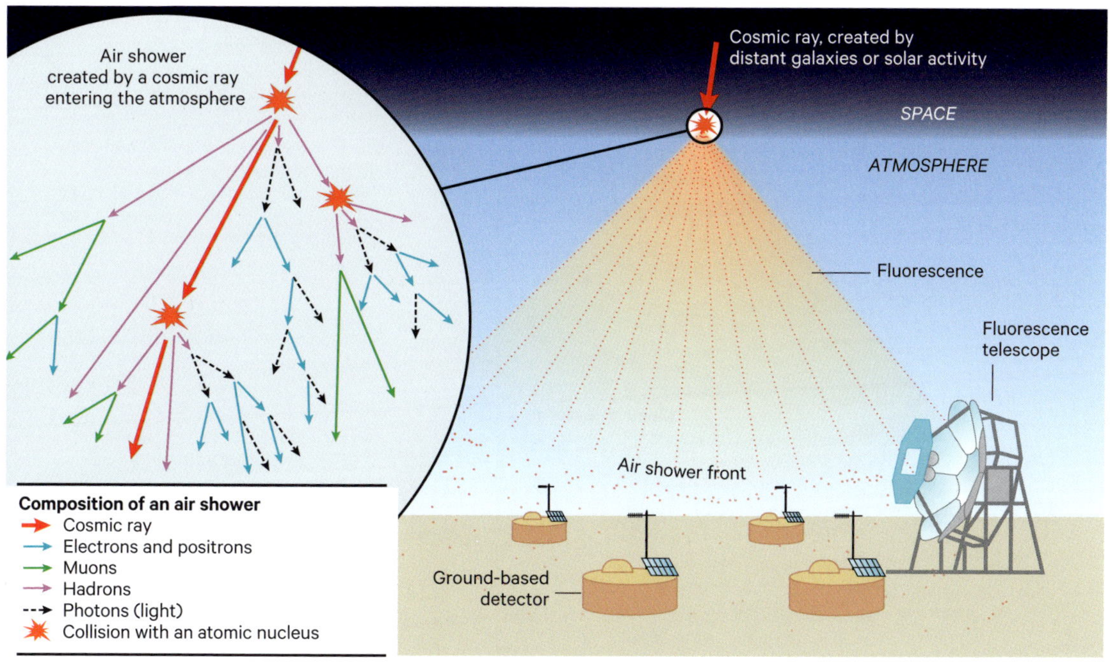

Gravitational wave observatories

In 2015, observations made with a detector that did not look very much like a traditional telescope confirmed the existence of gravitational waves. Their intensity, which is incredibly low in comparison to that of light, reflects the weakness of gravitational interaction. Their detection paved the way for a new astronomy of supermassive black holes.

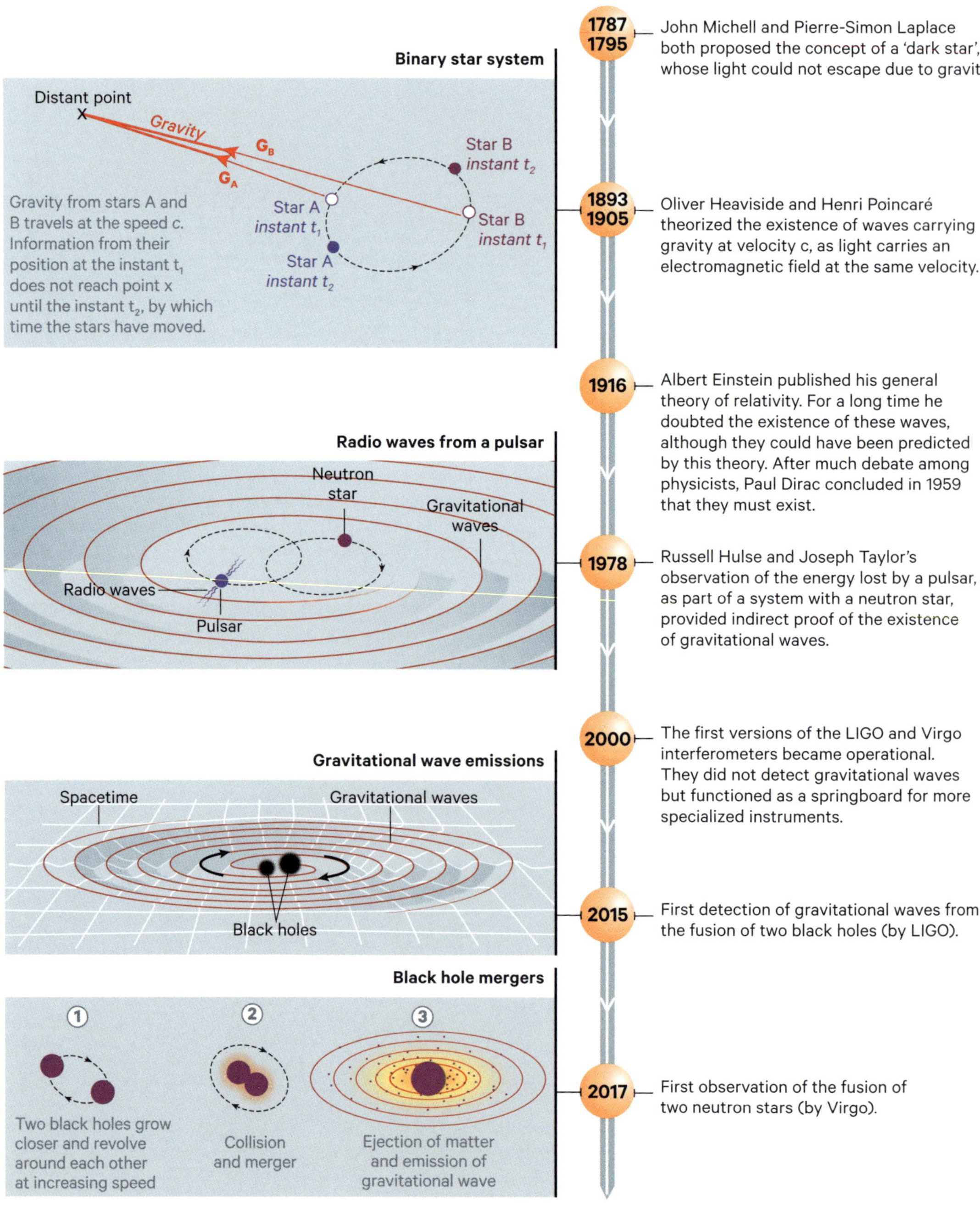

1787
1795 John Michell and Pierre-Simon Laplace both proposed the concept of a 'dark star', whose light could not escape due to gravity.

1893
1905 Oliver Heaviside and Henri Poincaré theorized the existence of waves carrying gravity at velocity c, as light carries an electromagnetic field at the same velocity.

1916 Albert Einstein published his general theory of relativity. For a long time he doubted the existence of these waves, although they could have been predicted by this theory. After much debate among physicists, Paul Dirac concluded in 1959 that they must exist.

1978 Russell Hulse and Joseph Taylor's observation of the energy lost by a pulsar, as part of a system with a neutron star, provided indirect proof of the existence of gravitational waves.

2000 The first versions of the LIGO and Virgo interferometers became operational. They did not detect gravitational waves but functioned as a springboard for more specialized instruments.

2015 First detection of gravitational waves from the fusion of two black holes (by LIGO).

2017 First observation of the fusion of two neutron stars (by Virgo).

Gravitational wave observatories worldwide

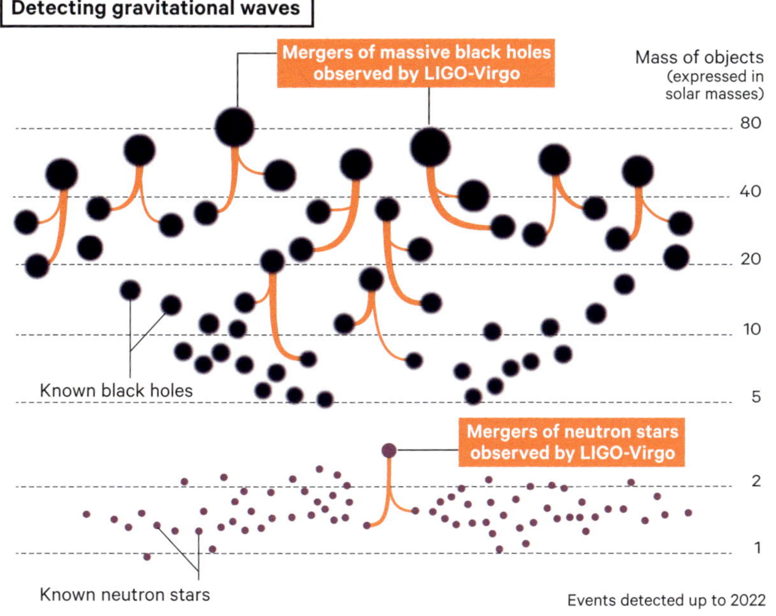

Laser Interferometer Gravitational Wave Observatory (LIGO)

Detecting gravitational waves

Decoding gravitational waves

'Spacetime tells matter how to move; matter tells spacetime how to curve.' This is how the physicist John Wheeler explained gravity in the context of Einstein's general relativity. General relativity states that there is a maximum velocity ($c \approx 300,000$ km/s) at which any kind of information can travel. Therefore there must be a wave that carries the gravitational field from one place to another. When a double star is the source of this gravitational field, the wave moves in space as the stars orbit. Compared to electromagnetism, gravity is an incredibly weak force. Detecting gravitational waves through experiments was a challenge, and it wasn't achieved until 2015.

The detector (a massive interferometer with a laser) measures the minuscule deformations of space caused by the passage of a wave. It is made up of two perpendicular arms, each of which is over a kilometre long. One extends by a few billionths of billionths of a millimetre, while the other shortens by the same amount. This difference between the arms is measured by perfectly stable mirrors, onto which the laser is reflected. Eventually, in 2015, a wave was detected that had been emitted by the fusion of two black holes.

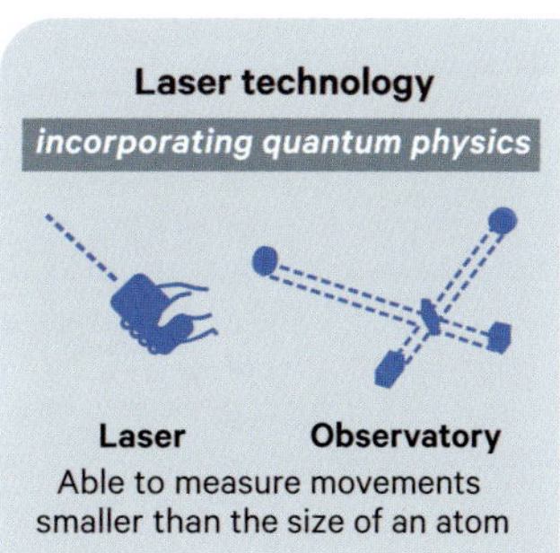

Laser technology
incorporating quantum physics

Laser   **Observatory**
Able to measure movements smaller than the size of an atom

Computer science and digital simulations

From the 1960s onwards, our knowledge of the sky was improved by the rapid growth of computer science and the construction of computers that drew on quantum physics. These were able to perform more complex calculations. Digital simulations have made it possible to demonstrate the evolution of phenomena from the distant past, and predict those that may occur in the future.

The evolution of supercomputers

Mflop: megaFLOP
Pflop: petaFLOP
Eflop: exaFLOP

Ever-improving processing power

Calculations have always been essential for accurately describing the sky. The advent of computers gave scientists access to greater calculating capabilities. A computer program is executed within its core, its central processing unit (CPU). The speed with which information is analysed by the core is measured in flops, the number of operations that can be carried out per second. One of the major advances of the 21st century was increasing the speed of these cores, and allowing many millions of them to work simultaneously. The amount of information that is fed into a computer, or that it produces after analysing it, requires huge amounts of rapid-access memory. Drawing on the physics of solids and material technology at a nanometric scale, processing power is constantly increasing, making it possible to carry out ever more complex scientific calculations.

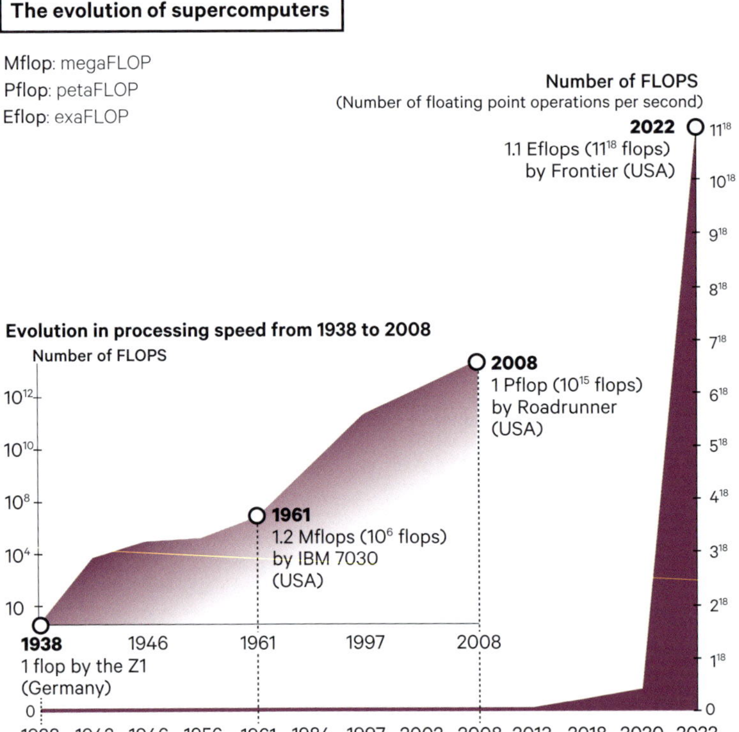

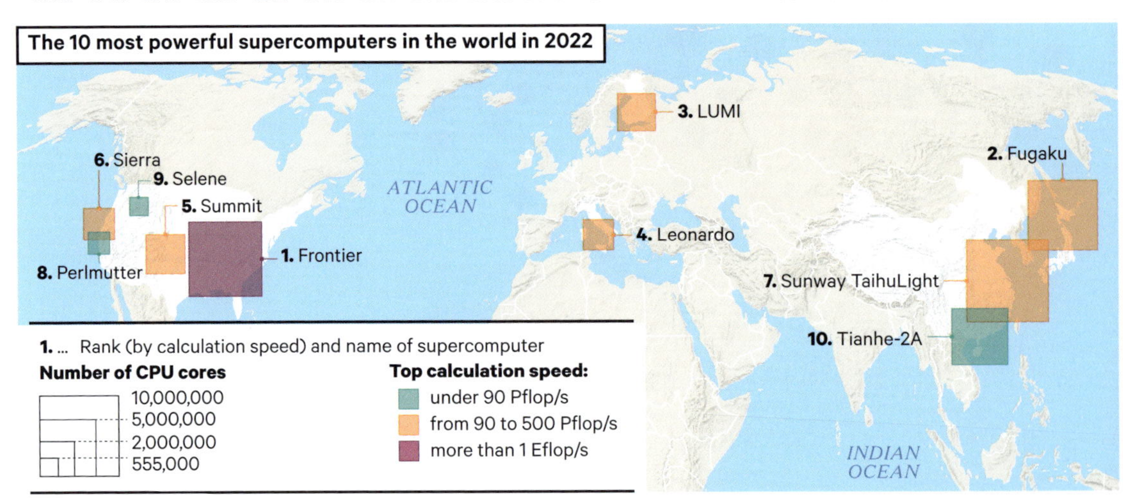

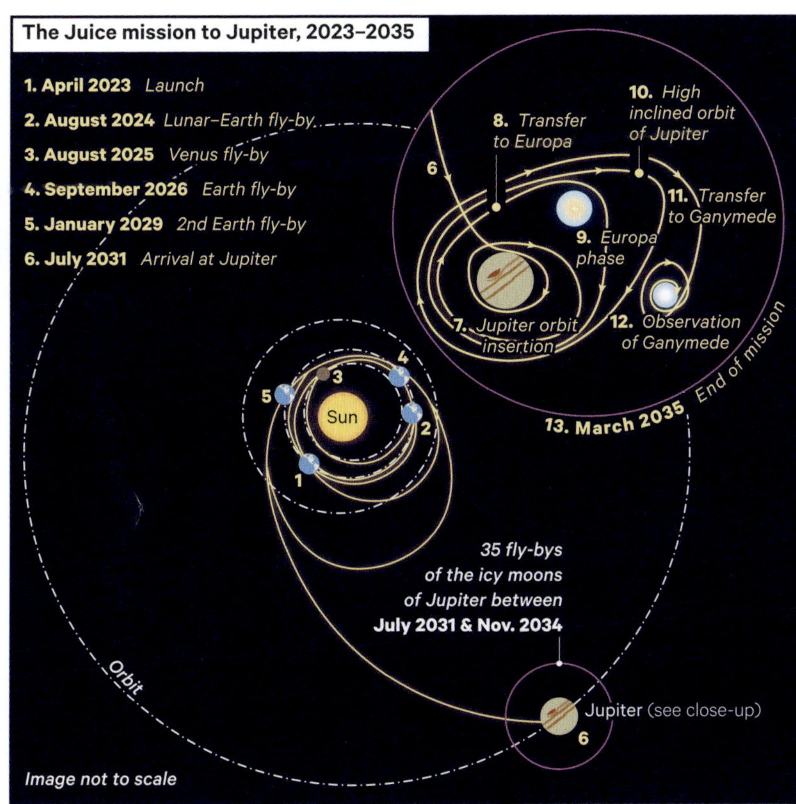

Navigating the Solar System

In 2023, the space probe Juice, equipped with a number of instruments, was launched by the European Space Agency towards the planet Jupiter and its moons. It is predicted to arrive after following a meandering path through the Solar System for eight years. Gravitational assistance, obtained by flying by Earth and Venus multiple times, will provide the probe with a large amount of energy that the launch rocket, Ariane 5, was not able to supply. The probe will then travel towards Jupiter, passing near Europa and Callisto before entering orbit around Ganymede, and eventually landing on it. None of this would be possible without powerful computers to perform the calculations required to plan the mission and follow its progress in real time.

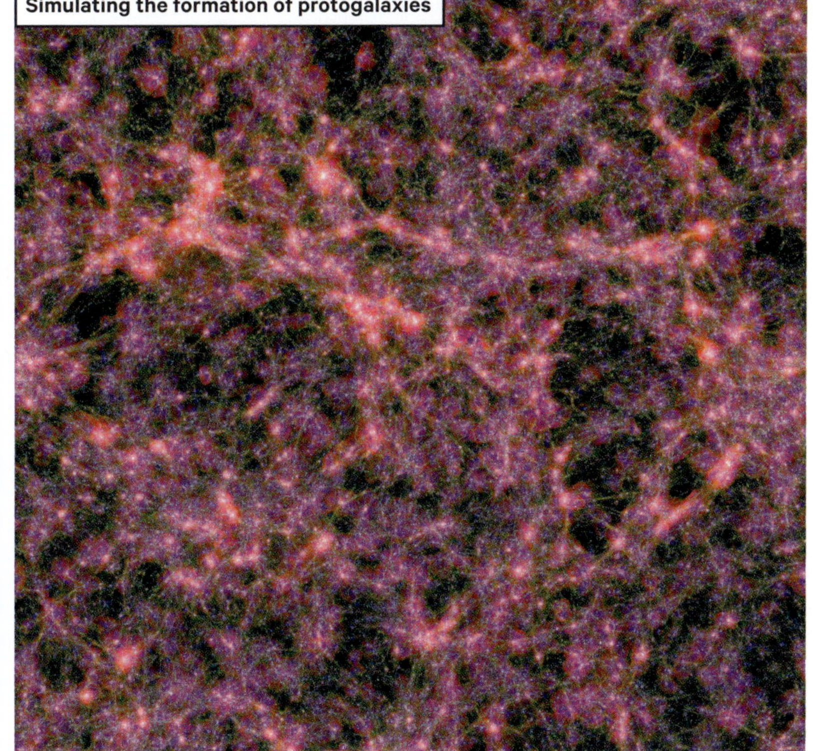

Simulating the evolution of matter

To understand how matter evolved, the computational astrophysics project Horizon used the laws of physics to calculate the composition of gas, made up mainly of atoms of hydrogen helium, oxygen and iron, over time in a vast cube of space with its sides measuring 300 million light-years. By accelerating time, a few weeks of calculations can simulate millions of years of evolution. After 6 million hours of calculations, the Jade supercomputer at CINES in France produced the image opposite, which shows the state of the universe just under 13.7 billion years ago. It shows filament-like structures overlapping each other, with areas of gas density in green, gas temperature in red and gas metallicity in blue.

Collaboration and competition

Advances in the field of astronomy are driven by the development of more powerful tools, which often require financial support that a single country alone cannot provide. This has led to international initiatives, although it has not meant the end of competition between scientists. Global participation in these initiatives varies a great deal, because it depends heavily on the wealth of the countries involved.

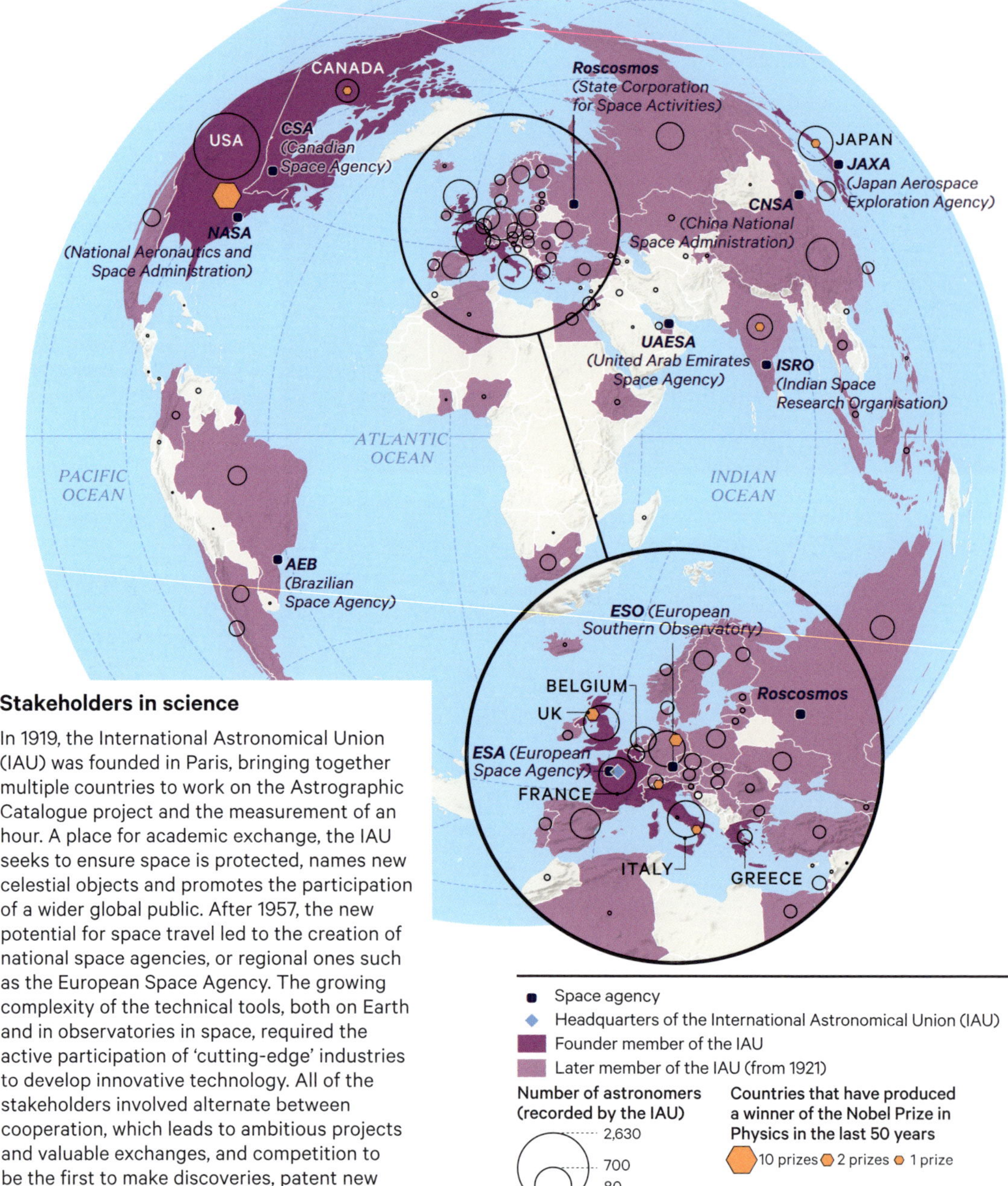

Stakeholders in science

In 1919, the International Astronomical Union (IAU) was founded in Paris, bringing together multiple countries to work on the Astrographic Catalogue project and the measurement of an hour. A place for academic exchange, the IAU seeks to ensure space is protected, names new celestial objects and promotes the participation of a wider global public. After 1957, the new potential for space travel led to the creation of national space agencies, or regional ones such as the European Space Agency. The growing complexity of the technical tools, both on Earth and in observatories in space, required the active participation of 'cutting-edge' industries to develop innovative technology. All of the stakeholders involved alternate between cooperation, which leads to ambitious projects and valuable exchanges, and competition to be the first to make discoveries, patent new processes or attract the best scientists.

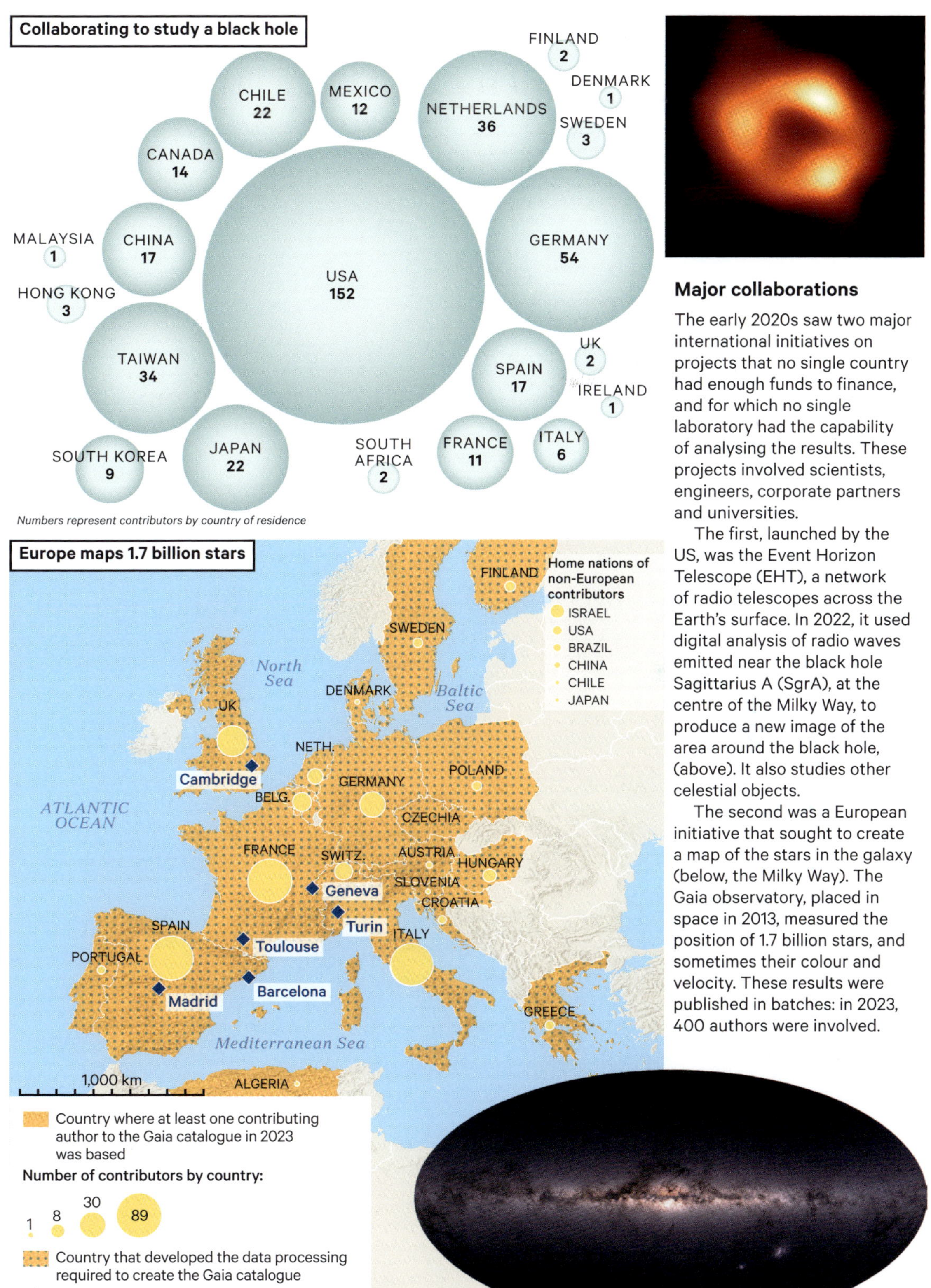

Major collaborations

The early 2020s saw two major international initiatives on projects that no single country had enough funds to finance, and for which no single laboratory had the capability of analysing the results. These projects involved scientists, engineers, corporate partners and universities.

The first, launched by the US, was the Event Horizon Telescope (EHT), a network of radio telescopes across the Earth's surface. In 2022, it used digital analysis of radio waves emitted near the black hole Sagittarius A (SgrA), at the centre of the Milky Way, to produce a new image of the area around the black hole, (above). It also studies other celestial objects.

The second was a European initiative that sought to create a map of the stars in the galaxy (below, the Milky Way). The Gaia observatory, placed in space in 2013, measured the position of 1.7 billion stars, and sometimes their colour and velocity. These results were published in batches: in 2023, 400 authors were involved.

5

Space exploration

After thousands of years in which the sky could only be seen from afar, humans can now finally explore it, in person or using probes and robots. Our knowledge of the Solar System has increased vastly. The Earth became a celestial object itself, and the humans that live on it could be viewed from a distance for the first time. The Moon and the planets in the Solar System have been mapped and their geography studied. Humans had conquered space and began to use it for their own ends, as it did not belong to anyone and was open to all those who had the means to reach it. The uses of satellites for observation, measurement and surveillance has gone hand in hand with globalization. Space law, still in its infancy, must now deal with issues such as militarization and attempts to lay claim to parts of space.

A timeline of flight

The origins of space exploration can be traced back to the technological advances made by Nazi Germany during the Second World War, with V2 missiles. The Cold War then brought fierce competition between the US and the USSR. Space technology was developed with the aim of bettering our scientific understanding of the Earth and the universe. Subsequently, it became a subject for both rivalry and collaboration on the international stage.

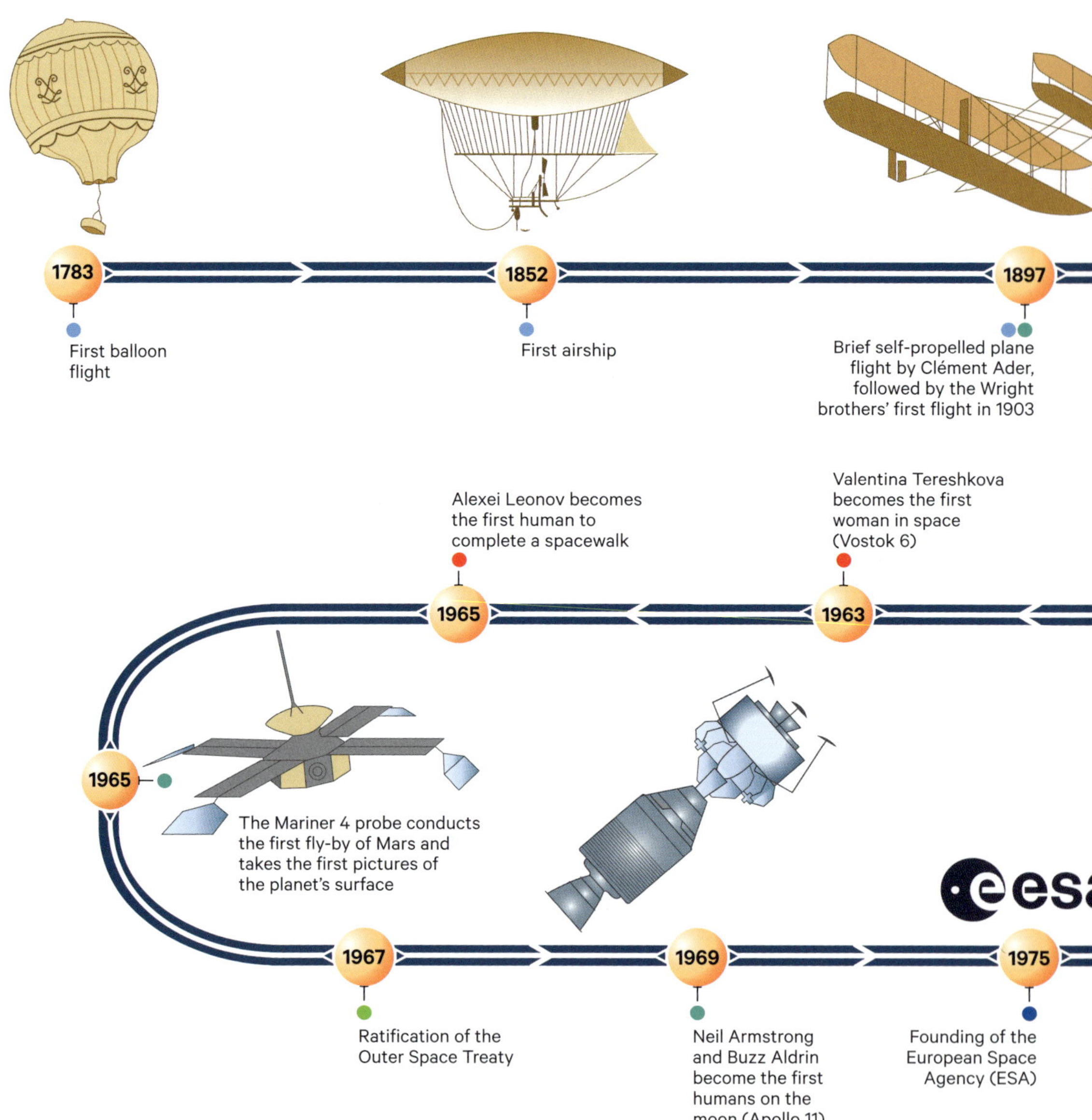

1783
First balloon flight

1852
First airship

1897
Brief self-propelled plane flight by Clément Ader, followed by the Wright brothers' first flight in 1903

1965
Alexei Leonov becomes the first human to complete a spacewalk

1963
Valentina Tereshkova becomes the first woman in space (Vostok 6)

1965
The Mariner 4 probe conducts the first fly-by of Mars and takes the first pictures of the planet's surface

1967
Ratification of the Outer Space Treaty

1969
Neil Armstrong and Buzz Aldrin become the first humans on the moon (Apollo 11)

1975
Founding of the European Space Agency (ESA)

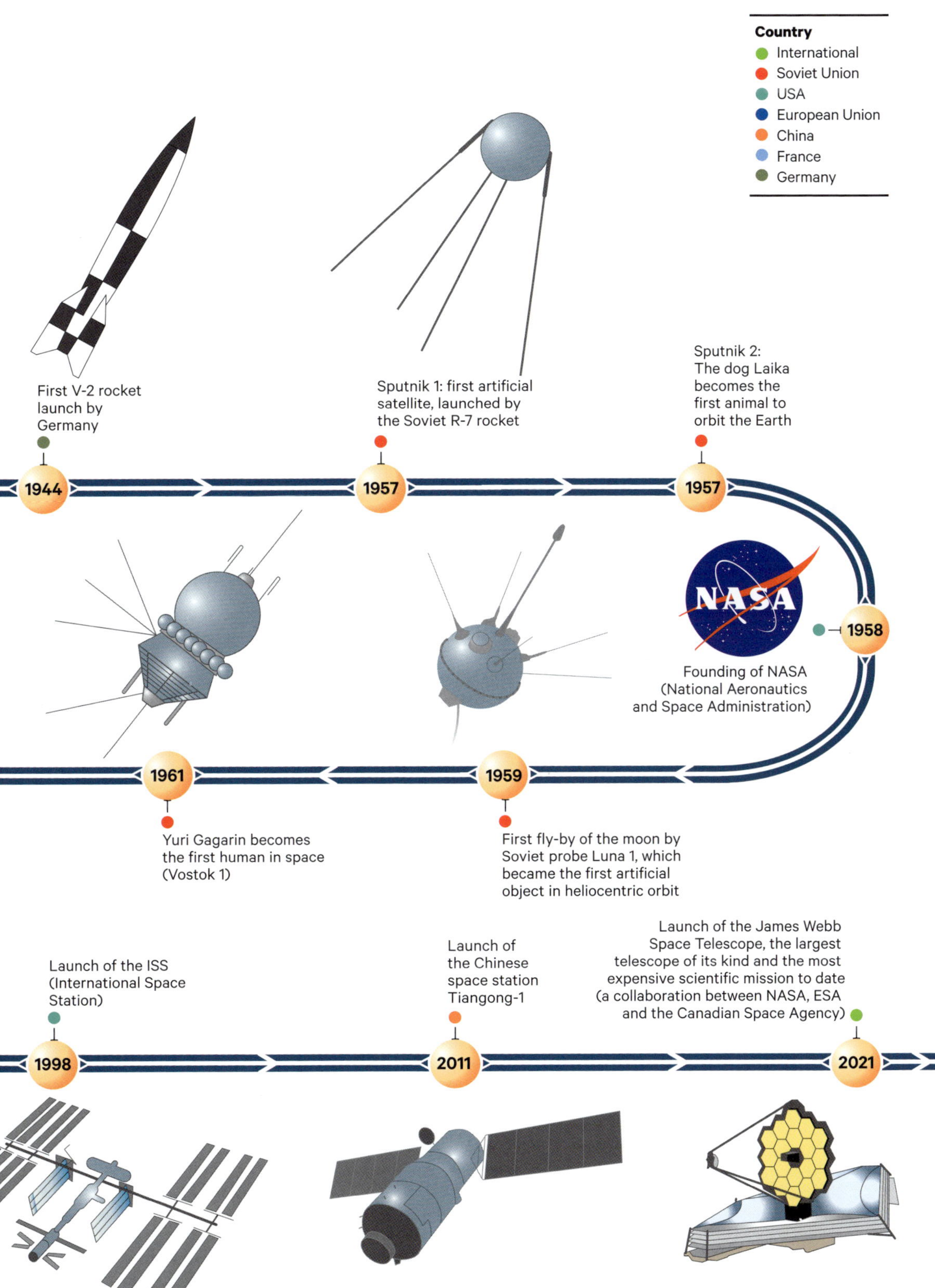

Country
International
Soviet Union
USA
European Union
China
France
Germany

First V-2 rocket
launch by
Germany

1944

Sputnik 1: first artificial
satellite, launched by
the Soviet R-7 rocket

1957

Sputnik 2:
The dog Laika
becomes the
first animal to
orbit the Earth

1957

NASA

1958

Founding of NASA
(National Aeronautics
and Space Administration)

1961

Yuri Gagarin becomes
the first human in space
(Vostok 1)

1959

First fly-by of the moon by
Soviet probe Luna 1, which
became the first artificial
object in heliocentric orbit

Launch of the ISS
(International Space
Station)

1998

Launch of
the Chinese
space station
Tiangong-1

2011

Launch of the James Webb
Space Telescope, the largest
telescope of its kind and the most
expensive scientific mission to date
(a collaboration between NASA, ESA
and the Canadian Space Agency)

2021

Humans in space

Since Yuri Gagarin first achieved the feat in 1961, more than 600 astronauts have travelled to space. While the USSR had the first successful missions, it was soon overtaken by the United States, who were the first to send crewed flights to the Moon, between 1969 and 1972. Since then, the US has remained the most powerful nation in space, but it is now facing competition from other nations, including China.

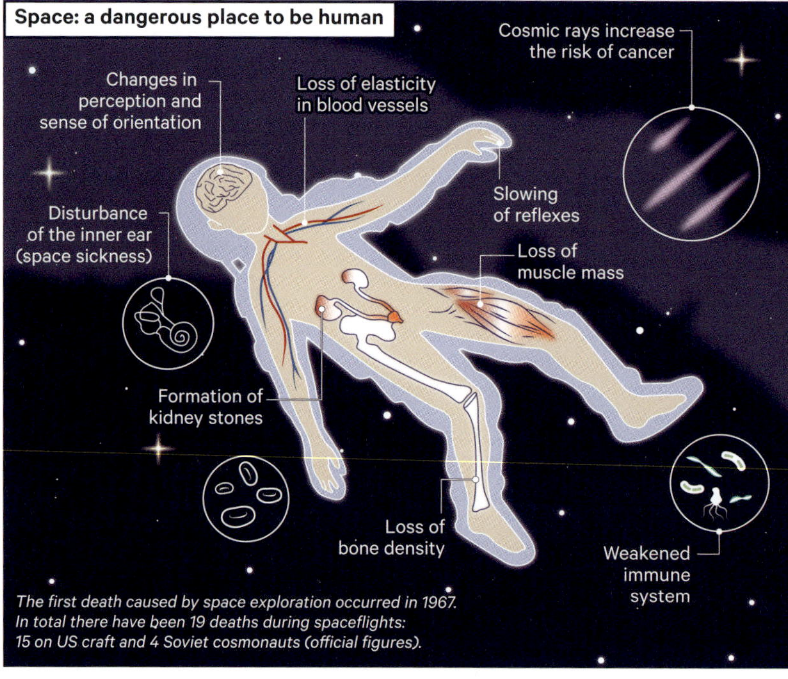

The first death caused by space exploration occurred in 1967. In total there have been 19 deaths during spaceflights: 15 on US craft and 4 Soviet cosmonauts (official figures).

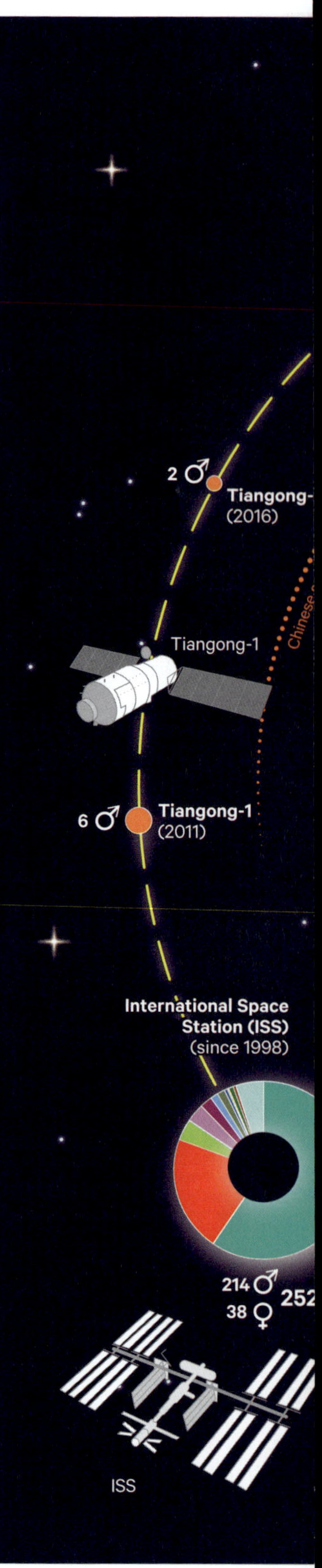

From ideological rivalry to international collaboration

The space race between the USSR and the United States after the Second World War was a reflection of the military, cultural, technological and ideological rivalry between these two superpowers. Successes were used as evidence to 'prove' the superiority of one political system over another. The Soviets achieved the first space flight and put the first space station in orbit. They were also the first to set up a programme that promoted international cooperation in space exploration. Named 'Interkosmos', it allowed astronauts from countries in the Soviet bloc, as well as the UK, France and India, to travel to space. As their investment started to bear fruit, the US caught up with and then overtook its competitor. Today, national rivalries coexist with international collaboration, with the private sector also playing an active part. Abandoned for many decades, plans to send humans to the Moon or Mars, or even to colonize them, are now back on the agenda. But in addition to financial concerns, these projects also raise the issue of the impact that spending an extended time in space may have on human health, both physical and mental.

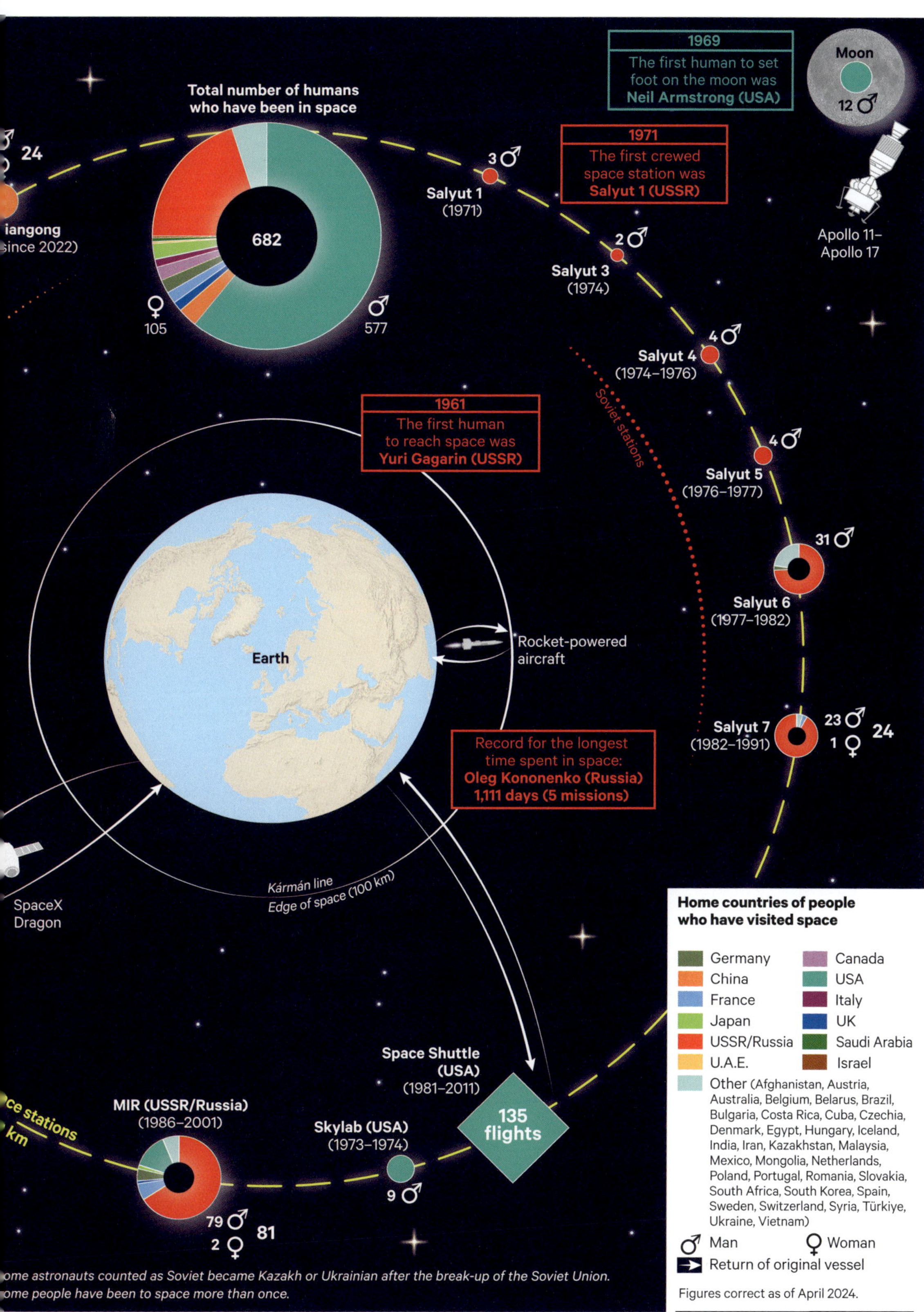

Some astronauts counted as Soviet became Kazakh or Ukrainian after the break-up of the Soviet Union.
Some people have been to space more than once.

Exploring the Moon

The conquest of the Moon was initially an US–Soviet race. While the USSR was the first country to land a probe there, the US was the first and so far only country to send humans. Then, in the 1970s, probe missions to the Moon petered out, before restarting in the 1990s, now with new stakeholders involved, particularly China. The Moon is now seen as a potential stopover point on the way to Mars.

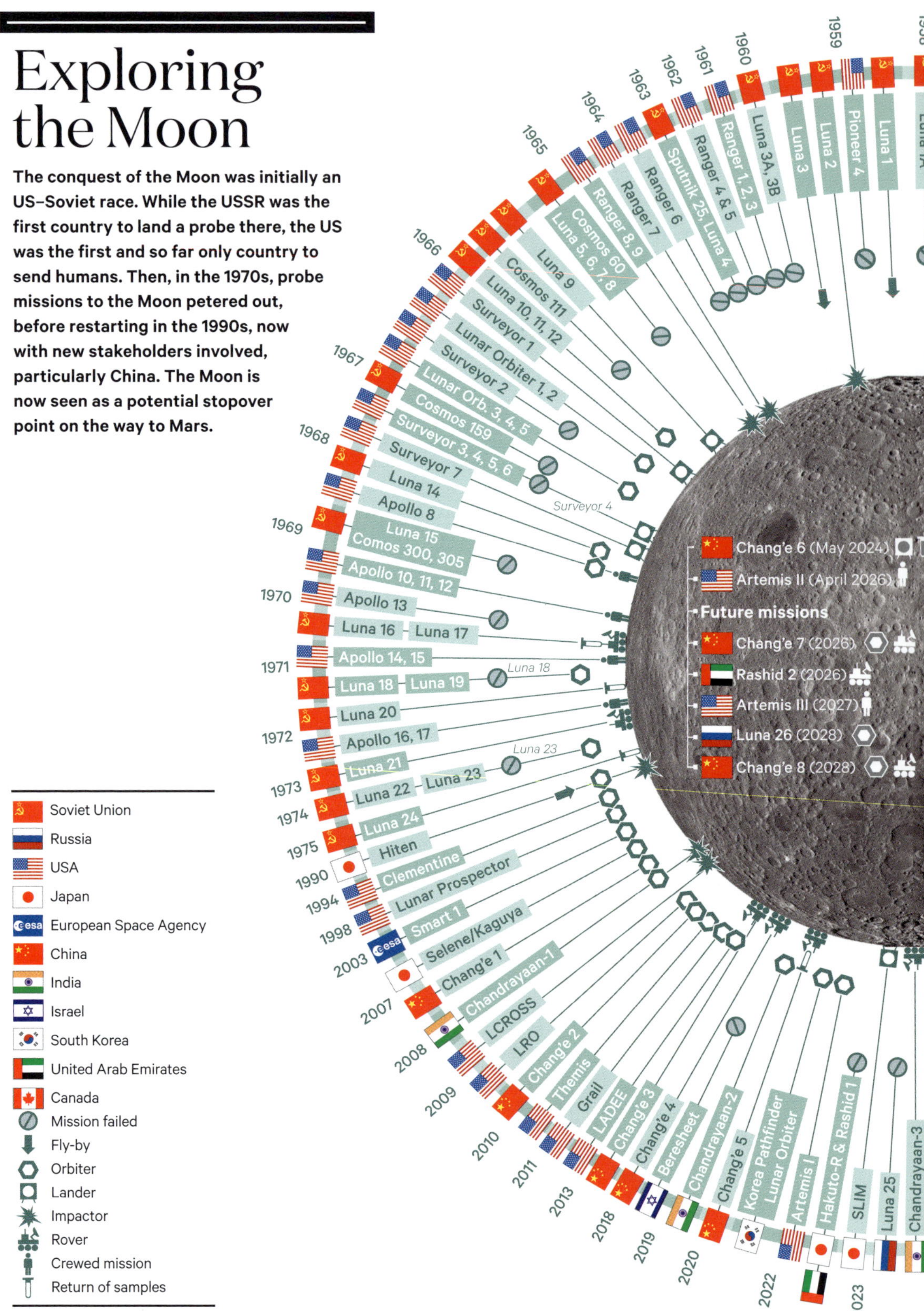

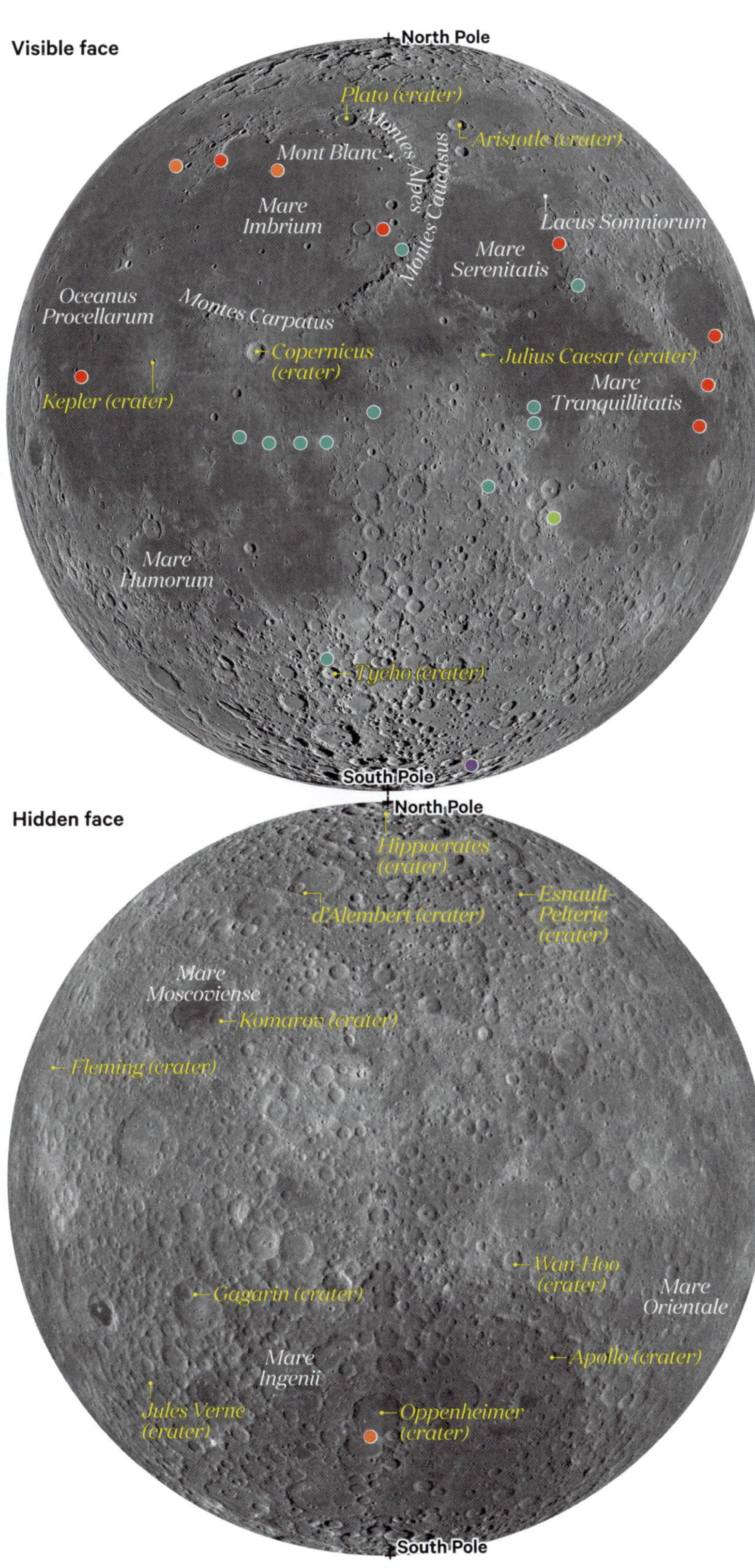

The hidden face of the Moon

Since the days of Galileo's telescope, the visible face of the Moon has been closely observed and described. Knowledge of its hidden face is much more recent and dates to the first time it was flown over by probes. Since then, the entire surface of the Moon has been mapped in detail. The Moon is often shown with its two faces side by side: the visible face, as it is seen from Earth, and the hidden face. This is easy for amateur astronomers to understand, as it reflects what can be seen through a telescope, but it gives a false impression that the two faces are somehow different in nature. This projection does, however, make it possible to see the differences in the naming of the Moon's geographical features. The features on the visible face were named in the 17th century by a number of scientists, including Johannes Hevelius, Giovanni Battista Riccioli and Francesco Maria Grimaldi. The 'seas' (marea) are named after human qualities or weather phenomena (the 'Sea of Rains', the 'Sea of Tranquillity'). The mountains are named after major mountain ranges on Earth (Alps, Caucasus) and the craters (created by the impact of meteorites) are named after scholars or philosophers (Plato, Copernicus). The features on the hidden face were named much more recently, as can be seen from the craters (Oppenheimer, Gagarin). The first landing on the hidden face, by an uncrewed Chinese probe, did not happen until 2024.

Will humans walk on Mars?

It is not easy to reach the red planet, as many failed projects since the 1960s can testify. While no humans have yet been sent there, a number of robot rovers have explored the planet's surface. Missions to return soil samples seek to answer a recurrent question: was there once life on Mars? NASA is planning a crewed mission to Mars in around 2040.

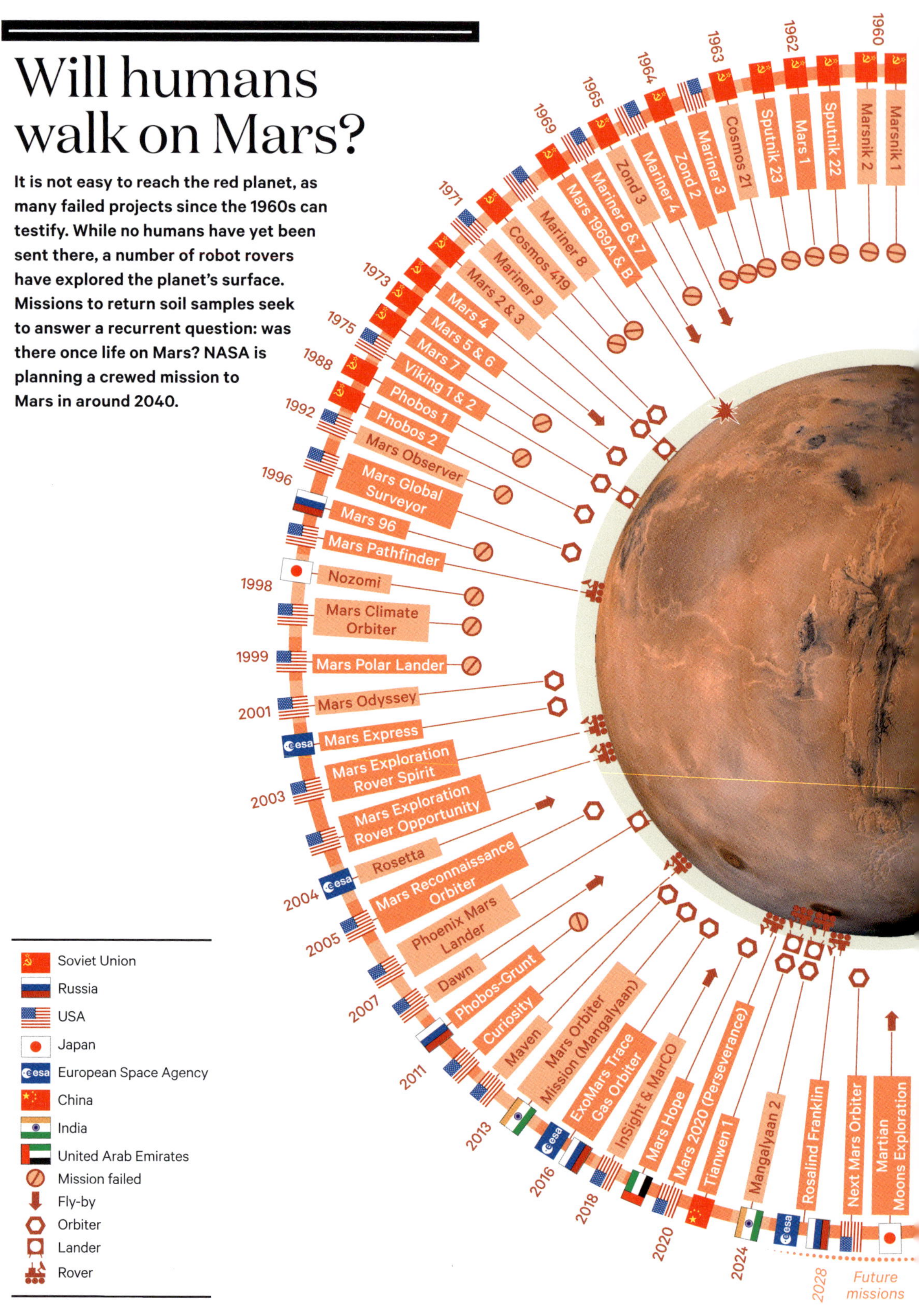

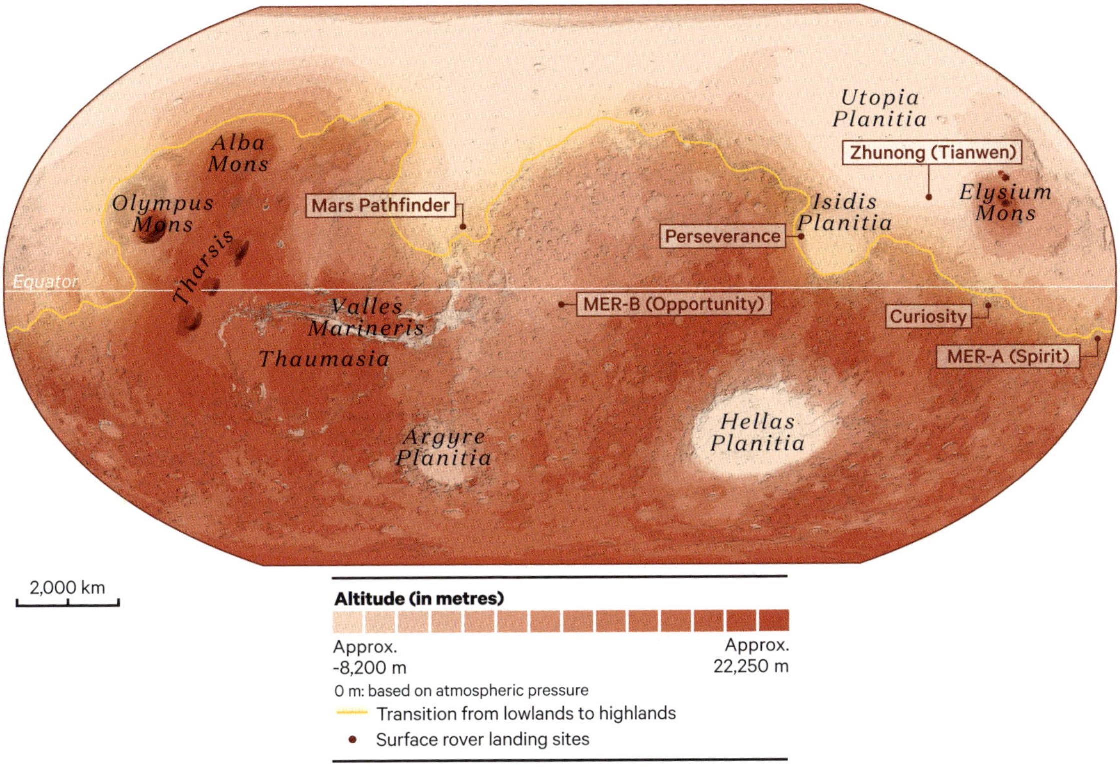

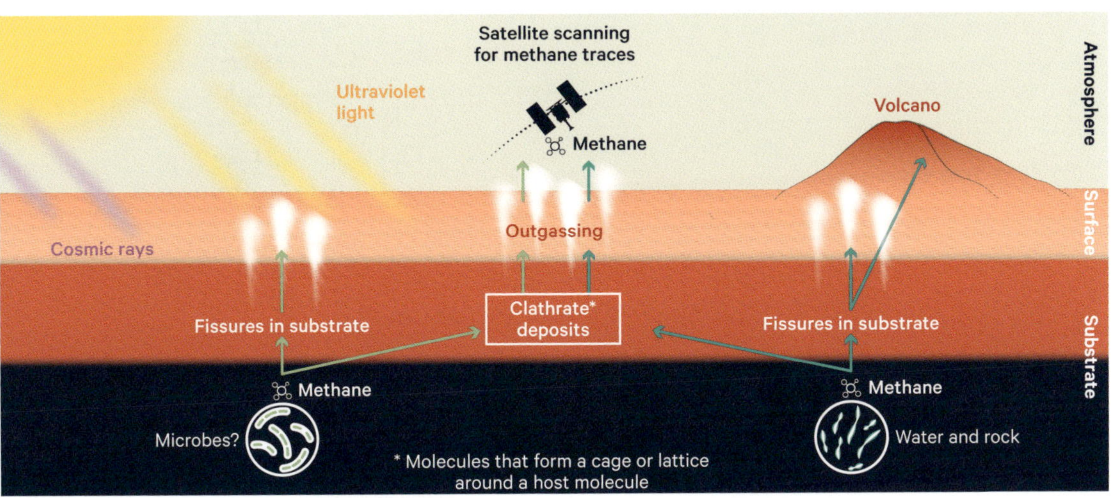

▲ North and south

The differences between the southern and northern hemispheres of Mars are clear at first glance. In the south, high plateaus cover around two thirds of the planet, while the north is made up of vast plains. The reason for this stark difference remains unknown. Another distinctive feature: the planet has the tallest volcanoes in the Solar System. The largest, Olympus Mons, is over 20 km in height.

▼ Is there life on Mars?

The soil on Mars contains frozen water and its atmosphere contains traces of methane. Probes in orbit and rovers on the ground have been using specialist equipment to analyse the concentrations of methane in Mars's atmosphere since 2010. The periodic detection of this unstable gas seems to suggest that there is an active source on the planet. Could it be geochemical processes or microbial life? This question so far remains unanswered.

The surface of Mars

The first fly-by of Mars by the space probe Mariner 4 in 1965 kicked off the study of its topography and geology. A number of missions to the red planet have revealed its huge volcanic massifs and canyons. Mars is a planet with a very geologically active past and it is possible that some of its volcanos might display occasional activity. Meanwhile, robot landers on the surface are searching for possible traces of life, past or present.

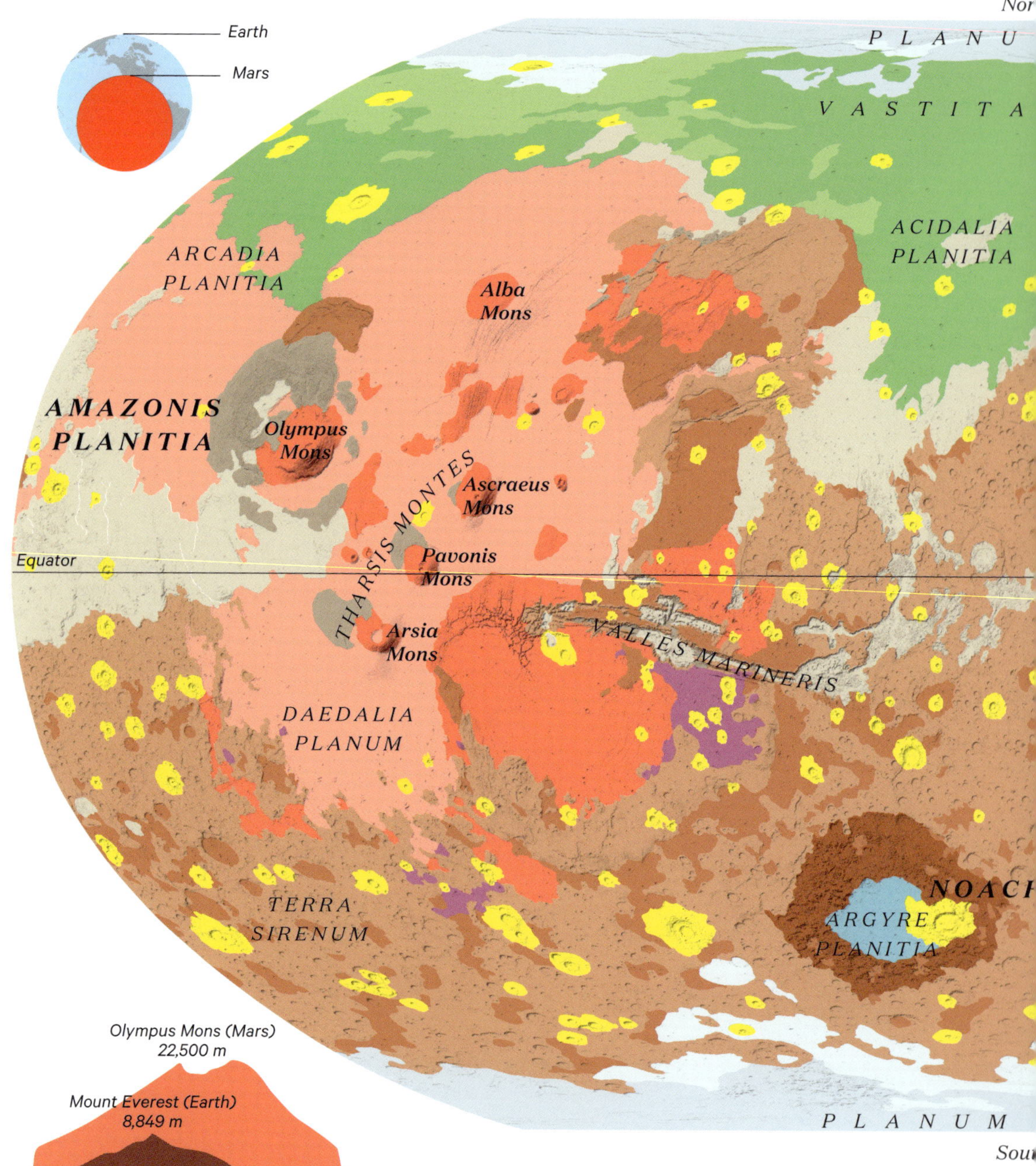

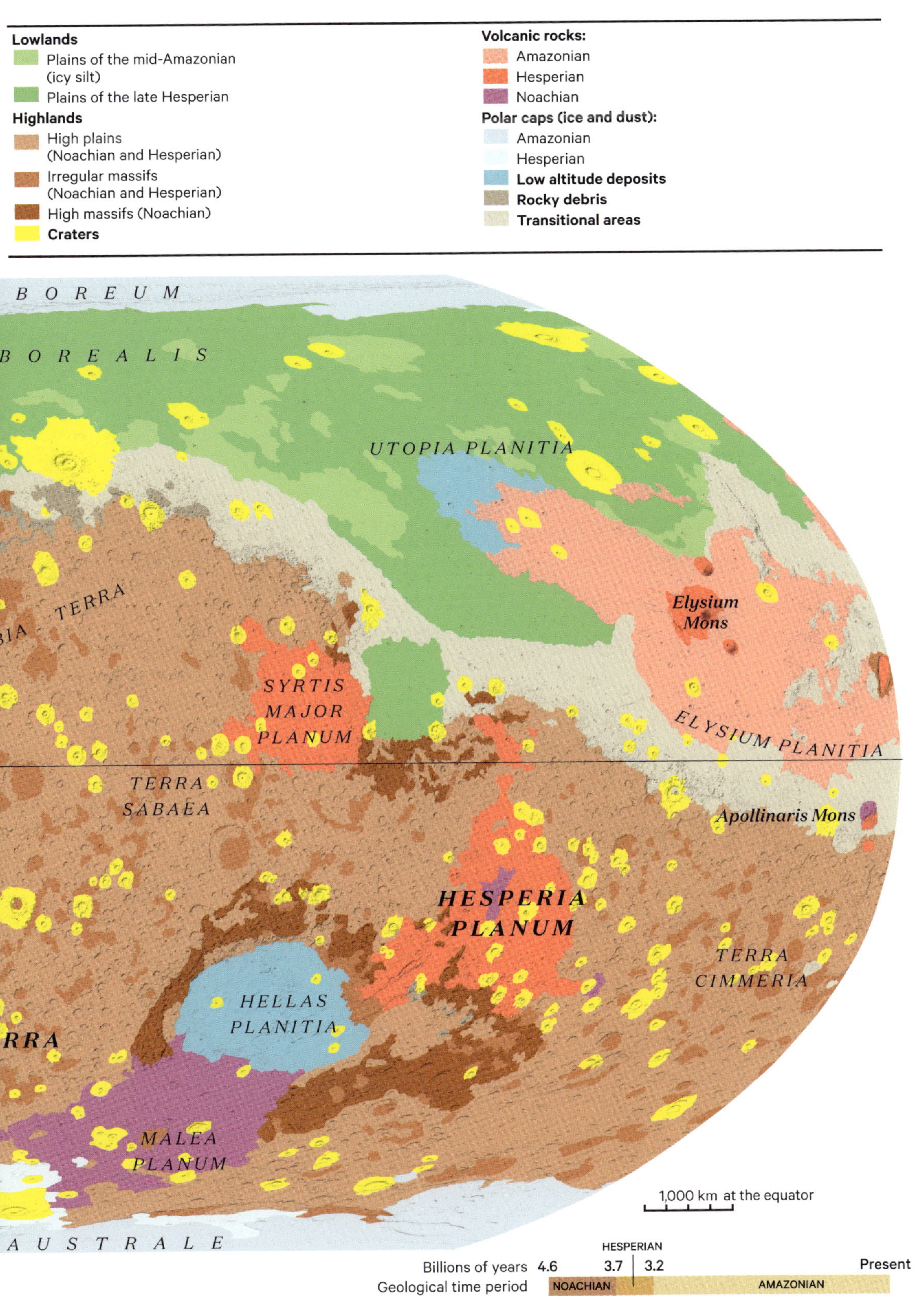
Lowlands
Plains of the mid-Amazonian (icy silt)
Plains of the late Hesperian
Highlands
High plains (Noachian and Hesperian)
Irregular massifs (Noachian and Hesperian)
High massifs (Noachian)
Craters
Volcanic rocks:
Amazonian
Hesperian
Noachian
Polar caps (ice and dust):
Amazonian
Hesperian
Low altitude deposits
Rocky debris
Transitional areas
BOREUM
BOREALIS
UTOPIA PLANITIA
BIA TERRA
Elysium Mons
SYRTIS MAJOR PLANUM
ELYSIUM PLANITIA
TERRA SABAEA
Apollinaris Mons
HESPERIA PLANUM
HELLAS PLANITIA
TERRA CIMMERIA
ERRA
MALEA PLANUM
AUSTRALE
1,000 km at the equator
HESPERIAN
Billions of years   4.6   3.7   3.2   Present
Geological time period   NOACHIAN   AMAZONIAN

Studying the Solar System

Despite two centuries of study, there is still much to be discovered about the Sun and its influence on the Earth, so this has been the subject of several recent missions. Other probes have passed close to the Sun, then moved on to observe other celestial bodies in the Solar System: planets, asteroids and comets. Over sixty years, more than 300 probes have provided data about these. Smaller objects that are further away remain relatively unexplored.

Mission: several months

1965 / 1969	Pioneer 6 to 9
1975 / 1976	Helios 1 & 2
1978	ISEE-3
1994 / 2009	Ulysses
1994 / 2026	WIND
1996 / 2026	SoHO
1997 / 2026	ACE
2001 / 2004	Genesis
2006	Stereo A & B
2015	DSCOVR
2018 / 2026	Parker Solar Probe
2020 / 2030	Solar Orbiter

Mission: several weeks

1962 / 1973	Mariner 2, 5 & 10
1967 / 1984	Venera 4 to 16
1978	Pioneer Venus
1985	Vega 1 & 2
1990	Galileo
1990 / 1994	Magellan
1998 / 1999	Cassini
2006 / 2014	Venus Express
2006 / 2007	Messenger
2015	Akatsuki
2018 / 2024	Parker Solar Probe
2020 / 2021	BepiColombo
2021 / 2026	Solar Orbiter

Earth, Moon & Mars pages 190, 19

Mission: several months

1974 / 1975	Mariner 10
2011 / 2015	Messenger
2025 / 2027	BepiColombo

NEPTUNE
URANUS
SATURN
and its moons
CERES &
ASTEROIDS
PLUTO

Mission: over 7 years
1979 Pioneer 11
1980 Voyager 1
1981 Voyager 2
2004 Cassini-
2017 Huygens

1989 Voyager 2
Mission: over 12 years

Mission: over 8 years
1986 Voyager 2

ROSETTA
CASSINI
NEW HORIZON
HAYABUSA 2

COMETS
(NON-EXHAUSTIVE)
1986 Giotto
      (Halley)
2001 Deep Space 1
      (Borrelly)
2004 Stardust
2006 (Wild)
2005 Deep Impact
      (Tempel 1)
2014 Rosetta
2016 (Churyumov–
      Gerasimenko)
2015 New Horizon

CERES
2015 Dawn
2018 (also orbited
      Vesta)

ASTEROIDS
1997 NEAR Shoemaker
2001 (253 Mathilde
      & 433 Eros)
2018 Osiris-REx
2019 (101955 Bennu)
2018 Hayabusa 2
2019 (162173 Ryugu)

Mission: from 5 to 10 years
1973 Pioneer 10
1974 Pioneer 11
1979 Voyager 1
1979 Voyager 2
1990 Galileo
1994
2000 Cassini
2007 New Horizon
2016 Juno
2025
2031 Juice
2035

Fly-by
Orbiter
Lander
Impactor
Returned samples
3 Number of fly-bys

Observing the Earth from space

Hundreds of satellites are dedicated to observing the Earth from orbit: modelling its weather patterns, monitoring natural disasters, and more. They analyse light (visible and infrared) emitted or diffused by the Earth's surface or its atmosphere.

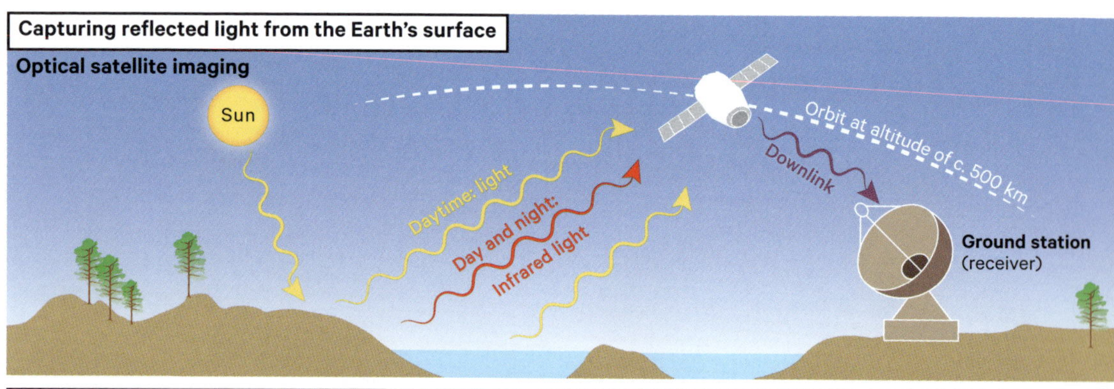

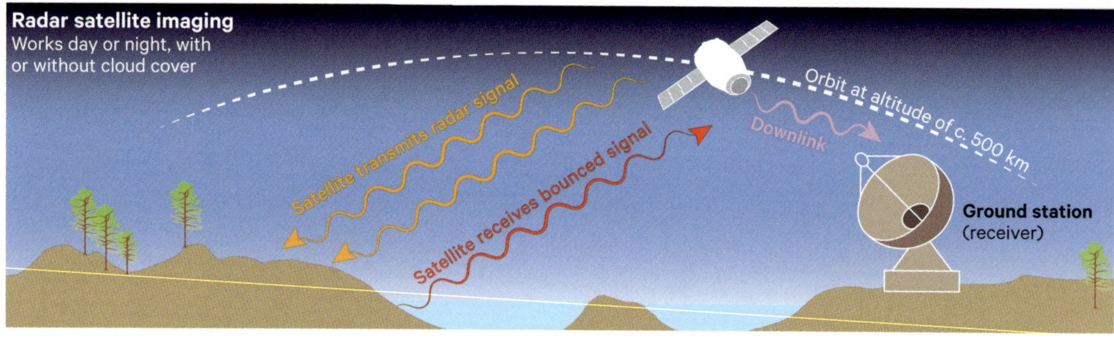

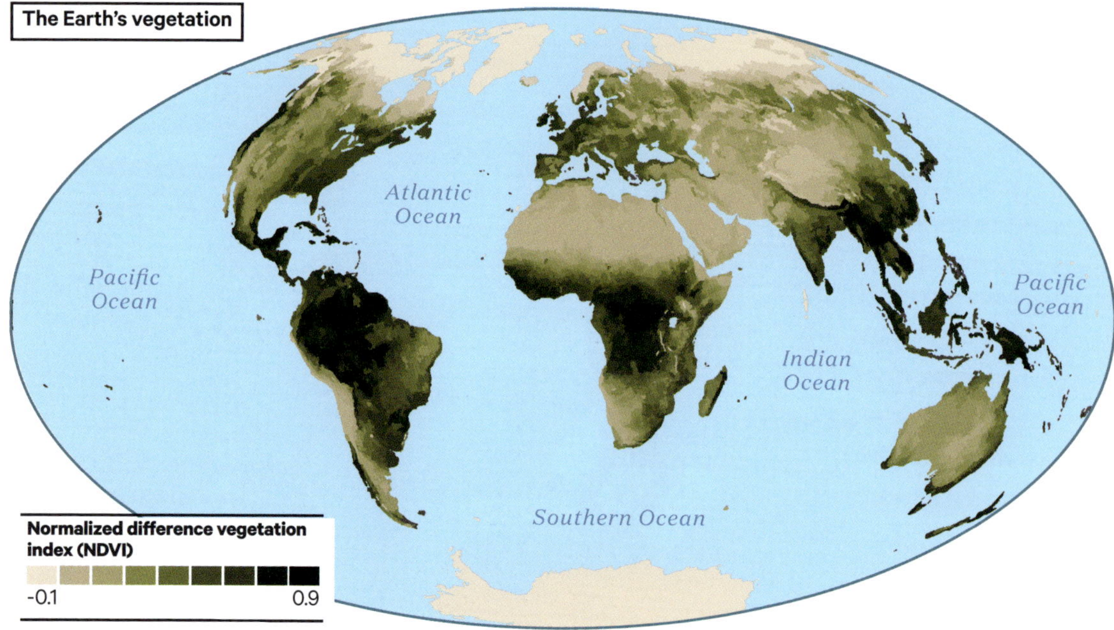

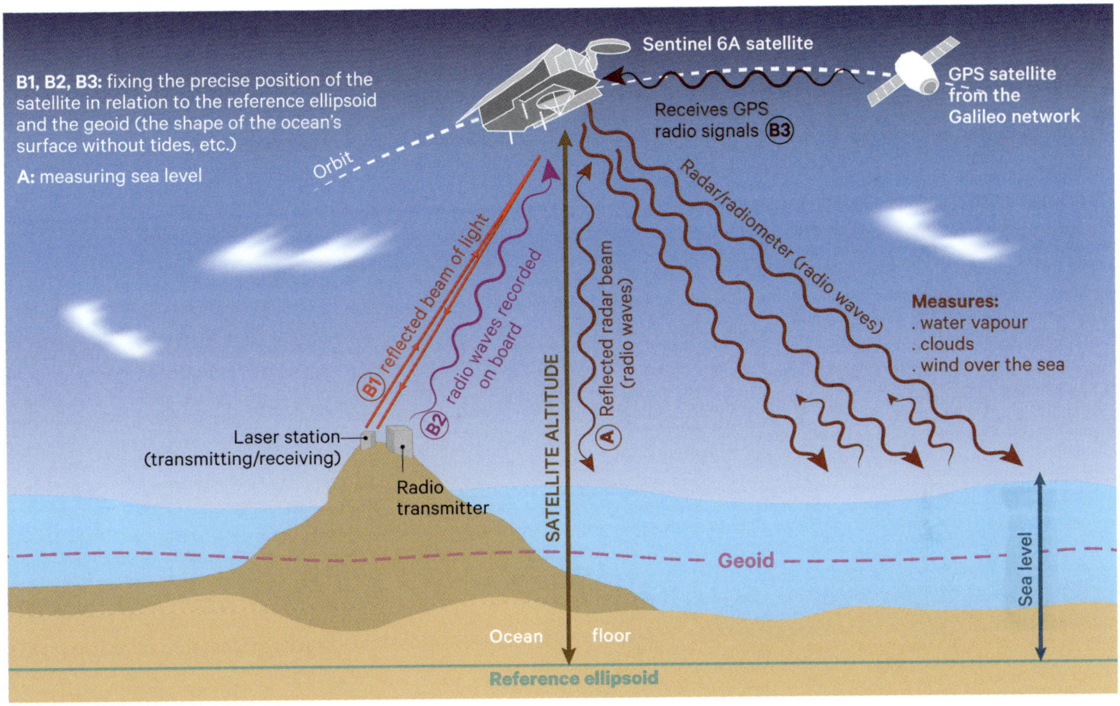

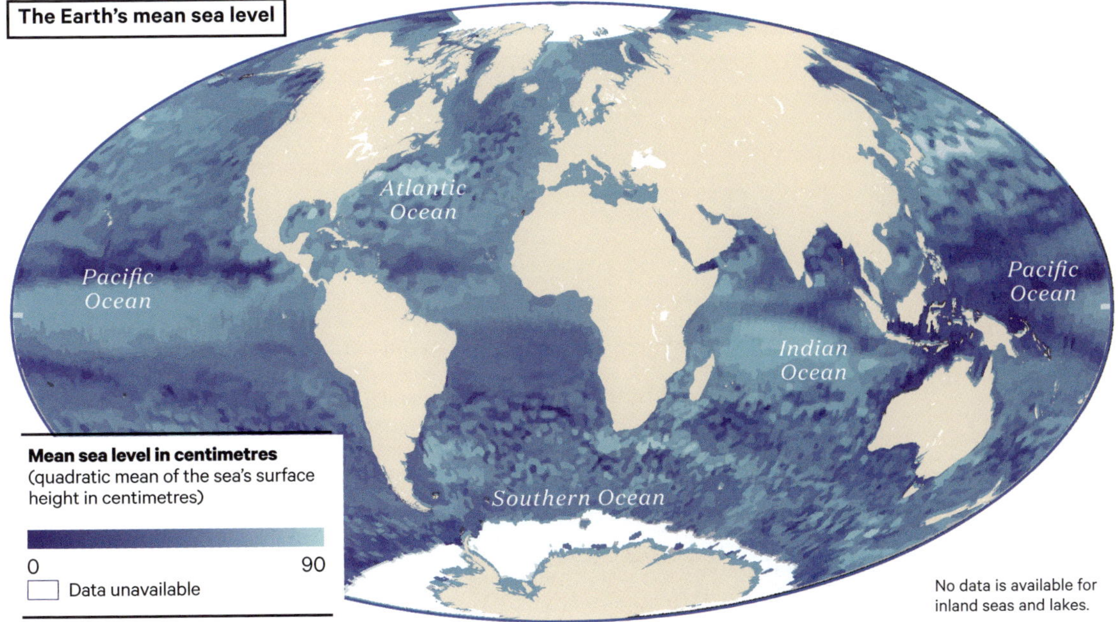

Satellite altimetry

Since the 1970s, satellite altimetry has been used to measure variations in the topography of the ocean's surface. A radio signal from a satellite is bounced off the water and returns to the satellite; the time it takes for the signal to travel there and back establishes the distance between the satellite and the surface. This can be used to deduce the sea level, relative to a simplified model of the Earth's surface, called the reference ellipsoid. Altimetry can also be used to map the topography of the ocean floor. A bump of 1 metre on the surface is caused by the gravitational attraction of an underwater mountain with a height of around 1,000 metres. The sea level is rising by an average of 4 mm a year, with local variations, due to climate change.

The Earth's mean sea level

GPS systems

First developed for military purposes in 1973 by the US Department of Defense, GPS (Global Positioning System) is now used everywhere in the world. Whether you are in the middle of the sea, on a mountain or in the desert, in good weather or during a storm, on the ground or in a plane, It establishes your geographical position, altitude and the time. The Russian equivalent GLONASS was developed in the late 1990s. There are also European and Chinese systems.

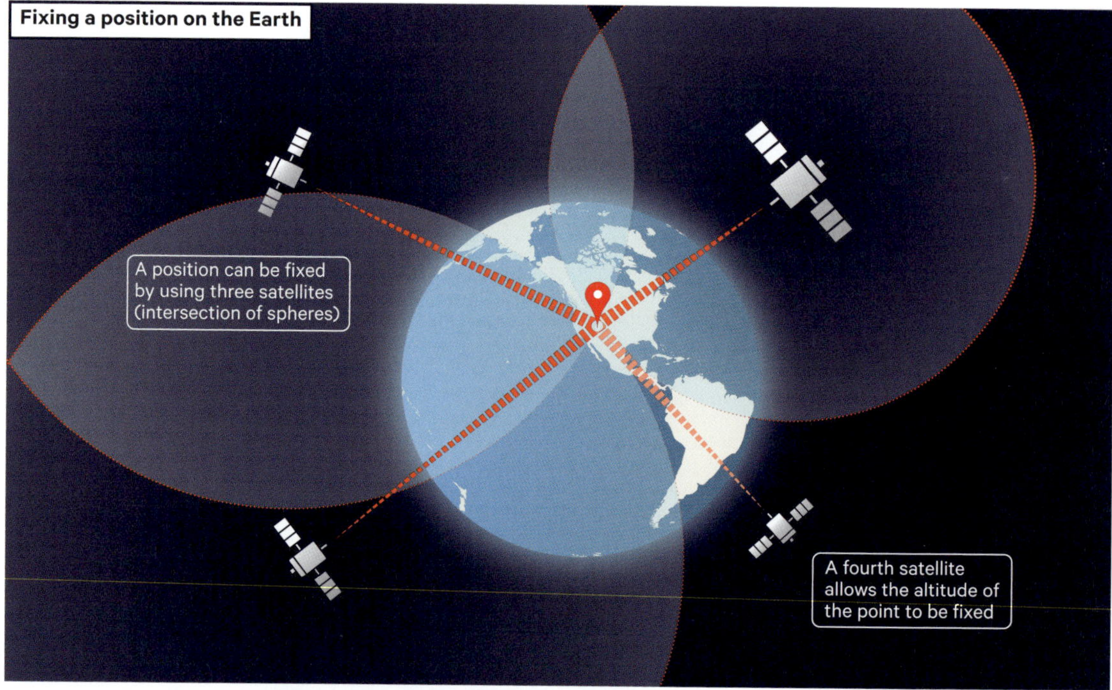

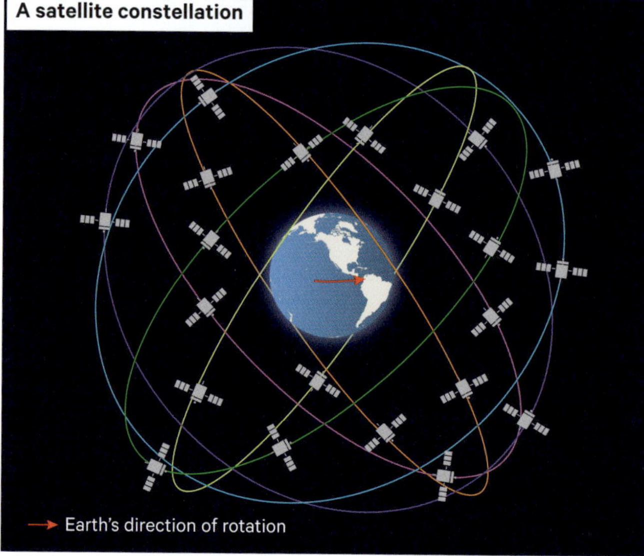

How GPS works

The user's position on Earth is determined using satellites that emit radiofrequencies. A satellite measures the distance between itself and the point in question. This means an imaginary sphere can be traced around each satellite. With two satellites, the position is reduced to a circle (the intersection of two spheres is a circle). With three satellites, there are only two possible points left. Then a fourth satellite makes it possible to determine the altitude and therefore the exact position. Navigation satellites orbit at an altitude of around 20,000 km. As four satellites are needed to determine each point, GPS systems need to have at least 25 operational satellites in order to pinpoint a position anywhere on the Earth's surface. These are distributed across six orbital planes.

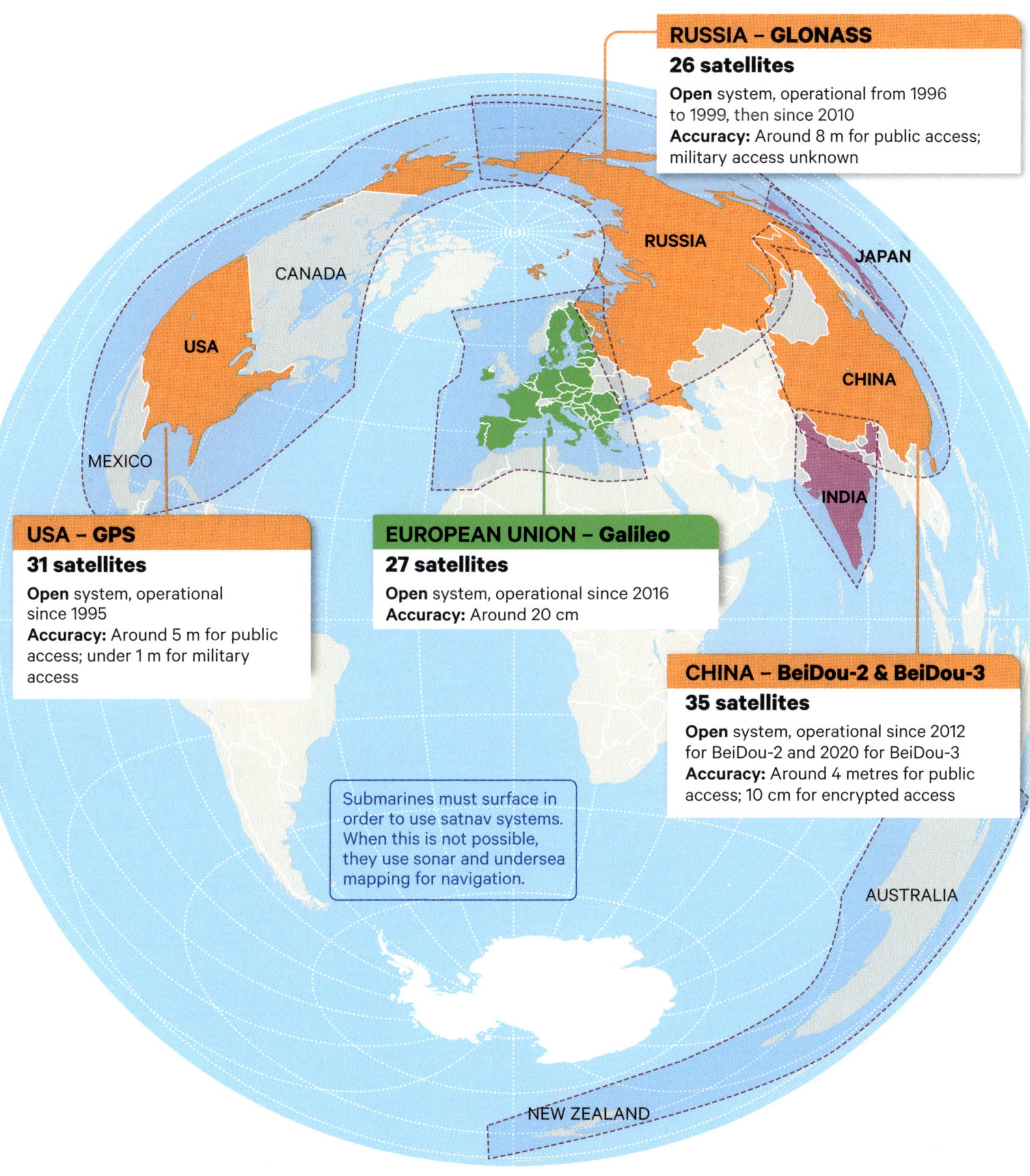

A difference in signals between two points can be used to establish distances with millimetric precision. This method has many potential uses, such as measuring the movement of tectonic plates.

Country or region with a global satellite navigation system:
- for civil and military use
- for civil use
- **Regional satellite navigation system**
- **Satellite-based augmentation system** (improves the precision and reliability of a satellite navigation system)

Competing systems

Fifty years after it was created, the American GPS system now has competition from three global systems (Russian, Chinese and European), which are at least as effective. These systems mean that those countries are not reliant on the US technology. A relatively inexpensive receiver makes it possible to determine your location and gives access to all these systems for free, although it doesn't provide the highest possible degree of precision. Two countries, India and Japan, have only regional systems. Many countries have developed increasingly precise satellite-based systems (SBAS), which correct any remaining errors in GPS measurements.

An increasingly full sky

The amount of satellites and debris in orbit is growing exponentially. This is an increasingly important issue, as it poses risks for space travel and interferes with astronomers' observations of the sky.

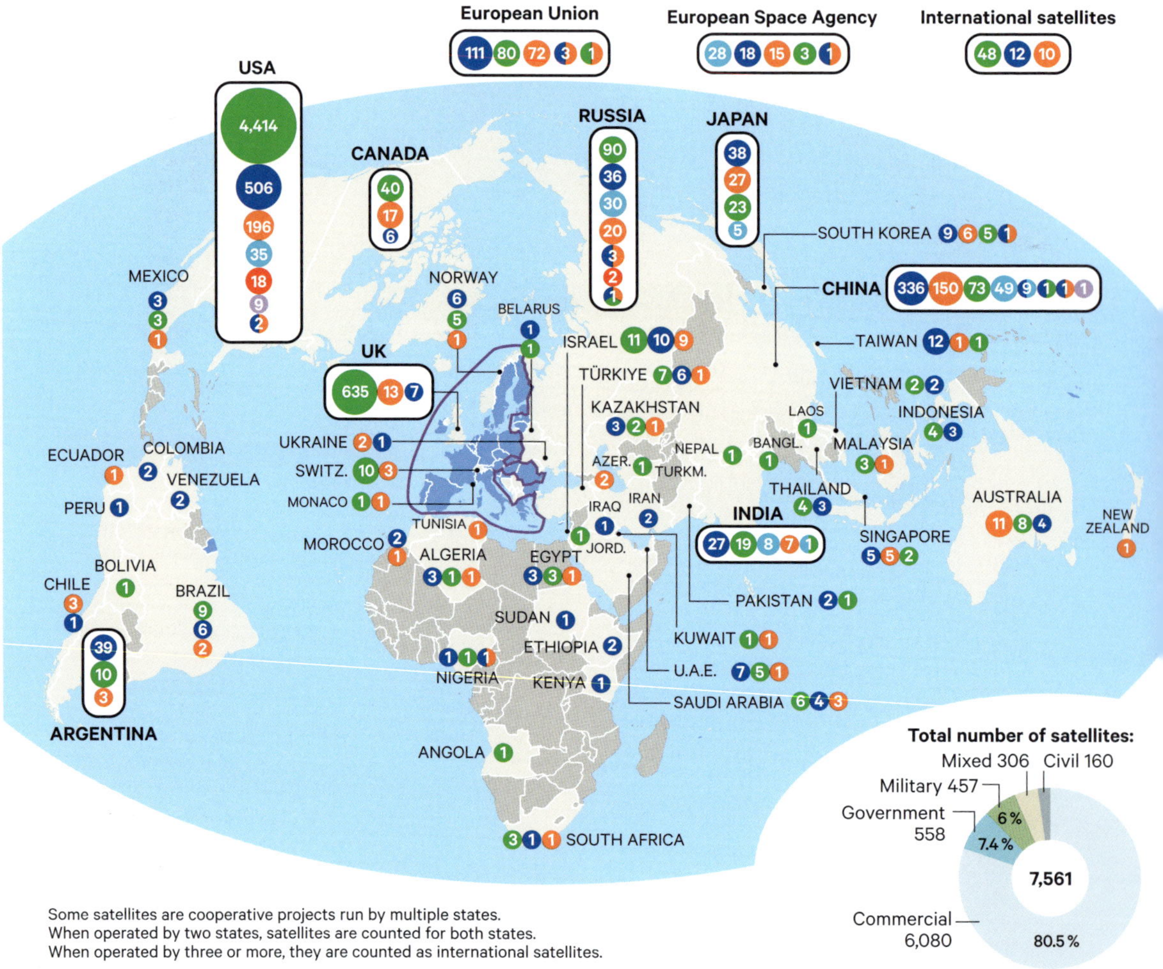

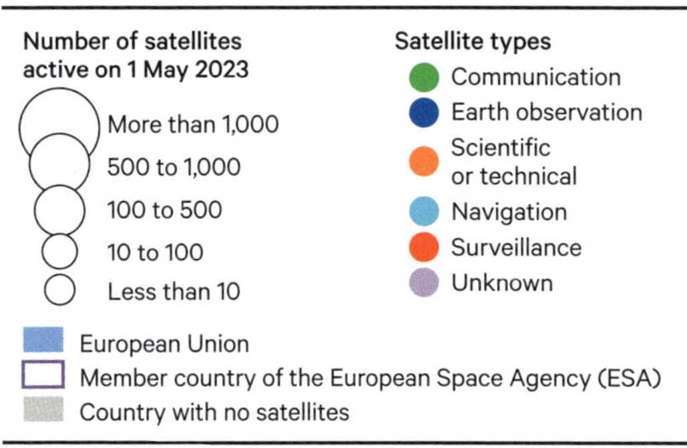

Some satellites are cooperative projects run by multiple states.
When operated by two states, satellites are counted for both states.
When operated by three or more, they are counted as international satellites.

Number of satellites active on 1 May 2023

- More than 1,000
- 500 to 1,000
- 100 to 500
- 10 to 100
- Less than 10

Satellite types

- Communication
- Earth observation
- Scientific or technical
- Navigation
- Surveillance
- Unknown

- European Union
- Member country of the European Space Agency (ESA)
- Country with no satellites

A myriad of satellites

The majority of satellites in orbit are used for communications. The Starlink constellation developed by Space-X (US), for example, contained more than 4,000 satellites in 2024 and may surpass 40,000 in the near future. A number of competing projects exist, including Amazon Leo (formerly Project Kuiper) and Eutelsat OneWeb. Some estimates predict there will be 100,000 satellites in orbit by the end of the 2020s. New satellites are therefore being developed that can detect small pieces of debris and analyse or predict the risk of collisions.

The white dots represent space debris of human origin. Most pieces are very small (under 1 cm). Debris accumulates in commonly used orbits and poses a potential threat to satellites and other spacecraft.

Dangerous debris

Since the dawn of the space age in 1957, thousands of launchers and satellites have been sent into space. Once their mission was completed, they were left in orbit and became space debris. But there is also other debris, varying in size, produced by explosions and collisions in space. Pieces can range in size from a fraction of a millimetre to the size of a bus. The Space Debris Office of the European Space Agency (ESA) monitors this debris constantly, but only objects over 10 cm in size in a low orbit can be detected by terrestrial radar or telescopes. In 2022, the European agency recorded 36,000 pieces of debris with a diameter of over 10 cm and 5,400 pieces of space debris larger than 1 metre. According to its statistical model, there are also 900,000 objects of more than 1 cm and 130 million objects over 1 mm. Travelling at an average speed of around 28,800 km per hour in low orbit, if a piece of space debris with a diameter of around a centimetre were to collide with a satellite, the impact would be the equivalent of being hit by a car at full speed. Any larger, and the satellite would almost certainly be destroyed. If the trajectory of a piece of catalogued space debris poses a potential danger, the orbit of the satellite in question is adjusted.

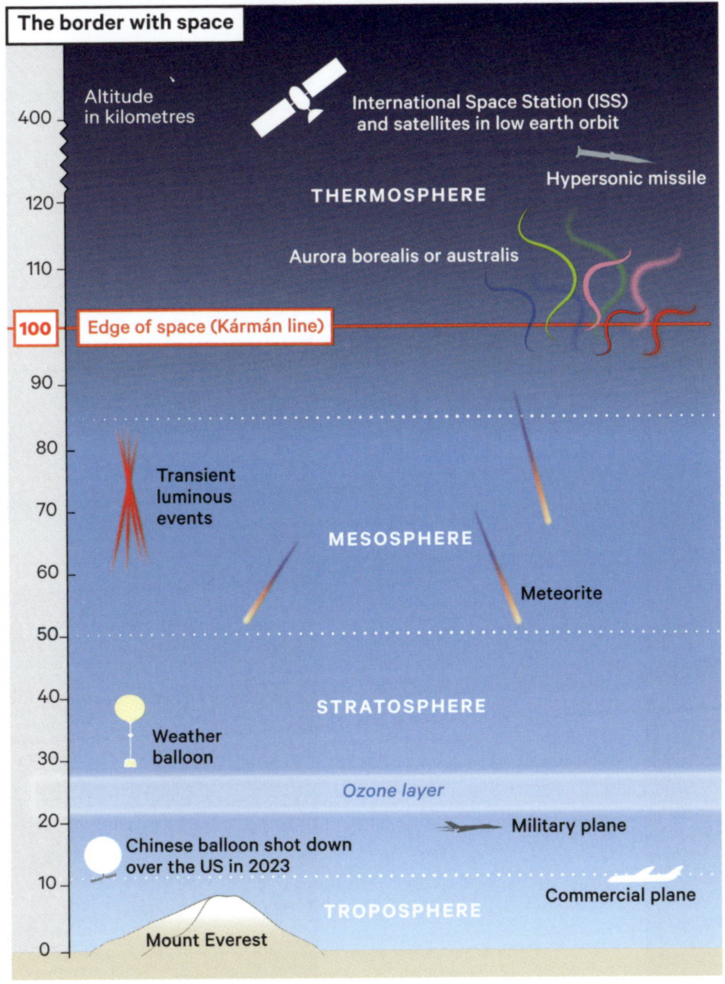

Cosmic fingerprints

Humans have now an impact on the cosmos: our electromagnetic signals travel at the speed of light, and our messages and probes go beyond the Solar System. In the future, it is possible that we may spread life to other planets; mining other planets is also being debated. All of these topics raise the question of our ethical responsibility towards the rest of the universe.

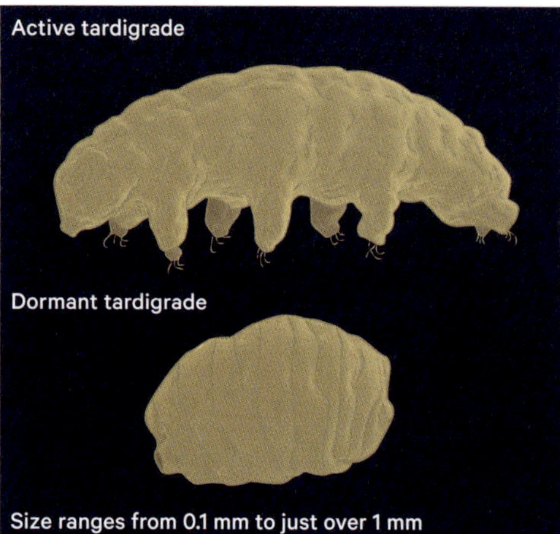

Living things

On 11 April 2019, the Israeli lander Beresheet, a mission led by a private company, crash-landed in the Mare Serenitatis on the Moon. It was carrying tardigrades, tiny animals capable of surviving in extreme conditions. Although space agencies try to protect celestial bodies from the risk of life being spread to them, the rising number of missions – some of which go beyond the Solar System – means that international regulation is required.

Objects

Within the Solar System, scientific equipment touches down, drills into the ground or crashes, deliberately or not, on the cores of comets and asteroids. On the Moon and Mars, plans to set up inhabited bases for tourism or industry raise the issue of humanity's ethical responsibility. Five probes and multiple pieces of debris are on trajectories to leave the Solar System. Launched from Earth, these missions have enough energy to keep travelling away from the Sun for millennia. Along with scientific instruments, they also carry messages for any extraterrestrial species that they may encounter.

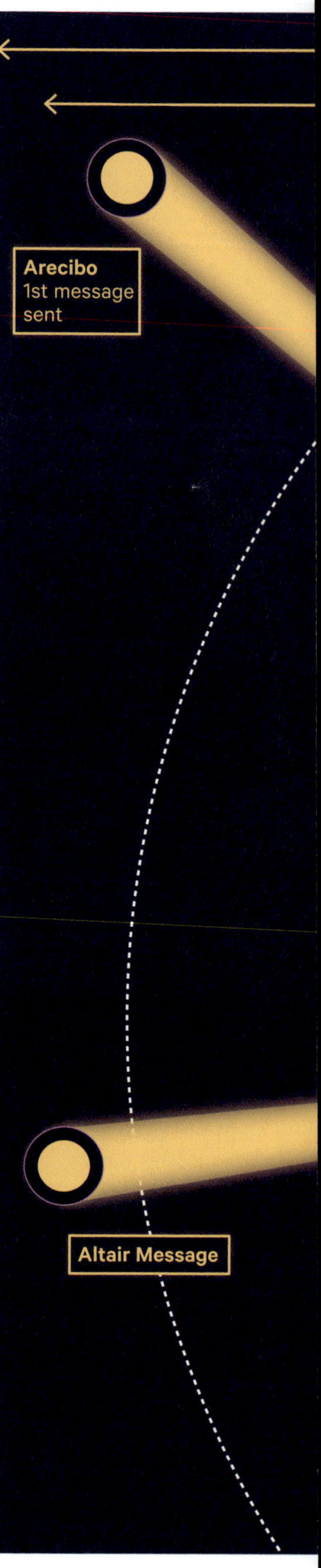

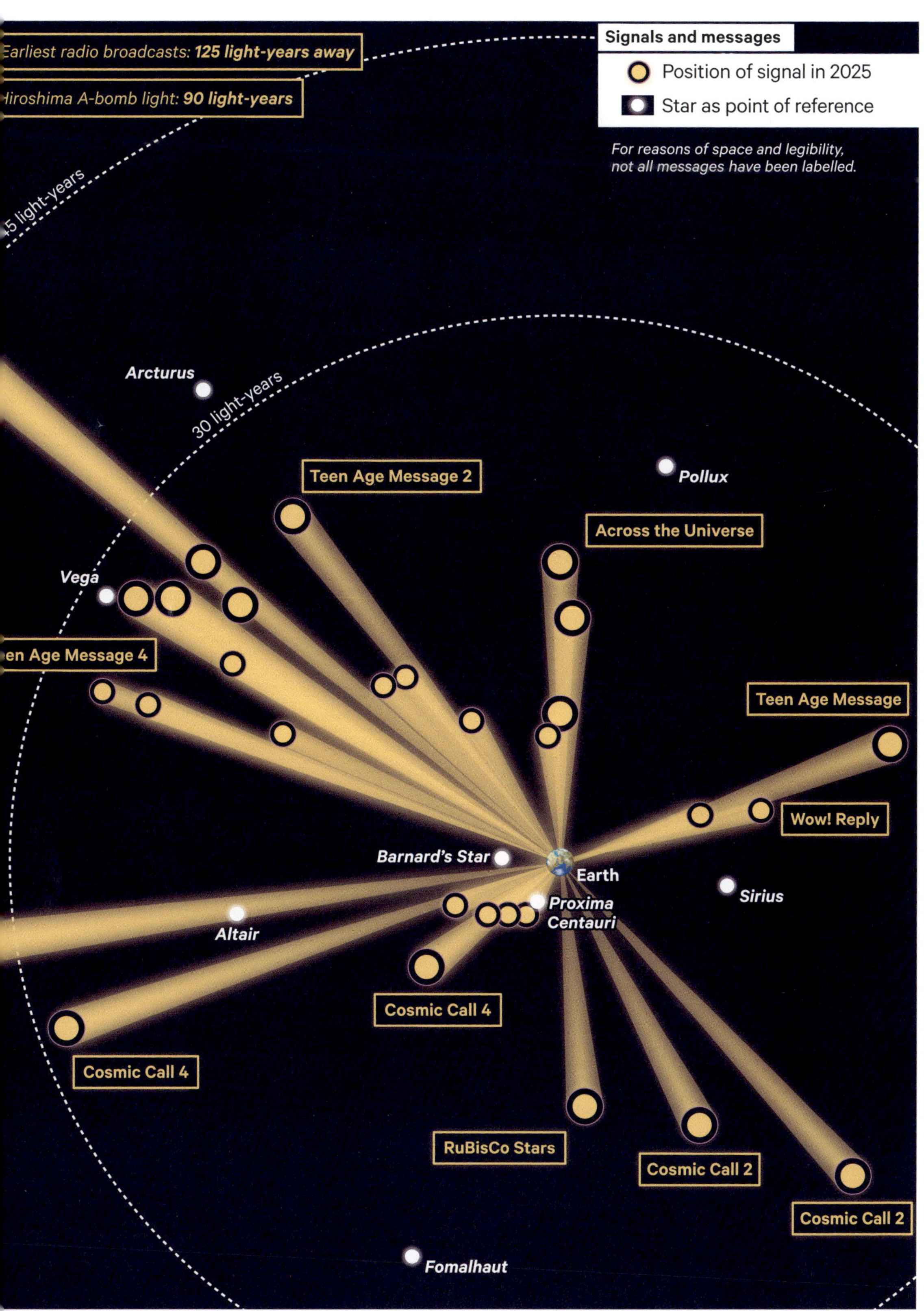
Earliest radio broadcasts: 125 light-years away
Hiroshima A-bomb light: 90 light-years
45 light-years
30 light-years
Signals and messages
Position of signal in 2025
Star as point of reference
For reasons of space and legibility, not all messages have been labelled.
Arcturus
Pollux
Teen Age Message 2
Across the Universe
Vega
Teen Age Message 4
Teen Age Message
Wow! Reply
Barnard's Star
Earth
Sirius
Proxima Centauri
Altair
Cosmic Call 4
Cosmic Call 4
RuBisCo Stars
Cosmic Call 2
Cosmic Call 2
Fomalhaut

The call of the sky

Christian Grataloup and Pierre Léna

Throughout this book, we have traced a centuries-long story and been amazed by what humans have managed to discover. The sky is sadly no longer a part of our daily lives because, in the modern world, wealth often leads to a specific kind of poverty: the majority of people in countries with a high standard of living are no longer able to see the stars. Bright city lights obscure the night sky and the motion of the planets, and our walls make us forget the movement of the Sun and our companion, the Moon. By contrast, in many developing countries, the starry night sky can still be seen clearly, and remains an endless source of fascination for travellers and curious tourists who come from urban worlds drowning in artificial light.

This book is not meant to be a substitute for the intense emotion we experience when looking up at the sky. It simply aims to outline the huge amount of research, experiments, calculations, dreams and fantastical tales that the sky has inspired in humans since the dawn of time. Throughout history, the interpretations that humans have projected onto the sky were as diverse as the societies from which they came. And then, in the west of the ancient world, from Sumerian scribes to Galileo, five centuries of study and reflection gave birth to the science of astronomy, which spread across the globalized world.

In the contemporary imagination, fed by science fiction, the sky may seem to have lost its mystery, now that the universe is finally within our reach. However, every day brings its share of discoveries, often demonstrated through staggering images, reminding us that there is still an infinite amount of knowledge to uncover. We only understand a tiny part of what makes up the universe. When we look to the past, as this book invites us to do, the distance we have travelled looks impressive, but it is almost impossible to imagine where this path will lead us in a century's time. Because in our quest for truth, the richness of the cosmos lies beyond our wildest imagination, the most abstract mathematics, the most elaborate physical or biological models, the most ingenious technology.

In the centuries to come, we may come to realize just how little we know. What will we discover about the origin of the universe when we are able to combine gravity and quantum physics, an advance as keenly awaited today as the discovery of the relationship between magnetism and electricity in the 19th century, which led to a new understanding of the nature of light? What oceans are hidden beneath the surface of the dozens of icy moons that orbit around Jupiter and Saturn? How many atmospheres and continents are there on the six thousand or so exoplanets in our galaxy that we know about, although it may contain many billions more? How many of them might be similar to Earth? When we look beyond this century, we may find ourselves confronting profound mysteries: whether matter has become life, either as we know it or perhaps in a very different form, spread throughout the billions of galaxies that we have photographed from our observatories in space; or we may discover, after our explorations, that life on Earth is unique, although it is impossible to be completely sure of this. Perhaps the reality of the universe is infinitely more complex.

We can draw inspiration from the rich history that we have traced. The moments when great minds have changed our understanding of the world are not simply the product of chance. The future of the sky is awaiting other giants, or perhaps simply humans standing on the shoulders of giants. For that to happen, the passion and tools required for scientific research must be shared more widely, offered to all young people, not only to those who have the good fortune to be born in the right place, both geographically and socially. This tough yet vital task will require all our strength and focus.

It was not so long ago that humans ventured beyond the Earth for the first time. What will they do with this new space, where they can indulge their curiosity, but also their battles and appetites? Will they come to see it as a commonwealth shared by all humanity, as was the case with Antarctica? Over the centuries, the creative power of the human mind has been expressed in science, while also drawing on many other social, ethical, philosophical and spiritual sources. In the future, these springs may be dried up by ignorance or violence, or they may feed our imaginations. But knowing this history can show us the way. Now, there is no doubt, the sky is calling.

Acknowledgments

This book would not have been possible without the aid and support of scientists, historians and researchers. Particular thanks go to Karine Chemla, Luc Dettwiller, Edward Duyker, Françoise Génova, Éric Lagadec, Jean-Loïc Le Quellec, Jean Mouette, Thierry Sarmant, Christophe Schmit, Pierre Teissier, David Valls-Gabaud and Zheng Ma-Cahuzac, as well as the library staff at the Observatoire de Paris, and Patrick Chevalley for his help with the Skychart software.

The dedicated involvement of Léna Hespel was crucial throughout this project. Many thanks for her enormous amount of work, her kindness and her generosity.

Thanks also to Philippe Pajot for his enthusiasm, his eye and his support.

Thanks to Marie-Sophie Putfin and Lucille Dugast, and the whole team at Légendes Cartographie for their work on the maps, for which they should be rightly proud.

This book was made possible by every department at the publishing house of Les Arènes. Thanks to all of them, especially Vincent Lever for the design, Sarah Ahnou for proofreading, Marie Baird-Smith and Anna Polonia. Making a book is one thing but letting people know about it is another. Many thanks to Isabelle Mazzaschi and Manon Lejeune in the press office and Salomé Bernard in bookstore publicity.

Thanks to Zoé Schwartz for her hard work and good humour during her internship. Thanks to Mahaut de Lataillade who was there when this book began its journey, and Oksana Pishko for her portraits of the great scientists of history.

Last of all, this book owes a great deal to Bertille Comar. She was the keystone of this project and the captain of our ship.

Jean-Baptiste Bourrat, Christian Grataloup and Pierre Léna

Bibliography

General works

James Evans, *The History & Practice of Ancient Astronomy*, Oxford & New York: Oxford University Press, 1998

Marc Lachièze-Rey & Jean-Pierre Luminet, *Celestial Treasury: From the Music of the Spheres to the Conquest of Space*, trans. Joe Laredo, Cambridge: Cambridge University Press 2001

Georges Minois, *L'Église et la science, Histoire d'un malentendu*, Paris: Fayard, 1991

Claude Boucher & Pascal Willis, *Les Références de temps et d'espace, Un panorama encyclopédique: histoire, présent et perspectives*, Paris: Hermann, 2017

René Taton, *A General History of the Sciences*, London: Thames & Hudson, 1966

James Poskett, *Horizons: A Global History of Science*, London: Penguin, 2021

Martin Harwit, *Cosmic Discovery: The Search, Scope, and Heritage of Astronomy*, Cambridge: Cambridge University Press, 2019

Ludwik Marian Celnikier, *Histoire de l'astronomie*, Cachan: Tec & Doc, 1996

Jean-Pierre Verdet, *Astronomie et astrophysique*, Paris: Larousse, 1993

Pierre Guillermier & Serge Koutchmy, *Total Eclipses: Science, Observations, Myths and Legends*, New York: Springer, 1999

Jean-Pierre Luminet, *The Big Bang Revolutionaries: The Untold Story of Three Scientists Who Reenchanted Cosmology*, Seattle: Discovery Institute Press, 2024

Nick Kanas, *Star Maps: History, Artistry, and Cartography*, Berlin: Springer, 2012

Robert Pansard-Besson, with Michel Serres & Pierre Léna, *Tours du Monde, tours du ciel, Une exploration de l'Univers à travers les âges*, Paris: EDP Science, 2009

Leïla Haddad & Guillaume Duprat, *Cosmos, Une histoire du ciel*, Paris: Éditions du Seuil, 2009

Didier Besset, *L'Univers, Une histoire humaine, De Lascaux à la découverte du Big Bang*, Lausanne: EPFL Press, 2023

Chapter 1

Daniel Kunth & Philippe Zarka, *L'astrologie est-elle une imposture?*, Paris: CNRS éditions, 2018

Yvon Georgelin, *Étoile des mages et astronomie chaldéenne*: http://astronomiechaldeenne.free.fr/

Jean-Loïc Le Quellec, *Avant nous le déluge: L'humanité et ses mythes*, Bordeaux: Éditions du Détour, 2021

African Cultural Astronomy: Current Archaeoastronomy and Ethnoastronomy Research in Africa, eds. Jarita C. Holbrook, Johnson O. Urama, R. Thebe Medupe, New York & Berlin: Springer, 2008

Macquarie Atlas of Indigenous Australia, eds. Bill Arthur & Frances Morphy, 2nd ed., Sydney: Macquarie, 2019

Madhuri Katti, 'Pathani Chandrasekhara Samanta: Traditional Eye Scanning the Sky', *Bhavana: The Mathematics Magazine*, vol. 7, issue 1, 2023

Serge Dunis, *Pacific Mythology, Thy Name is Woman: from Asia to the Americas in the Quest for the Island of Women*, Papeete, Tahiti: Haere Po, 2009

Jean Lefort, *La Saga des calendriers*, Paris: Belin, 1998

M. Heydari-Malayeri, *An Etymological Dictionary of Astronomy and Astrophysics: English–French–Persian*, L'Observatoire de Paris: https://dictionary.obspm.fr/

Chapter 2

Tang Yi Jie & Léon Vandermeersch, *Le Ciel*, Paris: Éditions Desclée De Brouwer, 2010

Joseph Needham, *Chinese Science*, London: Pilot Press, 1945

Isabelle Landry-Deron, *La Chine des Ming et de Matteo Ricci (1552–1610), Le Premier Dialogue des savoirs avec l'Europe*, Paris: Le Cerf/Institut Ricci, 2013

Jean-Baptiste Biot, *Précis de l'astronomie chinoise*, Paris: Gallica, 1861

Jean-Marc Bonnet-Bidaud, *Les Sciences de l'Empire du Milieu*, Paris: Belin, 2023.

Jean-Marc Bonnet-Bidaud, *4000 ans d'astronomie chinoise*, Paris: Belin, 2017

Ahmed Djebbar & Jean Rosmorduc, *Une histoire de la science arabe: Introduction à la connaissance du patrimoine scientifique des pays d'Islam*, Paris: Éditions du Seuil, 2001

Amir D. Aczel, *Finding Zero: A Mathematician's Odyssey to Uncover the Origins of Numbers*, New York: Palgrave Macmillan, 2015

Michel Serres, *Geometry: The Third Book of Foundations*, London: Bloomsbury, 2017

Henri Maspero, 'L'Astronomie chinoise avant les Han', *T'oung Pao*, vol. XXVI, Boston & Leiden, 1929

Léon Vandermeersch, *Les Deux Raisons de la Pensée Chinoise*, Paris: Gallimard 2013

Daniel Kunth & Elena Televich, *StarWords:*

The Celestial Roots of Modern Language, Berlin: Springer, 2024

Chapter 3
Bernhard R. Brandl, Remko Stuik & J. K. Katgert-Merkelijn (eds.), *400 Years of Astronomical Telescopes: A Review of History, Science and Technology,* New York & Berlin: Springer, 2010
Florence Trystram, *Le Procès des étoiles,* Paris: Payot, 2001
Dava Sobel, *Longitude: The True Story of a Lone Genius Who Solved the Greatest Scientific Problem of His Time,* London: Fourth Estate, 1995

Chapter 4
Jean-Pierre Luminet, *L'Invention du Big Bang,* Paris: Éditions du Seuil, 2014
Pierre Léna, *Astronomy's Quest for Sharp Images: From Blurred Pictures to the Very Large Telescope,* Berlin: Springer, 2020

Pierre Léna & Serge Koutchmy, *Eclipsed Suns, the Solar Corona and Exoplanets: From Concorde 001 to Telescopes in Space,* Berlin: Springer, 2025
Thérèse Encrenaz, *Planets: Ours and Others, from Earth to Exoplanets,* Hackensack, NJ: World Scientific, 2014
Dominique Lambert, *The Atom of the Universe: The Life and Work of Georges Lemaitre,* Krakow: Copernicus Center Press, 2016
Thierry Montmerle & Yi Zhou, *China and the International Astronomical Union: Divorce, Separation and Reconciliation (1958–1982),* Berlin: Springer, 2022

Chapter 5
Éric Chassefière, *Physics of the Terrestrial Environment, Subtle Matter and Height of the Atmosphere: Conceptions of the Atmosphere and the Nature of Air in the Age of Enlightenment,* Hoboken, NJ: John Wiley & Sons, 2021
Isabelle Sourbes-Verger, *Géopolitique du monde spatial,* Paris: Eyrolles, 2023
Xavier Pasco, *Le Nouvel âge spatial, De la Guerre froide au New Space,* Paris: CNRS éditions, 2017

Glossary

Note: Words in **bolder type** have separate glossary entries.

aberration of light
A phenomenon in which celestial objects seem to move slightly relative to their actual position. This effect is caused by the motion of the observer relative to the source of light, and is proportional to the ratio between the speed of the observer and the speed of light. If the observer is moving directly towards or away from the source, there is no aberration.

accretion
The process of a celestial object increasing in mass by accumulating matter (gaseous or solid) on or around itself, due to gravitational attraction.

aether
An invisible substance or 'fifth element' that was once believed to fill the interstellar void and form the **celestial spheres**. It was also believed to be the medium that light travels through, before it was known that light can pass through a vacuum.

arc-minute and **arc-second**
Units used to measure angles. A circle is divided into 360 degrees (°). One degree can be divided into 60 arc-minutes ('). One arc-minute can be divided into 60 arc-seconds ("). The terms 'minutes of arc' and 'seconds of arc' are also used.

armillary sphere
A three-dimensional object representing the apparent motion of the Sun, Moon and planets, as seen from Earth. Rings are used to show the paths of the **celestial equator**, the poles and the **ecliptic**. It also demonstrates the apparent motion of the **celestial sphere**, caused by the Earth's rotation around its axis. It is basically a simplified planetarium.

asterism
An identifiable pattern of stars in the sky. It is a looser term than **constellation**, which implies some kind of official recognition.

asteroids
A group of irregularly shaped rocky or metallic objects, ranging between a metre and several hundred kilometres in size, which orbit around the Sun in a belt located between Mars and Jupiter. Sometimes called minor planets.

astrobiology
Science that studies the potential existence of life elsewhere in the universe, past or present. Also called exobiology. **Exoplanets** are a particular source of interest to astrobiologists.

astrology
A set of principles based on the belief that the lives and fates of humans are affected by the positions of stars and planets in the Solar System. While it is true that humans are connected to the cosmos – life on Earth depends on light from the Sun, the atoms that make up human bodies come from the stars – astrology has no legitimate basis in science.

aurora
An atmospheric phenomenon that creates a pattern of moving coloured lights in the sky at night, most often close to the magnetic North or South Poles. Auroras are caused by charged particles in the Earth's atmosphere and are particuarly common during periods of solar activity. The resulting display of lights is named the aurora borealis in the northern hemisphere and the aurora australis in the southern hemisphere.

azimuth
The horizontal angle between a cardinal direction (usually north) and another direction.

Big Bang
A theory that describes how the universe was formed, expanding from a **singularity** that was extremely hot and dense. This expansion process is still ongoing and can be measured.

binary star
Star composed of two stars orbiting each other. Not to be confused with **double star**.

black hole
A celestial body so dense that nothing can escape from its **gravity**, not even light. A black hole is often formed when a large star reaches the end of its life and collapses.

celestial equator
A **great circle** that runs around the **celestial sphere** on the same plane as Earth's equator.

celestial sphere
A concept that visualizes the sky as a sphere with the Earth at its centre. This allows the positions of objects in the sky to be described without knowing their distance from the observer. See also **celestial spheres**.

celestial spheres
A set of imaginary nested spheres holding the stars and planets, which were once believed to rotate around an unmoving Earth. The spheres were thought to be composed of **aether**.

comet
A small celestial body made from ice and rock, which was originally formed on the far outer edges of our Solar System, and which follows a largely irregular orbit around the Sun. When it passes close the Sun, it releases gases, forming a bright glowing trail of light called a coma, and sometimes a tail of dust.

constellation
A shape formed by a group of stars, which suggests the form of an object or figure that it is named after. Most of these names are taken from mythological sources. In modern astronomy, the Earth's **celestial sphere** is divided into 88 regions, each bearing the name of a constellation.

corona
The outer atmosphere of the Sun, which extends into interplanetary space. The corona is made from low density gases. Its dim light is usually only visible from Earth during a total solar **eclipse**, when the bright light from the Sun is blocked by the Moon.

cosmic dust
Tiny particles of matter that exist in the **interstellar medium**. Also known as interstellar dust or extraterrestrial dust.

cosmic microwave background
A form of radiation that fills all space in the known universe, which can only be detected using particularly sensitive telescopes. Its variations in temperature and density can be measured, and provide vital evidence to support the theory of the **Big Bang**.

cosmic ray
A stream of high-energy particles (electrons, protons, neutrons, neutrinos, etc.) that generally comes from outside the Solar System. It is called a ray because it was originally believed to be a form of radiation.

cosmonaut
Russian term for an astronaut.

crater
A circular depression caused by the impact of a meteorite or other falling body on the surface of a planet or moon.

cycloid
A curve traced by a point on the circumference of a circle when the circle rolls along a straight line.

dark matter
A type of matter that doesn't interact with light or any other form of electromagnetic radiation, and is therefore invisible. However, its existence is implied by gravitational effects.

deferent
See **epicycle**.

diurnal motion
The apparent motion of the sky and celestial objects over the course of one day. This is caused by the Earth's rotation around its axis. For example, a star will seem to follow a circular path around the North or South Pole.

double star
Also called a visual double. Two stars that appear close to each other when viewed from Earth. Not to be confused with **binary star**.

escapement
A mechanism used to regulate the speed of clocks. The invention of the escapement led to greater accuracy in timekeeping, which allowed more accurate measurements of **longitude**.

eclipse
The temporary disappearance of a celestial body, as viewed by an observer on Earth. During a solar eclipse, the view of the Sun is blocked by the Moon. In a lunar eclipse, the Earth blocks the Sun's light from reaching the Moon, so it appears darker or redder in colour. An eclipse may be total or partial.

ecliptic
The Earth's **orbital plane**; in other words, the elliptical shape that the Earth follows as it orbits the Sun. It means that, viewed from the Earth, the Sun and other planets can be viewed in the same band of sky and the **constellations** move relative to this band over the course of a year.

epicycle
A circle that is part of a geometric model used in the **geocentric** theories of Hipparchus and Ptolemy to explain the apparently irregular motion of the planets around the 'fixed' Earth. A planet was thought to move in circles called epicycles, whose centre moved along an orbital path called a deferent.

equinox
The moment when the Sun appears directly over the Earth's equator.

Two of these occur each year, in March and September.

exoplanet
A planet that orbits a star other than the Sun.

fission
A phenomenon that occurs when the nucleus of an atom of element A is split into two, forming two elements, B and C. These two can then be subdivided again. Each split releases a vast amount of energy. This can be occur explosively (in nuclear weapons) or under controlled conditions (in a nuclear reactor).

galaxy
A group of many millions of stars held together by gravity, and also encompassing gas, dust and **dark matter**. The galaxy in which the Earth is located is known as the **Milky Way**.

geocentric
A term that describes any representation of the Solar System in which the Earth is positioned at the centre, with the Sun and planets orbiting around it.

geodesic
The straightest point between two points on a curved surface. Displayed on a flat surface, a geodesic appears as a curved line. On a sphere, the geodesic is the arc of a **great circle** passing through two points. In the field of **spacetime**, in which geometry is curved, light follows geodesics.

geoid
Representation of the Earth's surface, based on the shape that the oceans would take if they were subjected to the Earth's gravity and rotation but not to tides or wind.

gravitational wave
A ripple in **spacetime**, created when an object of large mass, such as a **black hole** or a **neutron star**, moves extremely fast. Waves of this kind travel at the speed of light.

gravity
The force of attraction that applies to any bodies with mass. Its laws were first described by Isaac Newton (universal gravitation).

great circle
A circle drawn around a sphere, which has the same diameter as the sphere and passes through its centre. This concept is used to establish a **meridian**.

heliocentric
A term that describes any representation of the Solar System in which the Sun is positioned at the centre, with the planets, including Earth, orbiting around it.

interstellar cloud
A region of the interstellar medium that contains an above-average density of matter.

interstellar medium
The 'empty' regions located between the stars in a galaxy. These areas contain gas in ionic, atomic and molecular form, as well as dust particles and crystals.

Kármán line
The boundary between the Earth's atmosphere and space, defined as lying at an altitude of 100 km above the Earth's surface.

Kuiper belt
An area of the Solar System that lies outside the **orbit** of Neptune. It is filled with scattered small objects composed of ice and rock, including remnants from when the Solar System first formed.

longitude
The geographical coordinate that measures how far east or west a point on the Earth's surface is.

mappa mundi
A medieval map of the world.

meridian
On the Earth's surface, a segment of a **great circle** that passes from the North Pole to the South Pole through a specific point is called that point's meridian. Midday and midnight occur at the same time at any point on that meridian.

meteor
Also known as a 'shooting star', this is a small piece of rock or debris that is heated as it enters the Earth's atmosphere, creating a streak of bright light as it burns up.

meteorite
A **meteor** that reaches the Earth's surface without entirely burning up.

Milky Way
The galaxy in which our Solar System is located. From Earth we view it in the direction of the galactic plane, so it appears as a broad band of hazy light across the night sky, composed of millions of stars.

nebula
(plural: **nebulae**) An interstellar cloud of gases and dust. Some are regions of space where stars are formed, while others are created by a **supernova**.

neutrino
Elementary particle with no mass and no electric charge. Neutrinos are produced by nuclear reactions and are difficult to detect because they interact very weakly with ordinary matter.

neutron star
A celestial body created when the core of a supermassive star reaches the end of its life and undergoes gravitational collapse. The resulting star is extremely dense, but not as dense as a **black hole**.

orbit
The curved path that one celestial object takes as it revolves around another celestial object of greater mass. The planets of the Solar System orbit the Sun, for instance. An orbit is generally round or elliptical in shape.

orbital plane
The flat disc of space drawn by the **orbit** of one celestial body around another. The Earth's orbital plane around the Sun is known as the **ecliptic**.

parallax
An apparent change in the position of a nearby object relative to distant objects when the observer changes position. The angle of this apparent displacement is known as the parallax and can be used to measure the distance of a celestial body, by observing it from the Earth at two different points in its **orbit**.

photon
A fundamental particle that transmits light. In a vacuum, it moves at the speed of light.

planet
A celestial body that orbits a star. Usually spherical in shape, it may be solid or gaseous, but is different from a star because it does not generate its own light or heat using nuclear reactions.

precession of the equinoxes
A slow shift in the orientation of Earth's axis of rotation, over a cycle of approximately 26,000 years. This shift means that the positions of the stars relative to Earth appear to move very slowly over the centuries.

probe
See **space probe**.

pulsar
A **neutron star** that spins rapidly and gives off pulses of electromagnetic radiation along a narrow band. These pulses are fixed in frequence and can only be registered if Earth is in the correct position relative to the pulsar.

reference ellipsoid
A simplified figure that approximates the shape of the Earth's surface, being slightly flattened at the poles and bulging at the equator. It is a mathematical representation of the **geoid**.

satellite
Term describing a small object

that orbits an object of greater mass due to the effects of **gravity**. A moon is an example of a natural satellite, while artificial satellites are human-made.

sidereal time
A time scale used by astronomers, based on the Earth's rotation relative to the 'fixed' stars. A sidereal day is around four minutes shorter than a solar day. See also **solar time**.

singularity
A point in space where gravity is so intense that space and time are affected. Normal observations are therefore impossible, so it can only be studied on a theoretical basis.

solar time
A time scale based on the position of the Sun in the sky. See also **sidereal time**.

solstice
When the Sun reaches its highest point in the sky over the North or South Pole. This event occurs twice a year, in winter and summer.

space probe
An uncrewed spacecraft designed to leave the Earth's atmosphere and send back information. Probes have been used to study the Moon, the planets, comets and deep space.

spacetime
A mathematical concept that combines the three dimensions of space with the fourth dimension of time to form a single continuum. This model allows a better understanding of relativity and its effects.

spectroscopy
A method of studing light by shining it through a solid, liquid or gas medium to produce a spectrum. This allows the light's intensity at different wavelengths to be measured. The instrument most commonly used to do this is called a spectrograph.

star
A ball of hot plasma held together by gravity. The mass of a star is so great that nuclear reactions occur inside it, giving off huge amounts of heat and light.

supernova
The explosion of a star. This event usually occurs late in the lifespan of a massive star. From Earth, a supernova can be seen as an extremely bright light that fades over months or years.

taikonaut
Chinese term for an astronaut.

vernal point
A point on the **celestial sphere** that marks one of the two places where the **celestial equator** crosses the **ecliptic**. These points are used to establish astronomical coordinates. The vernal point that marks the March equinox is also known as the first point of Aries.

zodiac
A belt-shaped area of the **celestial sphere** that passes north and south of the **ecliptic**. The Sun moves through this belt over the course of a year. The twelve **constellations** that occupy this belt were first identified by ancient civilizations and are linked with the practice of **astrology**.

zodiacal light
A long patch of glowing light that can sometimes be seen on the horizon before dawn or after sunset. It is caused by the Sun's light being scattered by interplanetary dust.

Picture credits

Sources:
p. 24: Observatoire de Paris website; p. 30: map based on research by Loïc Le Quellec; p. 36: Jean-Baptiste Biot, 'Précis de l'histoire de l'astronomie chinoise', *Journal des savants*, 1861; Sketches of Newark, Ohio, made in 1860 by David Wyrick en 1860; Keith Snedegar, 'Astronomy in Sub-Saharan Africa', *Encyclopaedia of the History of Science, Technology, and Medicine in Non-Western Cultures*, Dordrecht: Springer, 2008; p. 38: Denis Savoie, 'Les cadrans solaires', *Pour La Science*, no. 284, June 2001; p. 40: online date calculator: https://fr.planetcalc.com/; p. 42: Leïla Haddad and Guillaume Duprat, *Cosmos, Une histoire du Ciel*, Paris: Seuil, 2009; p. 44: Illustrations by Gary Urton, 'Animals and Astronomy in the Quechua Universe', *Proceedings of the American Philosophical Society*, University of Pennsylvania Press, vol. 125, 1981; p. 46: Keith Snedegar, 'Astronomy in Sub-Saharan Africa', *Encyclopaedia of the History of Science, Technology, and Medicine in Non-Western Cultures*, vol. 1, Dordrecht: Springer, 2008; p. 48: Polynesian compass by Hawaiian navigator Nainoa Thompson; Dominique Proust and Clotilde Proust, 'L'archéoastronomie à Rapa Nui', *Alliage: Culture, Science, Technique*, 2024; p. 52: F. C. Penrose, 'On the orientation of certain Greek temples and the dates of their foundation', *Philosophical Transactions of the Royal Society of London*, 1897; p. 58: James Evans, *The History & Practice of Ancient Astronomy*, Oxford: Oxford University Press, 1998; p. 68: http://serge.mehl.free.fr/anx/ epi_anim_pto.html; p. 70: Nick Kanas, *Star Maps: History, Artistry, and Cartography*, Dordrecht: Springer/Praxis Publishing, 2007; p. 74: James Evans, *The History & Practice of Ancient Astronomy*, Oxford: Oxford University Press, 1998; p. 76: http://www.ianridpath.com/startales/ startales2a.html#chinese; p. 78: https://www.scienceandsociety.co.uk/results. asp?image=10316284 – https://coron.et/new-1minute-reads/early-chinese-horology-su-sungs- water-clock; p. 80: Stefano Salvia, 'The Battle of the Astronomers: Johann Adam Schall von Bell

and Ferdinand Verbiest at the Court of the Celestial Emperors (1660–1670)', *Physics in Perspective*, 22, 81–109, 2020; p. 106: *400 Years of Astronomical Telescopes: A Review of History, Science and Technology*, New York: Springer, 2010; p. 108: sketch of observations of the moons of Jupiter after Galileo's *Sidereus nuncius*, published 1610; p. 124: *La Recherche*, no. 546, p. 35, April 2019; p. 152: Illustration from the National Astronomical Observatory of Japan; p. 162: Jessie L. Christiansen, 'Five thousand exoplanets at the NASA Exoplanet Archive', Nature Astronomy vol. 6, pp. 516–519, 2022; p. 164: F. W. Dyson, A. S. Eddington and C. Davidson, 'A Determination of the Deflection of Light by the Sun's Gravitational Field, from Observations Made at the Total Eclipse of May 29, 1919'; https://science.nasa.gov/resource/extrasolar-planet-detected-by-gravitational-microlensing; p. 170: *La Recherche*, no. 530, p. 25, December 2017; p. 176: Matthieu Quiret, 'À la pêche aux neutrinos en Méditerranée', *Les Échos*, 2007; p. 178: *La Recherche*, no. 551, p. 33, September 2019; p. 180: European Space Agency (ESA); p. 182: International Astronomical Union; Geoffrey C. Bower, 'Focus on First Sgr A* Results from the Event Horizon Telescope', *The Astrophysical Journal Letters*; Gaia Focused Product Release (FPR)/ESA; p. 190: *La Recherche*, no. 569, p. 118; © Bruno Bourgeois, 2022; p. 192: *La Recherche*, no. 569, p. 118; © Bruno Bourgeois, 2022; topographical map of Mars USGS 2014; 'How to create and destroy methane on Mars', ESA, 2019; p. 194: USGC; p. 196: *La Recherche*, no. 569, pp. 120–121; © Bruno Bourgeois, 2022; p. 198: NASA SWOT (Surface Water and Ocean Topography); NASA images produced by Reto Stockli and Jesse Allen, using data provided by the MODIS Land Science Team, caption by Michon Scott; NASA/JPL-Caltech: https://www.earthdata.nasa.gov/learn/articles/swot-data-release; p. 202: UCS Satellite Database; p. 204: SpaceEthics: https://spaceethics.vercel.app/.

Quiz answers

Question on page 41

What would the calendar be like for visitors to Mars?
Mars revolves at almost exactly the same speed as Earth, so the length of a Martian day (or 'sol') is only 2.75% longer than a solar day on Earth. The Martian sol is longer by 2,375.25 seconds (in atomic time), which is almost 40 minutes. The sol can be subdivided into Martian hours, minutes and seconds.

At the same time, however, the average year on Mars is 668.6 sols, because Mars is further from the Sun than Earth and so takes much longer to complete an orbit.

Question on page 85

An algebra problem
The solution to Al-Khwarizmi's problem of the divided inheritance is as follows:

The stranger receives 1/8 + 1/7 = 15/56 of the estate

The family therefore gets 41/56 to share between them. Of this, the husband's share is 1/4 of 41/56 = 41/224

What remains is then divided between the children, with the son getting 2/5 and each daughter getting 1/5.

Therefore, if the total inheritance is 1120 coins:
The stranger gets 300 coins.
The husband gets 205 coins.
The son gets 246 coins.
The daughters each get 123 coins.

Authors and contributors

Pierre Léna is an astrophysicist, a member of the French Académie des Sciences, professor emeritus at Paris Observatory and at the Université Paris-Cité. With the La Main à la Pâte foundation, he has been involved with the promotion and improvement of science teaching in schools for more than thirty years. He is passionate about the passing on of knowledge and spent almost three years creating this book.

Christian Grataloup has been called 'the most historian of geographers'. A doctor of geography and former lecturer at the Université Paris-Cité, he is a specialist in geohistory. Since 2019, he has been editor-in-chief of the *Atlas Historiques* series published by Éditions Les Arènes.

Léna Hespel is a science journalist. Holder of a master's degree in the history of science and graduate of the école Supérieure de Journalisme (ESJ) in Lille, she has worked for many French science publications (*Pour la science, Science et Vie, La Recherche*) and professional journals.

Philippe Pajot is a qualified astrophysicist and editor-in-chief of the journal La Recherche. He is author of the book *Parcours de mathématiciens* (2011).

Légendes Cartographie is an agency founded in 1996, specializing in designing maps for the fields of Earth sciences and social sciences. They created all the maps featured in this book.

Faouzia Charfi, physicist and historian, member of the Beit al-Hikma Foundation (the Tunisian Academy of Sciences, Letters and Arts)

Éric Chassefière, historian of science, research director at CNRS

Simonetta Cheli, **Gaia Maggi** & **Robert Meisner**, Earth Observation Programme Directorate at the European Space Agency (ESA)

Patrick Chevalley, creator of the software SkyChart/Cartes du Ciel

Pascal Crozet, historian of science, research director at the Université Paris-Cité

Olivier Darrigol, historian of science, research director at CNRS, team member of Sphere, research director at the Université Paris-Cité

Ewine van Dishoeck, astrophysicist, professor at Leiden University (Netherlands), president of the International Astronomical Union (2018–2021)

Serge Dunis, anthropologist, former lecturer at the universities of Wellington, Hawaii and French Polynesia

David Elbaz, astrophysicist, research director of the astrophysics department at the Atomic Energy Commission, Saclay, editor of the journal *Astronomy and Astrophysics*

Yvon Georgelin, astronomer, former director of Marseille Observatory

Étienne Ghys, lecturer at the École Normale Supérieure de Lyon, mathematician, permanent secretary of the Académie des Sciences

Matthieu Husson, CNRS researcher in the history of astronomy team at Paris Observatory, specializing in the Middle Ages

Bruno Jacomy, engineer, former deputy director of the Musée des Arts et Métiers and honorary chief curator of heritage at the French Ministry of Education

Michael Joalland, lecturer at CY Cergy Paris Université, historian specializing in the manuscripts of Isaac Newton

Francette Joanne, historian of Christian art

Étienne Klein, physicist and philosopher of science

Héloïse Kolebka, editor-in-chief of the journal *L'Histoire*

Serge Koutchmy (†2023), astrophysicist, research director at CNRS and the Institut d'Astrophysique, Paris

Daniel Kunth, astrophysicist, research director at CNRS and the Institut d'Astrophysique, Paris, secretary general of theUnion Rationaliste

Éric Lagadec, astrophysicist, teacher/researcher at the Côte-d'Azur Observatory, president of the Société Française d'Astronomie et d'Astrophysique

Anne-Marie Lagrange, astrophysicist, research director at CNRS and Paris Observatory, member of the Académie des Sciences

Colette Le Lay, research associate at the Centre François-Viète at the Université de Nantes, specializing in the history of science

James Lequeux, astronomer emeritus at Paris Observatory, historian of science

Jean-Pierre Luminet, astrophysicist, specializing in black holes and cosmology

François Mignard, astronomer at the Côte-d'Azur Observatory, member of the Académie des Sciences, president of the Bureau des Longitudes

Mamadou N'Diaye, researcher at the Côte-d'Azur Observatory

Adrien Normier, airline pilot, doctorate of philosophy and the ethics of space

Erwan Penchèvre, historian of science, project engineer at the École Normale Supérieure/ Université Paris Sciences & Lettres

Oksana Pishko, graphic designer and illustrator

Serge Plattard, professor at University College London, former director of the European Space Policy Institute

Jean Paul Poirier, honorary member of the Institut de Physique du Globe, member of the Académie des Sciences, historian of science

Dominique Proust, astrophysicist at Paris Observatory, honorary citizen of Easter Island

Brigitte Rocca, professor emeritus of astrophysics at the Université Paris-Saclay, specializing in galaxies

Patrick Rocher, astronomer at Paris Observatory and IMCCE (the Institute of Celestial Mechanics)

Shaohua Sang, archaeologist

Jean-Pierre Sarmant, honorary inspector general, physicist

Denis Savoie, research associate with the SYRTE team at Paris Observatory, scientific advisor at Universcience

Christophe Schmit, adjunct researcher at CNRS and Paris Observatory, SYRTE, historian of science

Isabelle Sourbès-Verger, geographer, research director at CNRS

Yi Zhou, writer and journalist at Radio France International

Index

Translated from the French *Atlas historique du ciel* by Bethany Wright

Cover designed by Steve O Connell

First published in the United Kingdom in 2026 by
Thames & Hudson Ltd, 6–24 Britannia Street, London WC1X 9JD

First published in the United States of America in 2026 by
Thames & Hudson Inc., 500 Fifth Avenue, New York, New York 10110

EU Authorized Representative: Interart S.A.R.L.
19 rue Charles Auray, 93500 Pantin, Paris, France
productsafety@thameshudson.co.uk
www.interart.fr

A CIP catalogue record for this book is available from the British Library

Library of Congress Control Number 2026939698

ISBN 978-0-500-03059-2
01

Printed and bound in Slovenia by DZS-Grafik d.o.o.

Be the first to know about our new releases,
exclusive content and author events by visiting
thamesandhudson.com
thamesandhudsonusa.com
thamesandhudson.com.au